LONDON MATHEMATICAL SOCIETY LECTURE NOTE SERIES

Managing Editor: Professor J.W.S. Cassels,
Department of Pure Mathematics and Mathematical Statistics,
16 Mill Lane, Cambridge CB2 1SB.

1. General cohomology theory and K-theory, P.HILTON
4. Algebraic topology, J.F.ADAMS
5. Commutative algebra, J.T.KNIGHT
8. Integration and harmonic analysis on compact groups, R.E.EDWARDS
9. Elliptic functions and elliptic curves, P.DU VAL
10. Numerical ranges II, F.F.BONSALL & J.DUNCAN
11. New developments in topology, G.SEGAL (ed.)
12. Symposium on complex analysis, Canterbury, 1973, J.CLUNIE & W.K.HAYMAN (eds.)
13. Combinatorics: Proceedings of the British Combinatorial Conference 1973, T.P.McDONOUGH & V.C.MAVRON (eds.)
15. An introduction to topological groups, P.J.HIGGINS
16. Topics in finite groups, T.M.GAGEN
17. Differential germs and catastrophes, Th.BROCKER & L.LANDER
18. A geometric approach to homology theory, S.BUONCRISTIANO, C.P. ROURKE & B.J.SANDERSON
20. Sheaf theory, B.R.TENNISON
21. Automatic continuity of linear operators, A.M.SINCLAIR
23. Parallelisms of complete designs, P.J.CAMERON
24. The topology of Stiefel manifolds, I.M.JAMES
25. Lie groups and compact groups, J.F.PRICE
26. Transformation groups: Proceedings of the conference in the University of Newcastle-upon-Tyne, August 1976, C.KOSNIOWSKI
27. Skew field constructions, P.M.COHN
28. Brownian motion, Hardy spaces and bounded mean oscillations, K.E.PETERSEN
29. Pontryagin duality and the structure of locally compact Abelian groups, S.A.MORRIS
30. Interaction models, N.L.BIGGS
31. Continuous crossed products and type III von Neumann algebras, A.VAN DAELE
32. Uniform algebras and Jensen measures, T.W.GAMELIN
33. Permutation groups and combinatorial structures, N.L.BIGGS & A.T.WHITE
34. Representation theory of Lie groups, M.F. ATIYAH et al.
35. Trace ideals and their applications, B.SIMON
36. Homological group theory, C.T.C.WALL (ed.)
37. Partially ordered rings and semi-algebraic geometry, G.W.BRUMFIEL
38. Surveys in combinatorics, B.BOLLOBAS (ed.)
39. Affine sets and affine groups, D.G.NORTHCOTT
40. Introduction to Hp spaces, P.J.KOOSIS
41. Theory and applications of Hopf bifurcation, B.D.HASSARD, N.D.KAZARINOFF & Y-H.WAN
42. Topics in the theory of group presentations, D.L.JOHNSON
43. Graphs, codes and designs, P.J.CAMERON & J.H.VAN LINT
44. Z/2-homotopy theory, M.C.CRABB
45. Recursion theory: its generalisations and applications, F.R.DRAKE & S.S.WAINER (eds.)
46. p-adic analysis: a short course on recent work, N.KOBLITZ
47. Coding the Universe, A.BELLER, R.JENSEN & P.WELCH
48. Low-dimensional topology, R.BROWN & T.L.THICKSTUN (eds.)
49. Finite geometries and designs, P.CAMERON, J.W.P.Hirschfield & D.R.Hughes (eds.)

50. Commutator calculus and groups of homotopy classes, H.J.BAUES
51. Synthetic differential geometry, A.KOCK
52. Combinatorics, H.N.V.TEMPERLEY (ed.)
53. Singularity theory, V.I.ARNOLD
54. Markov processes and related problems of analysis, E.B.DYNKIN
55. Ordered permutation groups, A.M.W.GLASS
56. Journëes arithmëtiques 1980, J.V.ARMITAGE (ed.)
57. Techniques of geometric topology, R.A.FENN
58. Singularities of smooth functions and maps, J.MARTINET
59. Applicable differential geometry, M.CRAMPIN & F.A.E.PIRANI
60. Integrable systems, S.P.NOVIKOV et al.
61. The core model, A.DODD
62. Economics for mathematicians, J.W.S.CASSELS
63. Continuous semigroups in Banach algebras, A.M.SINCLAIR
64. Basic concepts of enriched category theory, G.M.KELLY
65. Several complex variables and complex manifolds I, M.J.FIELD
66. Several complex variables and complex manifolds II, M.J.FIELD
67. Classification problems in ergodic theory, W.PARRY & S.TUNCEL
68. Complex algebraic surfaces, A.BEAUVILLE
69. Representation theory, I.M.GELFAND et al.
70. Stochastic differential equations on manifolds, K.D.ELWORTHY
71. Groups - St Andrews 1981, C.M.CAMPBELL & E.F.ROBERTSON (eds.)
72. Commutative algebra: Durham 1981, R.Y.SHARP (ed.)
73. Riemann surfaces: a view towards several complex variables, A.T.HUCKLEBERRY
74. Symmetric designs: an algebraic approach, E.S.LANDER
75. New geometric splittings of classical knots (algebraic knots), L.SIEBENMANN & F.BONAHON
76. Linear differential operators, H.O.CORDES
77. Isolated singular points on complete intersections, E.J.N.LOOIJENGA
78. A primer on Riemann surfaces, A.F.BEARDON
79. Probability, statistics and analysis, J.F.C.KINGMAN & G.E.H.REUTER (eds.)
80. Introduction to the representation theory of compact and locally compact groups, A.ROBERT
81. Skew fields, P.K.DRAXL
82. Surveys in combinatorics: Invited papers for the ninth British Combinatorial Conference 1983, E.K.LLOYD (ed.)
83. Homogeneous structures on Riemannian manifolds, F.TRICERRI & L.VANHECKE
84. Finite group algebras and their modules, P.LANDROCK
85. Solitons, P.G.DRAZIN
86. Topological topics, I.M.JAMES (ed.)
87. Surveys in set theory, A.R.D.MATHIAS (ed.)
88. FPF ring theory, C.FAITH & S.PAGE
89. An F-space sampler, N.J.KALTON, N.T.PECK & J.W.ROBERTS
90. Polytopes and symmetry, S.A.ROBERTSON
91. Classgroups of group rings, M.J.TAYLOR
92. Simple Artinian rings, A.H.SCHOFIELD
93. Aspects of topology, I.M.JAMES & E.H.KRONHEIMER (eds.)
94. Representations of general linear groups, G.D.JAMES
95. Low dimensional topology 1982: Proceedings of the Sussex Conference, 2-6 August 1982, R.A.FENN (ed.)
96. Diophantine equations over function fields, R.C.MASON
97. Varieties of constructive mathematics, D.S.Bridges & F.RICHMAN
98. Localization in Noetherian rings, A.V.JATEGAONKAR
99. Methods of differential geometry in algebraic topology, M.KAROUBI & C.LERUSTE
100. Stopping time techniques for analysts and probabilists, L.EGGHE

London Mathematical Society Lecture Note Series: 100

Stopping time techniques for analysts and probabilists

L. EGGHE
Limburgs Universitair Centrum
Universitaire Campus
B-3610 Diepenbeek
Belgium

The right of the University of Cambridge to print and sell all manner of books was granted by Henry VIII in 1534. The University has printed and published continuously since 1584.

CAMBRIDGE UNIVERSITY PRESS
Cambridge
London New York New Rochelle
Melbourne Sydney

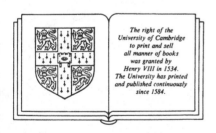

Published by the Press Syndicate of the University of Cambridge
The Pitt Building, Trumpington Street, Cambridge CB2 1RP
32 East 57th Street, New York, NY 10022, USA
296 Beaconsfield Parade, Middle Park, Melbourne 3206, Australia

First published 1984

Library of Congress catalogue card number: 84-45433

British Library cataloguing in publication data

Egghe, L.
Stopping time techniques for analysts and probabilists. - (London Mathematical Society lecture note series, ISSN 0076-0552; 100)
1. Functional analysis 2. Convergence
I. Title II. Series
515.7 QA320

ISBN 0 521 31715 0

Transferred to digital printing 2002

TABLE OF CONTENTS

PREFACE

Adapted sequences of integrable functions arise naturally in probability theory. Martingales, submartingales and supermartingales especially are very important to probabilists since they serve as mathematical models for many probabilistic phenomena. Consider for instance the fortune of a gambler. The martingale condition corresponds to the situation where this fortune remains constant in the sense of conditional mean. The supermartingale condition corresponds to the situation where at each play the game is unfavorable to the gambler in the same sense, while the submartingale condition corresponds to the situation where at each play the game is favorable in that sense. It is therefore clear that these notions are extremely important in probability theory, and so they have been heavily studied. One of the most interesting questions is when (and to what) does such an adapted sequence converge almost everywhere?

Such classes of adapted sequences do not only have interest in probability theory. They have also been used in other branches of mathematics such as potential theory, dynamical systems and many others.

However it is my feeling that not many analysts are used to dealing with martingales. That is even more the case with extensions of the martingale notion, involving stopping times. Nevertheless stopping time techniques do have many applications in real or functional analysis. This is what this book is about : to be of use to probabilists (of course) but also to analysts, by introducing them to the most important stopping time techniques. To look at a problem in analysis (real or functional) bearing in mind stopping time results may illuminate it and sometimes yield a surprisingly simple and elegant solution. As an example in real analysis there is the easy proof of the Radon-Nikodym theorem using stopping time techniques, given by Edgar and Sucheston. Also a surprising relationship between convergence in probability (i.e. in measure) and almost everywhere convergence can be described using stopping time techniques.

Furthermore, stopping times are very important in the study of topological or geometrical properties in Banach spaces. Many of them can be characterized by using stopping times. For a short description of the most important applications, see further on in this preface.

It is this double importance of the theory - for probabilists as well as for analysts - that encouraged me to study the convergence theory of generalized martingales in Banach spaces (including many new results in $\mathbb{R}$, the real line) and of extensions of generalized sub- or supermartingales in Banach lattices (also including many new results in $\mathbb{R}$).

In this book only some familiarity with measure theory as well as with functional analysis is presupposed. Concerning convergence of adapted sequences we only presuppose knowledge of the following theorem, to be found in Neveu [1975], theorem IV-1-2, p.62-64 and theorem II-2-9, p.26-29.

Theorem (J.Doob) : Let $(X_n, F_n)_{n \in \mathbb{N}}$ be a submartingale such that

$$\sup_{n \in \mathbb{N}} \int X_n^+ < \infty$$

then there exists $X_\infty \in L^1$ such that

$$\lim_{n\to\infty} X_n = X_\infty, \text{ a.e. } .$$

In this theorem obviously the following results are included :

a) Every real L^1-bounded martingale converges a.e. to a function $X_\infty \in L^1$.

b) Every positive supermartingale converges a.e. to a function $X_\infty \in L^1$ (but it must be emphasized that in fact (b) is used in the proof of the theorem above).

For new proofs of the real martingale convergence theorem, see Isaac [1965] or Lamb [1973].

Strictly speaking the above results and their proofs appear in this book independently : see theorem II.2.4.3, theorem III.2.2, theorem VII.2.12, theorem VIII.1.7, remark VIII.3.1, which yield several independent proofs of the results above, but this is more a curiosity and is not exactly the way in which to introduce Doob's result in classical probability courses.

So, apart from some elementary facts, the present book is completely self-contained.

Throughout the book E will stand for a Banach space with dual E' and $(\Omega, \mathcal{F}, P)$ will be a complete probability space.

Chapter I repeats some basic notions which are constantly used throughout the book. Classical measure theoretic notions and results are mentioned : we have first scalar and strong measurability and their relation expressed in the theorem of Pettis. Concerning integrability, the Bochner integral is introduced and the spaces L_E^p ($p \in [1, +\infty]$). On L_E^1 not only is the $\|\cdot\|_1$-norm considered but also the Pettis-norm of the form, where $f \in L_E^1$

$$\|f\|_{Pe} = \sup_{\substack{x' \in E' \\ \|x'\| \leq 1}} \int_\Omega |x'(f)| dP$$

Relations between $\|\cdot\|_1$ and $\|\cdot\|_{Pe}$ are given with a quick proof. Then we give the definition of an adapted sequence, which is the real subject of the book. By an adapted sequence we mean a sequence $(X_n, \mathcal{F}_n)_{n \in \mathbb{N}}$, where $(\mathcal{F}_n)_{n \in \mathbb{N}}$ is an increasing sequence of sub-σ-algebras of the σ-algebra $\mathcal{F}$ and where $X_n \in L_E^1$ for every $n \in \mathbb{N}$, such that every X_n is $\mathcal{F}_n$-measurable. For these processes we define the notion of stopping times, and for bounded stopping times, the elementary properties are derived, such as f.i. the "localization property". The existence of conditional expectations of a function $f \in L_E^1$ w.r.t. a sub-σ-algebra $\mathcal{G}$ of $\mathcal{F}$, i.e. a function $E^{\mathcal{G}} f$ such that

$$\int_A E^{\mathcal{G}} f = \int_A f$$

for every $A \in \mathcal{G}$, is proved. Also some elementary properties of $E^{\mathcal{G}}$ are indicated. Much of our attention in this book is focused on the problem of convergence of adapted sequences. Therefore, different types of convergence are discussed : pointwise almost everywhere (a.e.), $\|\cdot\|_1$-convergence, $\|\cdot\|_{Pe}$-convergence, convergence in probability. This last notion is studied for $(X_n)_{n \in \mathbb{N}}$ as well as for $(X_\tau)_{\tau \in T}$, where τ runs through the set of bounded stopping times and where

$$X_\tau = \Sigma X_n \chi_{\{\tau = n\}} \quad ;$$

here the sum ranges over all possible values τ can take (there are only finitely many). We have here the important result of Millet-Sucheston, which says that if $(X_\tau)_{\tau \in T}$ converges in probability, then it converges a.e., and hence also $(X_n)_{n \in \mathbb{N}}$. This elementary but important result was only proved in 1979. In 1980, Bellow-Egghe gave a simpler proof for it (including a small correction). It is this proof that is presented here; the method is important for what follows later on in the book. We feel that this result belongs in every elementary course on probability and perhaps also on analysis.

We have given a rather elaborate description of the introductory chapter I, to make this book more accessible for the non-expert, to whom this book is certainly directed. But I also hope that the expert will find the book useful since it unifies the material of numerous articles, selects the most elegant proofs, as we shall indicate later on, and since it contains a lot of very recent results. The book can be used as a seminar text in probability as well as in functional analysis. It is also hoped that it will encourage analysts to consider stopping time techniques occasionally and also that probabilists will understand the power of functional analysis and more specifically of Banach spaces, in probability theory. Let us now continue to describe the contents of the book.

Chapter II proves martingale convergence theorems in the framework of Banach spaces with the Radon-Nikodym property (RNP). The maximal lemma is very important in this connection as well as later on. It is however simple to prove. The special relation of martingale convergence theory and the geometry of Banach spaces is illustrated several times and, more specifically, the geometric notion of "dentability" in a Banach space is proved to be equivalent with the martingale convergence theorem : theorem of Rieffel, Huff et al.. The central result of this chapter is the a.e. convergence of martingales in Banach spaces with (RNP), known usually as Chatterji's result. In our book we call it the result of A. and C. Ionescu-Tulcea since they were the first to prove it. However in our book the proof of Chatterji is presented since it is short, elegant and independent of the real martingale convergence theorem of Doob. Two other proofs of this fundamental theorem are presented. One is based on the Kadec-Klee renorming theorem of Banach spaces which is, roughly speaking, a method of reducing the vector-valued convergence problem to the real, already known, case. The other proof, that of Chacon-Sucheston, uses an optimal stopping method by which the problem is reduced to that of martingales of the form $(E^{F_n}X, F_n)_{n \in \mathbb{N}}$, the convergence of which is fairly easy to prove. All these proofs are given because of the importance of the various methods for what follows in the book. The chapter closes with some martingale convergence properties in general Banach spaces, possibly without (RNP) : results of Korzeniowski, Chatterji and Burkholder-Shintani. It is my feeling that these results are not so well-known to the mathematical public and since the results and their proofs are nice, they are interesting enough to be included in this book. Especially the result of Korzeniowski-relating martingale convergence with weak convergence of the associated sequence of probability distributions-uses results a little beyond the scope of this book; f.i. the measurable selection theorem of Kuratowski and Ryll-Nardzewski is used in the proof.

Chapter III contains proofs of convergence theorems for sub- and supermartingales with values in Banach lattices. We prove first the theorem of Heinich on convergences a.e. of positive submartingales, with the original proof as well as the proof by using the renorming theorem of Davis-Ghoussoub-Lindenstrauss, in the same way as in the previous chapter the Kadec-Klee renorming theorem was used. Then the results of Szulga-Woyczynski on convergence of general submartingales are presented, with some new proofs. Also the result of Benyamini-Ghoussoub on the convergence of supermartingales is included. A result of Ghoussoub-Talagrand on weak convergence a.e. of class (B) supermartingales is postponed until chapter V, since for the proof more machinery is needed, which is available in chapter V. Also this chapter closes with a submartingale convergence theorem of Davis-Ghoussoub-Lindenstrauss, with values in general Banach spaces.

Chapter IV contains the basic inequalities - proved by Bellow-Egghe - upon which chapter V and VII are based to yield extensions of the convergence results of chapter II. It describes the validity of inequalities of the following form, where $(X_n, F_n)_{n \in \mathbb{N}}$ is an adapted sequence with values in a Banach space :

$$\limsup_{m,n \in \mathbb{N}} \|X_n(\omega) - X_m(\omega)\| \leq C. \limsup_{\substack{n,m \in \mathbb{N} \\ n \geq m}} \|E^{F_m}X_n(\omega) - X_m(\omega)\|$$

valid a.e..The norm can also be replaced by a continuous seminorm. Also stopping times instead of natural numbers can be involved. So if the righthand side is zero, i.e. a generalization of the martingale concept, then the left hand side is zero, i.e. convergent a.e.. The validity of such inequalities is studied and, as can be remarked, the results are very sharp : first they yield in chapter V and VII all known convergence results of extensions of martingales, thus giving a new unified proof for them, and secondly it is shown by Edgar and Mc Cartney-O'Brien that the conditions which were needed in the proof of these inequalities are really needed : the above inequality fails in certain (RNP) spaces which do not embed into a separable dual.

Chapter V contains the theory of uniform amarts, amarts, weak sequential amarts, weak amarts and (uniform) semiamarts. The class of uniform amarts comprises the class of quasi-martingales, yet the martingale convergence theorem of A. and C. Ionescu-Tulcea remains true for uniform amarts without additional assumptions. This was proved by Bellow, but becomes trivial using the Bellow-Egghe inequalities in chapter IV. Quasi-martingales arise in a natural way in the geometric theory of Banach spaces. However uniform amarts are easier to study than quasi-martingales. An application of the uniform amart a.e. convergence theorem to the geometric theory of Banach spaces is given. Next amarts - a contraction of the term asymptotic martingales - are studied. It is proved that convergence in the $\|\cdot\|_{Pe}$-norm obtains for uniformly integrable amarts in Banach spaces with (RNP), which is a result of Uhl but which is proved here by using again the Bellow-Egghe inequalities of chapter IV. Using amarts, a probabilistic characterization of the spaces E for which every operator from L^1 into E has a compact restriction to L^∞, is given by Egghe, based on a result of Uhl and another of Diestel. For applications of amarts : see chapter VI. Weak convergence a.e. is obtained for weak sequential amarts of class (B) if and only if E and E' have (RNP), again using the Bellow-Egghe inequalities of chapter IV. This is a result of Brunel-Sucheston. Another result of Brunel-Sucheston, giving a characterization of the separability of the dual E' in terms of a class of weak sequential amarts, is given by presenting the new proof of Edgar, which also gives an extension of this result, by characterizing Asplund operators using a class of weak sequential amarts. In chapter V the result of Ghoussoub-Talagrand, mentioned in chapter III, is also proved. Next a characterization of reflexivity in terms of weak convergence of weak amarts is given, being another result of Brunel-Sucheston. A study of (uniform) semiamarts concludes this chapter. These arise naturally in chapter IV, showing their importance in probability theory as well as in the geometric theory of Banach spaces. In chapter V we give some oscillatory results on semiamarts, using the notion of Banach limit.

Chapter VI deals with some applications of the amart convergence theory, as developed in chapter V. They were not included in chapter V since the applications are in the field of adapted nets, indexed by Z × Z or even more generally. By making a separate chapter, we can introduce these notions properly, and the results are separated from the general theory. Also, chapter VI can be skipped by the reader who is not interested in these topics, without losing continuity. We present some results of Millet-Sucheston and another of Astbury. In this last result an old problem of Krickeberg concerning the characterization of the Vitali condition for $(F_i)_{i \in I}$ in terms of convergence of adapted nets $(X_i, F_i)_{i \in I}$ is solved, using amarts (as was shown previously by Millet-Sucheston, martingales can not be used here!). In the notes and remarks of chapter VI, we also reveal a didactical application of amarts, mentioned by Edgar-Sucheston.

Chapter VII continues chapter V and yields another bunch of theorems which are applications of the inequalities proved in chapter IV. The chapter opens with some negative results on vector amart convergence theory. It shows that amarts never converge strongly a.e., except in the trivial case when E is finite dimensional (results of Bellow with proofs of Egghe). New extensions of the martingale concept are defined : pramarts, mils, and it is shown that they have better strong convergence properties than amarts (results of Bellow-Dvoretzky, Millet-Sucheston). Also, if $E = \mathbb{R}$, they generalise amarts and Doob's theorem extends to these new concepts without additional hypothesis (theorem of Mucci).

Chapter VIII studies extensions of sub- and supermartingales : subpramarts and superpramarts. For $E = \mathbb{R}$ it is shown that the sub- and supermartingales a.e. convergence theorems extend to sub- and superpramarts (Millet-Sucheston). The theorem of Heinich in chapter III extends to subpramarts without any additional hypothesis. This is a result of Słaby, solving a problem of Egghe. A second proof of Frangos is also presented, based on Talagrand's theorem that a Banach lattice with (RNP) is a separable dual. Both proofs of Słaby and Frangos are very recent and based on lemmas of Egghe. This result is applied by Słaby to yield a partial result on Sucheston's problem on pramartconvergence in Banach spaces. Frangos did the same, obtaining another partial result on the same problem. We now have that in a Banach space E with (RNP), every pramart with an L^1_E-bounded subsequence converges strongly a.e. provided that E is weakly sequentially complete (Słaby) or if E is isomorphic with a subspace of a separable dual (Frangos). Positive superpramarts converge a.e. in a Banach lattice only in the case when E is a sublattice of $\ell^1(\Gamma)$ (Egghe). Finally a mean convergence result of Egghe is proved for games which become worse (better) with time, extending a result of Subramanian, without actually using it.

Chapter IX closes the study developed in this book by making some remarks. First scalar convergence results are proved for the earlier introduced types of adapted sequences. Then a summary of the most important convergence results proved in this book is given. Finally the concept of Pettis integrable function - extending the notion of Bochner integrable function - is defined and some properties are mentioned. Also some results and literature on convergence of adapted sequences of Pettis integrable functions are mentioned, indicating that there is still a lot which has to be done in this area.

Chapter I : TYPES OF CONVERGENCE

I.1. Introduction

In this work, E will always stand for a real Banach space unless otherwise stated. Most of the results in this book are true however in complex Banach spaces without any real change. The norm on E is denoted by $\|\cdot\|$. When we say that F is a subspace of E we shall always mean a Banach subspace of E. The dual of E is denoted by E'. $(\Omega,\mathcal{F},P)$ will denote a complete probability space. This means that $\mathcal{F}$ is a σ-algebra of subsets of the set Ω and that P is a countably additive measure with $P(\Omega) = 1$ such that if $A \in \mathcal{F}$, $P(A) = 0$ and $B \subset A$ imply $B \in \mathcal{F}$.

I.1.1. Measurable functions

Let χ_A denote the characteristic function of $A \in \mathcal{F}$, i.e. the function

$$\chi_A(\omega) \begin{cases} = 0 & \text{if } \omega \notin A \\ = 1 & \text{if } \omega \in A \end{cases}$$

If $A_i \in \mathcal{F}$ for each $i = 1,\ldots,n$ and if $x_i \in E$ for each $i = 1,\ldots,n$, we say that the function

$$f = \sum_{i=1}^{n} x_i \chi_{A_i}$$

is a stepfunction. Let a function $f : \Omega \to E$ be the almost everywhere (a.e.) limit of f_n, i.e. : there exists $N \in \mathcal{F}$ such that $P(N) = 0$ and such that for each $\omega \in \Omega \setminus N$ and each $\varepsilon > 0$ there exists $n_o(\varepsilon) \in \mathbb{N}$ such that for each $n \geqslant n_o(\varepsilon)$:

$$\|f(\omega) - f_n(\omega)\| \leqslant \varepsilon$$

Then we say that f is <u>strongly measurable</u> or <u>measurable</u>, in case there cannot be any confusion. If $x'(f)$ is measurable for each $x' \in E'$, then we say that f is <u>scalarly measurable</u>.

Hence, in the definitions of scalar measurability, the nullset is dependent on the dual element x'. We obviously have that a strongly measurable function is scalarly measurable. The converse is not true, see Diestel and Uhl [1977], p.43. We shall often be dealing with scalar measurability as a working tool, but in this book only sequences of strongly measurable functions will be studied. A connection between strong and scalar measurability is given by Pettis' theorem.

<u>Theorem I.1.1.1</u> (B.J. Pettis) : Let $f : \Omega \to E$ be a function. The following two assertions are equivalent :

(i) f is strongly measurable.

(ii) f is scalarly measurable and f is <u>essentially separably valued</u>. This means : there exists $A \in F$, $P(A) = 0$ such that $f(\Omega \setminus A)$ is separable.

For a proof, see Diestel and Uhl [1977], p.42. Hence in a separable Banach space E the two notions coincide.

I.1.2. <u>Integrable functions</u>

Let $f : \Omega \to E$ be a function. We say that f is <u>Bochner-integrable</u> (w.r.t.P), or simply <u>integrable</u>, if there exists a sequence $(f_n)_{n\in \mathbb{N}}$ of stepfunctions such that

(i) $f_n \to f$, P-a.e., for $n \to \infty$

(ii) $\int_\Omega \|f_n - f_m\| dP \to 0$, for $m,n \to \infty$.

Instead of (ii) we might have required :

(ii)' $\int_\Omega \|f_n - f\| dP \to 0$, for $n \to \infty$, as is easily seen. Note that (i) implies that f is measurable. From (ii) (or (ii)') we can define the <u>Bochner integral</u> of f over $A \in F$ (or simply <u>integral of f over A</u>) :

$$\int_A f\, dP = \lim_{n\to\infty} \int_A f_n\, dP$$

where, if f is a stepfunction, say

$$f = \sum_{i=1}^{n} x_i \chi_{A_i}$$

we define

$$\int_A f\, dP = \sum_{i=1}^{n} x_i\, P(A_i \cap A)$$

Theorem I.1.2.1 (S. Bochner) : Let $f : \Omega \to E$ be a function. Then f is integrable iff f is measurable and $\int_\Omega \|f\| dP < \infty$.

We also have the following easily proved result :

Theorem I.1.2.2 : Suppose that $f : \Omega \to E$ is integrable. Then $f = 0$, P-a.e. iff $\int_\Omega \|f\| dP = 0$.

Denote by $\mathcal{L}^1_E(\Omega, F, P)$, or $\mathcal{L}^1_E$ if there is no confusion, the set of all integrable functions. The natural operations (defined pointwise) make $\mathcal{L}^1_E$ into a vector space. It is semi-normed by the function $f \to \int_\Omega \|f\| dP$. Hence $L^1_E(\Omega, F, P) = \mathcal{L}^1_E(\Omega, F, P)|_N$, or L^1_E for short, is a normed space with

$$N = \{f \in \mathcal{L}^1_E | f = 0, \text{ P-a.e.}\} .$$

This follows trivially from theorem I.1.2.2. We denote the norm on L^1_E by $\|\cdot\|_1$.

Furthermore, for $1 < p < \infty$, define L^p_E to be the space of those functions f in L^1_E such that $\|f\|_p = (\int_\Omega \|f\|^p dP)^{1/p} < \infty$. We put on L^p_E the norm $\|\cdot\|_p$. Also, for $p = \infty$, we define L^∞_E to be the space of those functions f in L^1_E such that $\|f\|_\infty < \infty$. Here

$$\|f\|_\infty = \inf \{\lambda | P(\|f\| > \lambda) = 0\}$$

Also $L^\infty_E, \|\cdot\|_\infty$ is a normed space. If $1 \leq p \leq \infty$ and $E = \mathbb{R}$ we denote $L^p_{\mathbb{R}} = L^p$. We have, as is well known :

Theorem I.1.2.3 : L_E^p is a Banach space ($1 \leqslant p \leqslant \infty$), for every Banach space E. As in the scalar case we recall that simple functions are dense in L_E^p ($1 \leqslant p < \infty$).

We note also the following result (see f.i. Dinculeanu [1967], p.187).

Theorem I.1.2.4 : Suppose that $f \in L_E^1$. Then $f = 0$, a.e. iff $\int_A f \, dP = 0$, for each $A \in \mathcal{F}$.

Let $(f_n)_{n \in \mathbb{N}}$ be a sequence in L_E^1. We say that $(f_n)_{n \in \mathbb{N}}$ is uniformly integrable if

$$\lim_{\lambda \to \infty} \sup_{n \in \mathbb{N}} \int_{\{\|f_n\| > \lambda\}} \|f_n\| dP = 0$$

The following result is well known (see f.i. Chung [1974], p.96-97).

Theorem I.1.2.5 : Let $(f_n)_{n \in \mathbb{N}}$ be a sequence in L_E^1. $(f_n)_{n \in \mathbb{N}}$ is uniformly integrable iff the following two conditions are satisfied :

(i) $$\sup_{n \in \mathbb{N}} \int_\Omega \|f_n\| dP < \infty$$

(ii) $$\lim_{P(A) \to 0} \sup_{n \in \mathbb{N}} \int_A \|f_n\| dP = 0,$$ i.e. : For each $\varepsilon > 0$, there is a $\delta(\varepsilon) > 0$ such that for any $A \in \mathcal{F}$ with $P(A) \leqslant \delta(\varepsilon)$ we have :

$$\sup_{n \in \mathbb{N}} \int_A \|f_n\| dP \leqslant \varepsilon .$$

The following result is well known (see f.i. Breiman [1978], p.91-92).

Theorem I.1.2.6 : Let $(f_n)_{n \in \mathbb{N}}$ be uniformly integrable. Suppose $f_n \to f$ a.e.. Then

$$\lim_{n \to \infty} \int_\Omega \|f_n - f\| dP = 0 .$$

From this result, the Lebesgue dominated convergence theorem is an obvious corollary.

On L^1_E we consider a weaker norm, called the Pettis norm :

$$\|\cdot\|_{Pe} : L^1_E \to \mathbb{R}^+$$

$$f \to \|f\|_{Pe} = \sup_{\substack{x' \in E' \\ \|x'\| \leq 1}} \int_\Omega |x'(f)|dP$$

We have $\|f\|_{Pe} \leq \|f\|_1$, but $\|\cdot\|_{Pe}$ and $\|\cdot\|_1$ are not equivalent norms, as we shall see later. That $\|\cdot\|_{Pe}$ is indeed a norm on L^1_E is seen from the following result and theorem I.1.2.4 :

Theorem I.1.2.7 : $\|\cdot\|_{Pe}$ is equivalent to the seminorm on $\mathcal{L}^1_E$ (or norm on L^1_E)

$$f \to \sup_{A \in F} \|\int_A f\, dP\|$$

Indeed :

$$\sup_{A \in F} \|\int_A f\, dP\| \leq \|f\|_{Pe} \leq 2 \sup_{\substack{x' \in E' \\ \|x'\| \leq 1}} \sup_{A \in F} |\int_A x'(f)dP|$$

$$= 2 \sup_{A \in F} \|\int_A f\, dP\|$$

The second inequality is a standard inequality for scalar measures; see f.i. Dunford and Schwartz [1957] , p.97. We have said already that $\|\cdot\|_{Pe}$ and $\|\cdot\|_1$ are not equivalent norms. In fact more is true :

Theorem I.1.2.8 : The following assertions are equivalent :

(i) $\|\cdot\|_{Pe}$ on L^1_E is complete.

(ii) $\|\cdot\|_{Pe}$ and $\|\cdot\|_1$ are equivalent.

(iii) $\dim E < \infty$.

Proof : (iii) ⇒ (ii) ⇒ (i) is obvious.

(i) ⇒ (ii) follows from the closed graph theorem.

(ii) ⇒ (iii). Suppose $\dim E = \infty$. From the well-known Dvoretzky-Rogers theorem there exists a sequence $(x_n)_{n \in \mathbb{N}}$ in E such that Σx_n is unconditionally convergent and such that $\Sigma\|x_n\| = \infty$; see e.g. Lindenstrauss and Tzafriri [1979], p.16 or Dvoretzky and Rogers [1950]. From the unconditional convergence, it follows that :

$$\sup_{\substack{x' \in E' \\ \|x'\| \leq 1}} \Sigma |x'(x_n)| < \infty$$

Take now $\Omega = (0,1]$ with Lebesgue measure, and take for $n = 1,2,\ldots$

$$A_n = (\frac{1}{2^n}, \frac{1}{2^{n-1}}]$$

Put, for every $n \in \mathbb{N}$

$$f_n = \sum_{k=1}^{n} 2^k x_k \chi_{A_k}$$

We have

$$\sup_{n \in \mathbb{N}} \|f_n\|_{Pe} = \sup_{n \in \mathbb{N}} \sup_{\substack{\|x'\| \leq 1 \\ x' \in E'}} \int_0^1 |x'(f_n)| =$$

$$= \sup_{n \in \mathbb{N}} \sup_{\substack{\|x'\| \leq 1 \\ x' \in E'}} \sum_{k=1}^{n} |x'(x_k)| < \infty$$

while

$$\sup_{n \in \mathbb{N}} \|f_n\|_1 = \sum_{n=1}^{\infty} \|x_n\| = \infty ,$$

contradicting (ii). □

This method of proof will also be used in a more intricate way later on in this book.

I.2. Adapted sequences

I.2.1. Definitions

Suppose that $(F_n)_{n \in \mathbb{N}}$ is an increasing sequence of sub-σ-algebras of the σ-algebra F. $(F_n)_{n \in \mathbb{N}}$ is called a stochastic basis. Suppose that $(X_n)_{n \in \mathbb{N}}$ is a sequence of integrable functions. Hence each X_n belongs to $L^1_E(\Omega,F,P)$. If every X_n is F_n-measurable, i.e., is the a.e. limit in E of a sequence of stepfunctions with steps in F_n, we say that

$(X_n, F_n)_{n\in\mathbb{N}}$ is an adapted sequence or a stochastic process. We remark that any sequence $(X_n)_{n\in\mathbb{N}}$ in L^1_E can be considered as an adapted sequence w.r.t. $(F_n)_{n\in\mathbb{N}}$, where

$$F_n = \sigma(X_1,\dots,X_n)$$

for every $n\in\mathbb{N}$. Here $\sigma(X_1,\dots,X_n)$ denotes the smallest σ-algebra keeping every $X_1,\dots,X_n$ measurable.

A function $\tau : \Omega \to \mathbb{N}\cup\{+\infty\}$ is called a stopping time w.r.t. $(F_n)_{n\in\mathbb{N}}$ if for each $n\in\mathbb{N}$, $\{\tau = n\}\in F_n$. Here we denote $\{\tau = n\} = \{\omega\in\Omega \mid \tau(\omega) = n\}$. The set of all stopping times is denoted by T^*. We order T^* as follows : for $\sigma,\tau\in T^*$ we denote $\sigma\leqslant\tau$ if $\sigma(\omega)\leqslant\tau(\omega)$ for each $\omega\in\Omega$. The subset of T^* consisting of all bounded stopping times is denoted by T. So $\tau\in T$ if $\tau\in T^*$ and $\tau(\Omega)$ is a finite subset of $\mathbb{N}$. The order induced by T^* on T has the property that $\mathbb{N}$ is cofinal in T. For $\tau\in T$, denote

$$F_\tau = \{A\in F \mid A\cap\{\tau = n\}\in F_n, \text{ for each } n\in\mathbb{N}\}$$

and

$$X_\tau = \sum_{n=\min\tau}^{\max\tau} X_n\, \chi_{\{\tau=n\}} .$$

This means, for each $\omega\in\Omega$: $(X_\tau)(\omega) = X_{\tau(\omega)}(\omega)$. It is obvious that X_τ is F_τ-measurable and that X_τ is integrable. If $\sup_{\tau\in T}\int_\Omega \|X_\tau\| < \infty$ then $(X_n,F_n)_{n\in\mathbb{N}}$ is said to be of class (B).

Stopping times satisfy a "localization property" : Let $\{\sigma_1,\dots,\sigma_n\}\subset T$ and $A_i\in F_{\sigma_i}$, for each $i = 1,\dots,n$ such that $(A_i)_{i=1,\dots,n}$ is a partition of Ω. Then $\tau\in T$ if $\tau =: \sigma_i$ on A_i. Indeed, for each $k\in\mathbb{N}$:

$$\{\tau = k\} = \bigcup_{i=1}^{n} [A_i\cap\{\sigma_i = k\}]$$

$$\in F_k$$

For the same reason we have that, if σ and $\tau\in T$, then $\{\sigma=\tau\}\in F_\sigma\cap F_\tau$. We denote $F_\infty = \sigma(\bigcup_n F_n)$, the σ-algebra generated by $(F_n)_{n\in\mathbb{N}}$. If $\sigma\in S$, a subset of T, denote $S(\sigma) = \{\tau\in S \,\|\, \tau\geqslant\sigma\}$.

All these notions can be generalized to adapted nets $(X_i,F_i)_{i\in I}$, where I is a directed index set, $F_i\subset F_j\subset F$ if $i\leqslant j$, $i,j\in I$ and X_i is F_i-measurable; for this, see chapter VI.

I.2.2. Conditional expectations

Let $\mathcal{G}$ be a sub-σ-algebra of the σ-algebra $\mathcal{F}$. We state and prove the following existence theorem for conditional expectations of functions in L^1_E.

Theorem I.2.2.1 : There exists a unique map :

$$E^{\mathcal{G}} : L^1_E(\Omega,\mathcal{F},P) \to L^1_E(\Omega,\mathcal{G},P/\mathcal{G})$$

such that for each $A \in \mathcal{G}$ and each $f \in L^1_E(\Omega,\mathcal{F},P)$ we have :

$$\int_A f\, dP = \int_A E^{\mathcal{G}}(f)\, dP$$

Furthermore for each $p \in [1,+\infty)$, $E^{\mathcal{G}} \in \mathcal{L}(L^p_E(\Omega,\mathcal{F},P), L^p_E(\Omega,\mathcal{G},P/\mathcal{G}))$ and $\|E^{\mathcal{G}}\|_p \leq 1$, where $\mathcal{L}(\cdot,\cdot)$ denotes the space of the continuous linear operators between two Banach spaces and $P/\mathcal{G}$ denotes the restriction of P to the sub-σ-algebra $\mathcal{G}$.

Proof : Fix $A \in \mathcal{F}$. Define

$$\mu : \mathcal{G} \to [0,1]$$

by $\mu(B) = P(A \cap B)$ for $B \in \mathcal{G}$. Obviously $\mu \ll P$ on $\mathcal{G}$. So, by the Radon-Nikodym theorem, there exists a function $\varphi_A \in L^1(\Omega,\mathcal{G},P/\mathcal{G})$ such that :

$$\mu(B) = P(A \cap B) = \int_B \varphi_A\, dP$$

for each $B \in \mathcal{G}$. Let f be a stepfunction of the form

$$f = \sum_{i=1}^{n} x_i \chi_{A_i}$$

where $x_i \in E$ and $A_i \in \mathcal{F}$. Define

$$E^{\mathcal{G}}(f) = \sum_{i=1}^{n} x_i \varphi_{A_i} .$$

Then we see that the conditional expectation property is fullfilled on stepfunctions. Extension to $L^1_E(\Omega,\mathcal{F},P)$ is possible since stepfunctions are dense in this space. The uniqueness of $E^{\mathcal{G}}$ is obvious from theorem I.1.2.4. Linearity is obvious too. So we only have to show that

$$\|E^{\mathcal{G}}\|_p \leq 1$$

for each $p \in [1,+\infty)$. Suppose first that $f \in L^p$ and let $at+b$ be a support line of $\Phi(t) = t^p$. Then

$$aE^{\mathcal{G}}(f) + b = E^{\mathcal{G}}(af+b) \leq E^{\mathcal{G}}(|f|^p)$$

Hence $|E^{\mathcal{G}}f|^p \leq E^{\mathcal{G}}(|f|^p)$, by taking suprema over the support lines. This inequality is called Jensen's inequality and is valid for any convex function Φ. From this :

$$\int_\Omega |E^{\mathcal{G}}(f)|^p \leq \int_\Omega |f|^p \tag{1}$$

Now let f be an E-valued stepfunction $f = \sum_{i=1}^n x_i \chi_{A_i}$. Then :

$$\|E^{\mathcal{G}}(f)\|_p \leq (\int_\Omega (\sum_{i=1}^n \|x_i\| E^{\mathcal{G}}(\chi_{A_i}))^p \, dP)^{1/p}$$

$$= (\int_\Omega (E^{\mathcal{G}} (\sum_{i=1}^n \|x_i\| \chi_{A_i}))^p \, dP)^{1/p}$$

$$\leq (\int_\Omega \|f\|^p \, dP)^{1/p} = \|f\|_p ,$$

the last inequality by (1). This finishes the proof since stepfunctions are dense in L^p_E. □

$E^{\mathcal{G}}(f)$ is called the conditional expectation of f w.r.t. $\mathcal{G}$.

Example I.2.2.2 : If we start with $(F_n)_{n\in\mathbb{N}}$ as in I.2.1 and if we take $f \in L^1_E(\Omega,F,P)$, the sequence $(E^{F_n}(f),F_n)_{n\in\mathbb{N}}$ is an adapted sequence. In fact it is a martingale, converging to f a.e. and in the $\|\cdot\|_1$ sense as we shall see later in chapter II.

Properties I.2.2.3 : We have the following properties of conditional expectations $E^{\mathcal{G}}$:

(i) $E^G E^G = E^G$

(ii) $E^G(x) = x$ where $x \in E$

(iii) E^G is linear.

(iv) E^G is a contraction.

It is interesting to note that these properties characterize conditional expectation operators in $\mathcal{L}(L^1_E(\Omega,F,P), L^1_E(\Omega,F,P))$ if E has a strictly convex norm, i.e. : the unit sphere $\{x \in E \mid \|x\| = 1\}$ contains no straight lines. For the proof, see Landers and Rogge [1981] . For properties of strict convexity, see Diestel [1975b] , from p.23 on.

I.3. Convergence

In this section we shall discuss some types of convergence which will be studied later on in the book. $(X_n,F_n)_{n \in \mathbb{N}}$ denotes an adapted sequence with values in the Banach space E. One might as well take any sequence $(X_n)_{n \in \mathbb{N}}$ in L^1_E, but we study only adapted sequences in this book. On the other hand $(X_n)_{n \in \mathbb{N}}$ is always adapted w.r.t. $(F_n)_{n \in \mathbb{N}}$, where $F_n = \sigma(X_1,\ldots,X_n)$. Relations between different types of convergence will only be stated if they are well known. For comparatively new results we shall provide a proof.

I.3.1. Pointwise convergence

$(X_n,F_n)_{n \in \mathbb{N}}$ is said to be strongly convergent a.e. to a (measurable) function X_∞ if there is a nullset N such that for each $\omega \in \Omega \setminus N$ and each $\varepsilon > 0$, there exists an $n_o \in \mathbb{R}$ such that for each $n \in \mathbb{N}(n_o)$, $\|X_\infty(\omega) - X_n(\omega)\| \leq \varepsilon$. $(X_n,F_n)_{n \in \mathbb{N}}$ is said to be weakly convergent a.e. to a (measurable) function X_∞ if there is a nullset N such that for each $\omega \in \Omega \setminus N$ and each weak-zero-neighbourhood V, there exists an $n_o \in N$ such that for each $n \in \mathbb{N}(n_o)$, $X_\infty(\omega) - X_n(\omega) \in V$. $(X_n,F_n)_{n \in \mathbb{N}}$ is said to be scalarly convergent a.e. to a (not necessarily measurable) function X_∞ if for every $x' \in E'$, $(x'(X_n))_{n \in \mathbb{N}}$ converges a.e. to $x'(X_\infty)$. Note that X_∞ is always scalarly measurable.

I.3.2. Mean convergence

$(X_n, F_n)_{n\in\mathbb{N}}$ is said to be mean convergent to X_∞ (or $\|\cdot\|_1$-convergent) if

$$\lim_{n\to\infty} \|X_n - X_\infty\|_1 = 0$$

$(X_n, F_n)_{n\in\mathbb{N}}$ is said to be mean convergent to X_∞ of order p ($p > 1$) or $\|\cdot\|_p$-convergent) if

$$\lim_{n\to\infty} \|X_n - X_\infty\|_p = 0$$

One supposes here of course that $X_n \in L_E^p$ for each $n \in \mathbb{N}$. It is easily seen that $\|\cdot\|_p$-convergence ($p > 1$) implies $\|\cdot\|_1$-convergences. Also if $(X_n)_{n\in\mathbb{N}}$ is $\|\cdot\|_1$-convergent to X_∞, there exists a subsequence $(n_k)_{k\in\mathbb{N}}$ such that $(X_{n_k})_{k\in\mathbb{N}}$ converges a.e. to X_∞. This result however can be generalized (see I.3.4). It is also worth noting that for each $f \in L_E^\infty$: $\|f\|_\infty = \lim_{p\to\infty} \|f\|_p$.

I.3.3. Pettis-convergence

$(X_n, F_n)_{n\in\mathbb{N}}$ is said to be Pettis-convergent to X_∞ if

$$\lim_{n\to\infty} \|X_n - X_\infty\|_{Pe} = 0$$

It is obvious that $\|\cdot\|_1$-convergence implies Pettis-convergence, but from theorem I.1.2.8 not conversely in infinite dimensional Banach spaces. $(X_n, F_n)_{n\in\mathbb{N}}$ is said to be Pettis-Cauchy if $(X_n)_{n\in\mathbb{N}}$ is a $\|\cdot\|_{Pe}$-Cauchy sequence. Also from I.1.2.8 we see that this property is strictly weaker than Pettis-convergence.

I.3.4. Convergence in probability

$(X_n, F_n)_{n\in\mathbb{N}}$ is said to be convergent to X_∞ in probability if for every $\varepsilon > 0$, there exists an $n_0 \in \mathbb{N}$ such that if $n \in \mathbb{N}(n_0)$

$$P(\{\|X_n(\omega) - X_\infty(\omega)\| \geqslant \varepsilon\}) \leqslant \varepsilon$$

Pointwise convergence implies convergence in probability, but not conversely. However, if $(X_n)_{n\in\mathbb{N}}$ convergesto X_∞ in probability, then $(X_n)_{n\in\mathbb{N}}$ has an a.e.-convergent subsequence. Also in L^1_E, the convergence in probability induces a metric on L^1_E, namely the metric, where $X,Y \in L^1_E$:

$$d(X,Y) = \int_\Omega \frac{\|X-Y\|}{1+\|X-Y\|}\,dP$$

which is complete on the space of measurable functions.
Also, $\|\cdot\|_1$-convergence implies convergence in probability. All the above results are well-known and can be found in the classical literature on measure theory or probability theory. They are valid for any sequence in L^1_E.

I.3.5. Convergence in probability in the stopping time sense

More specifically for adapted sequences we can define some other convergence properties : $(X_n,F_n)_{n\in\mathbb{N}}$ is said to converge to X_∞ in probability in the stopping time sense if for every $\varepsilon > 0$, there exists a $\sigma_o \in T$ such that if $\sigma \in T(\sigma_o)$

$$P(\{\|X_\sigma(\omega) - X_\infty(\omega)\| \geqslant \varepsilon\}) \leqslant \varepsilon \ .$$

$(X_n,F_n)_{n\in\mathbb{N}}$ is said to converge to X_∞ a.e. in the stopping time sense if there is a nullset N such that for every $\omega \in \Omega\setminus N$ and every $\varepsilon > 0$, there exists a $\sigma_o \in T$ such that if $\sigma \in T(\sigma_o)$

$$\|X_\sigma(\omega) - X_\infty(\omega)\| \leqslant \varepsilon$$

However this convergence is obviously seen to be equivalent to pointwise convergence to X_∞. Surprisingly however convergence in probability in the stopping time sense is also equivalent to pointwise convergence. This was proved by Millet and Sucheston in 1979, see Millet and Sucheston [1980a] . Later an easier proof was given by Bellow and Egghe [1982] (with a small correction). It is this proof we shall present here. Some preliminary results are needed. They have independent interest and they will also be used on some other occasions.

Lemma I.3.5.1 (Edgar and Sucheston [1976a]), Austin, Edgar and Ionescu-Tulcea [1974]) : Let $(X_n, F_n)_{n\in\mathbb{N}}$ be an adapted sequence with values in the Banach space E. Let X be an F_∞ measurable function.
Suppose also that for each $\omega \in \Omega$, $X(\omega)$ is a clusterpoint of $(X_n(\omega))_{n\in\mathbb{N}}$. In case $E = \mathbb{R}$, $X(\omega)$ may have values in $\overline{\mathbb{R}} = \mathbb{R} \cup \{-\infty, +\infty\}$). Then for each $\varepsilon > 0$, $\delta > 0$ and $m \in \mathbb{N}$, there exists $\tau \in T(m)$ such that

$$P(\{\omega \mid \|X_{\tau(\omega)}(\omega) - X(\omega)\| > \delta\}) \leqslant \varepsilon$$

Proof : Choose $N \in \mathbb{N}(m)$ and a F_N-measurable function X' such that

$$P(\{\omega \in \Omega \mid \|X(\omega) - X'(\omega)\| < \tfrac{\delta}{2}\}) > 1 - \tfrac{\varepsilon}{2}$$

Since $\{\omega \mid \|X(\omega) - X'(\omega)\| < \frac{\delta}{2}\} \subset \bigcup_{n=N}^{\infty} \{\omega \mid \|X_n(\omega) - X'(\omega)\| < \frac{\delta}{2}\}$ there exists $N' \in \mathbb{N}(N)$ such that

$$P(\bigcup_{n=N}^{N'} \{\omega \mid \|X_n(\omega) - X'(\omega)\| < \tfrac{\delta}{2}\}) > 1 - \tfrac{\varepsilon}{2}$$

Let $\tau(\omega)$ be the smallest integer n such that $N \leqslant n \leqslant N'$ and $\|X_n(\omega) - X'(\omega)\| < \frac{\delta}{2}$ if it exists. Otherwise let $\tau(\omega) = N'$. So

$$P(\{\omega \mid \|X_{\tau(\omega)}(\omega) - X'(\omega)\| < \tfrac{\delta}{2}\}) > 1 - \tfrac{\varepsilon}{2}$$

Hence , $\tau \in T(m)$ and

$$P(\{\omega \mid \|X_{\tau(\omega)}(\omega) - X(\omega)\| < \delta\}) > 1 - \varepsilon \qquad \square$$

Corollary I.3.5.2 : Let X and $(X_n, F_n)_{n\in\mathbb{N}}$ be as in I.3.5.1. Then there is a strictly increasing sequence $(\tau_n)_{n\in\mathbb{N}}$ in T with $\tau_n \geqslant n$, $\forall n \in \mathbb{N}$, such that

$$X(\omega) = \lim_{n\to\infty} X_{\tau_n}(\omega), \text{ a.e. .}$$

For instance, this is the case if $E = \mathbb{R}$ and for $X(\omega) = \limsup_{n\in\mathbb{N}} X_n(\omega)$.

From now on - whenever there cannot be any confusion - we write $P(X > \lambda)$ instead of $P(\{\omega \| X(\omega) > \lambda\})$, if X is a function on Ω.

Corollary I.3.5.3 : For each $\sigma \in T$, let $g(\sigma)$ be a real valued F_σ-measurable function. Assume that for every $n \in \mathbb{N}$:

$$\chi_{\{\sigma=n\}} g(\sigma) = \chi_{\{\sigma=n\}} g(n) \qquad (1)$$

Then for each $\lambda > 0$

$$P(\limsup_{\sigma \in T} g(\sigma) > \lambda) \leq \limsup_{\sigma \in T} P(g(\sigma) > \lambda)$$

Proof : From (1) we see that

$$\limsup_{\sigma \in T} g(\sigma) = \limsup_{n \in \mathbb{N}} g(n)$$

Fix $\sigma_o \in T$. From I.3.5.2, there exists an increasing sequence $(\sigma_n)_{n \in \mathbb{N}}$ in $T(\sigma_o)$ such that

$$\limsup_{n \in \mathbb{N}} g(n) = \lim_{n \to \infty} g(\sigma_n), \text{ a.e. } .$$

So, for each $\lambda > 0$:

$$\{\limsup_{\sigma \in T} g(\sigma) > \lambda\} \subset \liminf_{n \in \mathbb{N}} \{g(\sigma_n) > \lambda\} \text{ , a.e. } .$$

So, it follows from Fatou's lemma that

$$P(\limsup_{\sigma \in T} g(\sigma) > \lambda) \leq \liminf_{n \in \mathbb{N}} P(g(\sigma_n) > \lambda) \qquad (1)$$

$$\leq \sup_{\sigma \in T(\sigma_o)} P(g(\sigma) > \lambda)$$

Since $\sigma_o \in T$ was arbitrary, the corollary follows. □

Lemma I.3.5.4 : Let $\sigma \in T$ and $(g(\tau))_{\tau \in T(\sigma)}$ be a family of E-valued F_σ-measurable functions. Assume that

If $A \in F_\sigma$ and $\tau', \tau'' \in T(\sigma)$ are such that $\tau'(\omega) = \tau''(\omega)$ for $\omega \in A$, then

$$\chi_A\, g(\tau') = \chi_A\, g(\tau'') \qquad (2)$$

Then for each $\lambda > 0$ we have :

$$P(\sup_{\tau\in T(\sigma)} \|g(\tau)\| > \lambda) \leq \sup_{\tau\in T(\sigma)} P(\|g(\tau)\| > \lambda)$$

Proof : We shall show first that there exists a sequence $(\sigma_n)_{n\in\mathbb{N}}$ in $T(\sigma)$ such that

$$(*) \qquad \sup_{\tau\in T} \|g(\tau)\| = \sup_{n\in\mathbb{N}} \|g(\sigma_n)\|, \text{ a.e. }.$$

For this we can assume that all the functions $\|g(\tau)\|$ with $\tau \in T(\sigma)$ have values in $[0,1]$, since we can always use an increasing bijection between $\mathbb{R}$ and $[0,1]$. For every countable subset C of $T(\sigma)$, define $\sup_{\tau\in C} \|g(\tau)\| =: g_C^*$. Let Γ be the set of all countable subsets of $T(\sigma)$ and put $\alpha = \sup_{C\in\Gamma} \int g_C^*$. If $(C_n)_{n\in\mathbb{N}}$ is a sequence in Γ such that $\lim_{n\to\infty} \int g_{C_n}^* = \alpha$, then $\int g_{C_\infty}^* = \alpha$, where $C_\infty = \bigcup_{n\in\mathbb{N}} C_n \in \Gamma$. It is now clear that (*) holds where $\{\sigma_n | n\in\mathbb{N}\} = C_\infty$. Indeed, for every $\tau \in T(\sigma)$ we have $\alpha = \int g_{C_\infty}^* \leq \int g_{C_\infty}^* \vee \|g(\tau)\| \leq \alpha$; so $g_{C_\infty} \geq g(\tau)$, a.s., proving (*). From (*) we have :

$$\sup_{\tau\in T(\sigma)} \|g(\tau)\| = \lim_{n\to\infty} \sup_{k=1,\dots,n} \|g(\sigma_k)\| .$$

For each n, we can find a partition $\{A_1,\dots,A_n\}$ of Ω with $A_i \in F_\sigma$ for $1 \leq i \leq n$ and such that on A_i :

$$\|g(\sigma_i)\| = \sup_{k=1,\dots,n} \|g(\sigma_k)\|$$

It follows from the "localization property" in T - see I.2.1, and (2) that $\gamma_n \in T(\sigma)$ if we put $\gamma_n = \sigma_i$ on A_i. Clearly

$$\|g(\gamma_n)\| = \sup_{k=1,\dots,n} \|g(\sigma_k)\|$$

So, $(\|g(\gamma_n)\|)_{n\in\mathbb{N}}$ increases to $\sup_{\tau\in T(\sigma)} \|g(\tau)\|$. Hence, for each $\lambda > 0$:

$$P\left(\sup_{\tau \in T(\sigma)} \|g(\tau)\| > \lambda\right) = P\left(\lim_{n\to\infty} \|g(\gamma_n)\| > \lambda\right)$$

$$= \lim_{n\to\infty} P(\|g(\gamma_n)\| > \lambda) \qquad (1)$$

$$\leq \sup_{\tau \in T(\sigma)} P(\|g(\tau)\| > \lambda) . \ \square$$

We now have enough machinery to prove the main result of this section :

Theorem I.3.5.5 (Millet-Sucheston) : For each $\sigma \in T$ and $\tau \in T(\sigma)$, let $f(\sigma,T)$ be an E-valued, F_σ -measurable function. Assume that the family $(f(\sigma,\tau))_{\substack{\sigma \in T \\ \tau \in T(\sigma)}}$ satisfies :

(1) For every $\sigma \in T$, $\tau \in T(\sigma)$ and $n \in \mathbb{N}$

$$\chi_{\{\sigma=n\}} f(n,\tau) = \chi_{\{\sigma=n\}} f(\sigma,\tau)$$

(i.e. the localization property in the first variable).

(2) For every $\sigma \in T$, $A \in F_\sigma$ and $\tau',\tau'' \in T(\sigma)$ such that $\tau'(\omega) = \tau''(\omega)$ on A we have :

$$\chi_A f(\sigma,\tau') = \chi_A f(\sigma,\tau'')$$

(i.e. the localization property in the second variable).

If $(f(\sigma,\tau))_{\substack{\sigma \in T \\ \tau \in T(\sigma)}}$ converges to f_∞ in probability then $(f(\sigma,\tau))_{\substack{\sigma \in T \\ \tau \in T(\sigma)}}$ converges a.e. to f_∞ (both convergences are meant in $\sigma \in T$, uniformly in $\tau \in T(\sigma)$).

Proof : For each $\varepsilon > 0$, choose $n_o \in \mathbb{N}$ and an F_{n_o} -measurable function f_{n_o} such that

$$P(\|f_{n_o} - f_\infty\| > \tfrac{\varepsilon}{4}) \leq \tfrac{\varepsilon}{4}$$

By the convergence in probability of $f(\sigma,\tau)$ in f_∞ it follows that there is a $\sigma_o \in T(n_o)$ such that for each $\sigma \in T(\sigma_o)$ and $\tau \in T(\sigma)$:

$$P(\|f(\sigma,\tau) - f_\infty\| > \frac{\varepsilon}{4}) \leq \frac{\varepsilon}{4}$$

whence

$$P(\|f(\sigma,\tau) - f_{n_0}\| > \frac{\varepsilon}{2}) \leq \frac{\varepsilon}{2}$$

Using (1), (2), I.3.5.3 and I.3.5.4 we see :

$$P(\limsup_{\sigma \in T} (\sup_{\tau \in T(\sigma)} \|f(\sigma,\tau) - f_{n_0}\|) > \frac{\varepsilon}{2})$$

$$\leq \limsup_{\sigma \in T} P(\sup_{\tau \in T(\sigma)} \|f(\sigma,\tau) - f_{n_0}\| > \frac{\varepsilon}{2})$$

$$\leq \limsup_{\sigma \in T} \sup_{\tau \in T(\sigma)} P(\|f(\sigma,\tau) - f_{n_0}\| > \frac{\varepsilon}{2}) \leq \frac{\varepsilon}{2}$$

So

$$P(\limsup_{\sigma \in T} (\sup_{\tau \in T(\sigma)} \|f(\sigma,\tau) - f_\infty\|) > \varepsilon) \leq \varepsilon . \quad \square$$

Corollary I.3.5.6 : The above theorem can and will be applied in the following important cases. Here $(X_n, \mathcal{F}_n)_{n \in \mathbb{N}}$ is an E-valued adapted sequence and $\sigma \in T$ and $\tau \in T(\sigma)$:

(i) $f(\sigma,\tau) = X_\sigma$;

(ii) $f(\sigma,\tau) = E^{\mathcal{F}_\sigma} X_\tau$;

(iii) $f(\sigma,\tau) = E^{\mathcal{F}_\sigma} X_\tau - X_\sigma$.

Corollaries I.3.5.7 (Edgar [1979])

(i) Under the conditions of lemma I.3.5.3 we also have :

$$\int_\Omega \limsup_{\sigma \in T} g(\sigma) \leq \limsup_{\sigma \in T} \int_\Omega g(\sigma)$$

(ii) Under the conditions of lemma I.3.5.4 we also have :

$$\int_\Omega \sup_{\tau \in T(\sigma)} \|g(\tau)\| = \sup_{\tau \in T(\sigma)} \int_\Omega \|g(\tau)\|$$

Proof : (i)

This follows from (1) in the proof of lemma I.3.5.3, Fatou's lemma and the identity for positive integrable functions f :

$$\int_0^\infty P(f > \lambda)d\lambda = \int_\Omega f \qquad (1)$$

which can be proved using Fubini.

(ii)

This follows from the fact that, in the notation of the proof of lemma I.3.5.4, $(\|g(\gamma_n)\|)_{n\in\mathbb{N}}$ increases to $\sup_{\tau\in T(\sigma)} \|g(\tau)\|$, and the monotone convergence theorem. □

I.4. Notes and Remarks

I.4.1. In some publications, adapted sequences of vector measures are studied. In this case we are given $(\mu_n, F_n)_{n\in\mathbb{N}}$ where $(F_n)_{n\in\mathbb{N}}$ is an increasing sequence of sub-σ-algebras of F and where μ_n is an E-valued measure on F_n. For $\tau \in T$ one defines

$$\mu_\tau(A) = \sum_{n=\min\tau}^{\max\tau} \mu_n(A\cap\{\tau = n\}) \quad .$$

See the work of Bru and Heinich [1979a], [1979b], Heinich [1978b], Chatterji [1971], Schmidt [1979a] to [1982]. Adapted sequences of additive measures on algebras are studied by the last author as is the convergence of their generalized Radon-Nikodym derivatives.

I.4.2. Convergence a.e. (or in probability) in the stopping time sense is called by Dvoretzky [1976] : S-convergence.

I.4.3. Theorem I.1.2.8 in Fréchet spaces is true when (iii) is changed into "E is nuclear"; the same proof works. For the definition and properties of nuclear spaces : see Pietsch [1972].

I.4.4. Theorem I.3.5.5 can be proved if we start with a net $(F_i)_{i \in I}$ of sub-σ-algebras of F and with appropriate change in the definitions of T, convergence in probability and convergence a.e.. Here we need the "Vitali-condition" on $(F_i)_{i \in I}$; see Bellow and Egghe [1982] or Millet and Sucheston [1980a]. The interested reader may see chapter VI, for more information concerning the Vitali-condition V.

Chapter II : MARTINGALE CONVERGENCE THEOREMS

This chapter is concerned with martingale convergence theorems with values in a Banach space E.

We start with some elementary results, concerned with martingales of the type $X_n = E^{F_n} X_\infty$, where $X_\infty \in L^1_E(\Omega, F, P)$. We then proceed by stating and proving the main martingale convergence results in Banach spaces with the Radon-Nikodym-property. To this end, some results concerning the Radon-Nikodym-property are reviewed.

In this connection the link between the geometry of Banach spaces and Banach-valued probability theory is indicated.

The chapter closes with some martingale convergence results in Banach spaces which do not necessarily have the Radon-Nikodym-property.

II.1. Elementary results

In this section we prove some easy convergence theorems on martingales with values in a Banach space. They have interest on their own, but they are also used in the next sections. Thus they prove to be extremely important in the general theory of martingale convergence.

Definition II.1.1 : Let $(X_n, F_n)_{n \in \mathbb{N}}$ be an E-valued adapted sequence. We say that $(X_n, F_n)_{n \in \mathbb{N}}$ is a martingale if

$$E^{F_n} X_{n+1} = X_n, \text{ a.e.}$$

for each $n \in \mathbb{N}$. Otherwise stated, $(X_n, F_n)_{n \in \mathbb{N}}$ is a martingale if

$$\int_A X_n = \int_A X_{n+1} ,$$

for each $A \in F_n$ and each $n \in \mathbb{N}$.

Examples II.1.2

1) Let $X \in L^1_E(\Omega, F, P)$ and suppose that $(F_n)_{n\in\mathbb{N}}$ is an increasing sequence of sub-σ-algebras of F. Then $(E^{F_n}X, F_n)_{n\in\mathbb{N}}$ is a martingale.

2) For any sequence $(x_n)_{n\in\mathbb{N}}$ in E, it follows that

$$X_1 = x_1\chi_{[0,1)},$$

$$X_2 = (x_1 - x_2)\chi_{[0,\frac{1}{2})} + (x_1+x_2)\chi_{[\frac{1}{2},1)},$$

$$X_3 = (x_1-x_2-x_3)\chi_{[0,\frac{1}{4})} + (x_1-x_2+x_3)\chi_{[\frac{1}{4},\frac{1}{2})} + (x_1+x_2-x_3)\chi_{[\frac{1}{2},\frac{3}{4})}$$

$$+ (x_1+x_2+x_3)\chi_{[\frac{3}{4},1)}, \ \ldots$$

is a martingale w.r.t. $F_n = \sigma(X_1,\ldots,X_n)$, the smallest σ-algebra making $X_1,\ldots,X_n$ measurable (take $F_\infty = \mathcal{B}[0,1)$, the Borel-$\sigma$-algebra on $[0,1)$).

In II.1.3, II.1.4 and II.1.5 we suppose that $F = F_\infty$ for convenience.

Theorem II.1.3 : The martingale in example II.1.2(1) converges to X in the norm $\|\cdot\|_p$, for each $1 \leq p < \infty$ if $X \in L^p_E$.

Proof : The martingale is certainly $\|\cdot\|_p$-convergent in case X is a $\cup F_n$-stepfunction, since in this case $(E^{F_n}X, F_n)_{n\in\mathbb{N}}$ is eventually constant. Since these functions are dense in $L^p_E(\Omega,F,P)$ and since $\|E^{F_n}\|_p \leq 1$ for each $n \in \mathbb{N}$ and $p \in [1,+\infty)$, a limiting argument gives the result. □

From II.1.3 it follows already that there is a subsequence of $(E^{F_n}X)_{n\in\mathbb{N}}$ which converges to X a.e.. However, the martingale itself converges to X a.e.. This will be proved in theorem II.1.6.

We shall need a very important maximal inequality but first an elementary inequality is proved for martingales.

Lemma II.1.4 : Let $(X_n, F_n)_{n\in\mathbb{N}}$ be a martingale and let $\sigma \in T$ and $\tau \in T(\sigma)$ be arbitrary. Then :

(i) $$E^{F_\sigma} X_\tau = X_\sigma$$

(ii) For every $A \in F_\sigma$

$$\int_A \|X_\sigma\| \leq \int_A \|X_\tau\|$$

Proof : (i) In fact, if $A \in F_\sigma$, define

$$\tau' \begin{cases} = \tau & \text{on } A \\ = \max \tau & \text{on } \Omega \setminus A \end{cases}$$

$$\sigma' \begin{cases} = \sigma & \text{on } A \\ = \max \tau & \text{on } \Omega \setminus A \end{cases}$$

So $\sigma' \in T$, $\tau' \in T(\sigma')$ and we have :

$$\int_A X_\tau = \int_A X_{\tau'} = \int_\Omega X_{\tau'} - \int_{\Omega\setminus A} X_{\tau'} = \int_\Omega X_{\tau'} - \int_{\Omega\setminus A} X_{\max \tau} \qquad (1)$$

$$\int_A X_\sigma = \int_A X_{\sigma'} = \int_\Omega X_{\sigma'} - \int_{\Omega\setminus A} X_{\sigma'} = \int_\Omega X_{\sigma'} - \int_{\Omega\setminus A} X_{\max \tau} \qquad (2)$$

Finally :

$$\int_\Omega X_{\tau'} = \sum_{k=\min \tau'}^{\max \tau'} \int_{\{\tau'=k\}} X_k = \sum_{k=\min \tau'}^{\max \tau'} \int_{\{\tau'=k\}} X_{\max \tau}$$

$$= \int_\Omega X_{\max \tau} = \int_\Omega X_1 .$$

So :

$$\int_\Omega X_{\tau'} = \int_\Omega X_{\sigma'} .$$

Hence (1) = (2) and so $E^{F_\sigma} X_\tau = X_\sigma$.

(ii) Follows readily from (1) since

$$\int_A \|X_\sigma\| = \int_A \|E^{F_\sigma} X_\tau\| \leq \int_A E^{F_\sigma}\|X_\tau\| = \int_A \|X_\tau\|$$

since $A \in F_\sigma$. □

We now come to the "maximal lemma". We present it in its full generality since its proof is very simple and since we need the full generality later on in the book. The general form is due to Chacon and Sucheston [1975] , the special form to Neveu [1965] .

Lemma II.1.5 (Chacon and Sucheston)

(i) Let $(X_n, F_n)_{n \in \mathbb{N}}$ be an adapted sequence of measurable functions of class (B).
Then, for every $\lambda > 0$,

$$P(\sup_{n \in \mathbb{N}} \|X_n\| > \lambda) \leq \frac{1}{\lambda} \sup_{\tau \in T} \int_\Omega \|X_\tau\| \tag{1}$$

(ii) In case $(X_n, F_n)_{n \in \mathbb{N}}$ is also a martingale (1) reduces to, for each $\lambda > 0$,

$$P(\sup_{n \in \mathbb{N}} \|X_n\| > \lambda) \leq \frac{1}{\lambda} \sup_{n \in \mathbb{N}} \int_\Omega \|X_n\| \tag{2}$$

(iii) In case the martingale $(X_n, F_n)_{n \in \mathbb{N}}$ is of the form $X_n = E^{F_n} X$ with $X \in L^1_E$ for each $n \in \mathbb{N}$, (2) reduces to, for each $\lambda > 0$,

$$P(\sup_{n \in \mathbb{N}} \|X_n\| > \lambda) \leq \frac{1}{\lambda} \int_\Omega \|X\| \quad . \tag{3}$$

Proof : Let $N \in \mathbb{N}$ be fixed. Put

$$A = \{\sup_{n \leq \mathbb{N}} \|X_n\| > \lambda\}$$

Define σ as follows :

$$\sigma(\omega) \begin{cases} = \min \{n \in \mathbb{N} \mid n \leq N \text{ and } \|X_n(\omega)\| > \lambda\}, \text{ if } \omega \in A \\ = N, \text{ if } \omega \notin A \end{cases}$$

Then $\sigma \in T$ and

$$\sup_{\tau \in T} \int_\Omega \|X_\tau\| \geq \int_\Omega \|X_\sigma\| \geq \int_A \|X_\sigma\| \geq \lambda P(A) .$$

If we let N go to ∞, (1) follows.

(ii) If we use lemma II.1.4 we see that for each $\tau \in T$

$$\int_\Omega \|X_\tau\| \leq \int_\Omega \|X_{\max \tau}\|$$

where $\max \tau \in \mathbb{N}$. So (1) gives :

$$P(\{\sup_{n \in \mathbb{N}} \|X_n\| > \lambda\}) \leq \frac{1}{\lambda} \sup_{\tau \in T} \int_\Omega \|X_\tau\| = \frac{1}{\lambda} \sup_{n \in \mathbb{N}} \int \|X_n\|$$

(iii) This follows since for each $n \in \mathbb{N}$

$$\int_\Omega \|X_n\| = \int_\Omega \|E^{F_n}X\| \leq \int_\Omega \|X\| \qquad \square$$

<u>Theorem II.1.6</u> : $(E^{F_n}X)_{n \in \mathbb{N}}$ converges to X, a.e..

<u>Proof</u> : Again as in II.1.3 we see that if X is a $\bigcup_n F_n$-stepfunction, the result is trivial. Now let $X \in L^1_E(\Omega, F, P)$ and let $\varepsilon > 0$ and $\delta > 0$ be given. Choose a $\bigcup_n F_n$-stepfunction X' such that $\|X - X'\|_1 \leq \frac{\varepsilon\delta}{2}$. From

$$X_n - X_m = (E^{F_n}X' - E^{F_m}X') + E^{F_n}(X - X') - E^{F_m}(X - X')$$

where we denoted $X_n = E^{F_n}X$, for every $n \in \mathbb{N}$, we have a.e.

$$\limsup_{m,n \in \mathbb{N}} \|X_n - X_m\| \leq 2 \sup_{n \in \mathbb{N}} \|E^{F_n}(X - X')\|$$

Using lemma II.1.5(iii),

$$P(\limsup_{m,n \in \mathbb{N}} \|X_n - X_m\| > \varepsilon) \leq P(\sup_{n \in \mathbb{N}} \|E^{F_n}(X - X')\| > \frac{\varepsilon}{2})$$

$$\leq \frac{2}{\varepsilon} \|X - X'\|_1 \leq \delta$$

Hence, $(X_n)_{n \in \mathbb{N}}$ converges a.e., and by II.1.3, to X. $\square$

The above proof can be generalized; see theorem II.2.4.1.

II.2. Main results

II.2.1. The Radon-Nikodym-property

This property of Banach spaces has become widely considered. Therefore we shall indicate the main results only briefly, except for those related to martingale convergence (see following sections). We begin with some well-known definitions in vectormeasure theory.

Definitions II.2.1.1 : Let $\mu : \mathcal{F} \to E$ be a vectormeasure : i.e. a countably additive set function. We say that μ is P-continuous or continuous w.r.t. P if $\lim_{P(A)\to 0} \mu(A) = 0$. Since $\mathcal{F}$ is a σ-algebra and P and μ are countably additive this is equivalent to $P(A) = 0 \Rightarrow \mu(A) = 0$. We denote this by $\mu \ll P$. We say that μ is of bounded variation if

$$|\mu|(\Omega) < \infty$$

where

$$|\mu|(\Omega) = \sup_{\mathcal{A}} \sum_{A\in\mathcal{A}} \|\mu(A)\|$$

where $\mathcal{A}$ runs through the set of finite disjoint families in $\mathcal{F}$; see also Dinculeanu [1967], I.3. Define also, for $A \subset \Omega$

$$|\mu|(A) = \sup_{\mathcal{B}} \sum_{B\in\mathcal{B}} \|\mu(B)\|$$

where $\mathcal{B}$ runs through the set of finite disjoint families $(B_i)_{i\in\{1,\ldots,n\}}$ such that $B_i \subset A$ and $B_i \in \mathcal{F}$ for every $i = 1,\ldots,n$.
$|\mu|$ is called the variation of μ and is also a countably additive measure; see e.g. Dinculeanu [1967], I.3.

The above definitions extend naturally to the case where $\mathcal{F}$ is only an algebra and where μ is only additive. In this case the notation $\mu \ll P$ is reserved for the property $\lim_{P(A)\to 0} \mu(A) = 0$ and is not equivalent with the property $P(A) = 0 \Rightarrow \mu(A) = 0$, as is well-known.

A Banach space E is said to have the Radon-Nikodym-property, abbreviated (RNP) if for every probability space $(\Omega, \mathcal{F}, P)$ and every vectormeasure $\mu: \mathcal{F} \to E$ such that μ has bounded variation and that $\mu \ll P$, there exists $f \in L^1_E(\Omega, \mathcal{F}, P)$ such that for each $A \in \mathcal{F}$

$$\mu(A) = \int_A f \, dP$$

This is not a property that every Banach space has! The simplest example is the Banach space c_o. Indeed : consider the following vectormeasure μ on $([0,1], \mathcal{B}[0,1], \lambda)$, where $\mathcal{B}[0,1]$ denotes the Borel-sets in $[0,1]$ and λ is Lebesguemeasure

$$\mu : ([0,1], \mathcal{B}[0,1], \lambda) \to c_o$$

$$A \to \mu(A) = \left(\int_A \cos nt \, dt\right)_{n \in \mathbb{N}}$$

$\mu(A) \in c_o$ for every $A \in \mathcal{B}[0,1]$, by the lemma of Riemann-Lebesgue. Furthermore, that μ has bounded variation and that $\mu \ll \lambda$ is completely obvious. But μ cannot have a Radon-Nikodym-density in $L^1_{c_o}$, since $(\cos nt)_{n \in \mathbb{N}} \notin c_o$ for almost every $t \in [0,1]$. For another proof of the fact that c_o lacks (RNP), see the remarks after theorem II.2.2.1. Furthermore, we mention that $L^1(\Omega, \mathcal{F}, P)$ has (RNP) if and only if the space has the form $\ell^1(\Gamma)$, for any index set Γ; see Diestel and Uhl [1977].

The most important results concerning (RNP) can be found in Diestel and Uhl [1977]. We mention some of them without proof.

Theorem II.2.1.2 : If E is reflexive or if E is a separable dual then E has (RNP).

Theorem II.2.1.3 : Every subspace of a Banach space with (RNP) has (RNP). If every separable subspace of a Banach space E has (RNP), then E has (RNP).

So, from theorems II.2.1.2 and II.2.1.3 we see that every subspace of a separable dual Banach space has (RNP) (Uhl). The converse is not true as was shown by P. Mc Cartney and R. O'Brien [1980] and independently by J. Bourgain and F. Delbaen [1980]. The example of Mc Carntney and O'Brien has some probabilistic relevance. For this reason we shall discuss this proof in detail in chapter IV.

There exists a geometric characterization of (RNP). The underlying idea of the proof is martingale convergence. We shall give it in full detail in section II.2.3.
However, to demonstrate very simply the relation between (RNP) and martingale convergence we prove the mean convergence of martingales with values in a (RNP) Banach space E, yielding also a simple characterization of (RNP).

II.2.2. Mean convergence of martingales and (RNP)

Theorem II.2.2.1 : For a Banach space E the following assertions are equivalent :

(i) E has (RNP)

(ii) Every uniformly integrable martingale is $\|\cdot\|_1$-convergent

(iii) Every uniformly bounded martingale is $\|\cdot\|_1$-convergent.

Proof : (i) ⇒ (ii)
Let $(X_n, F_n)_{n\in\mathbb{N}}$ be the uniformly integrable martingale.
Define, for each $n \in \mathbb{N}$ and each $A \in F_n$:

$$\mu_n(A) = \int_A X_n \, dP$$

By the martingale property, $\lim\limits_{n\to\infty} \mu_n(A) =: \mu(A)$ exists for each $A \in \bigcup\limits_n F_n$.
Using uniform integrability we hence see that

$$\mu(A) = \lim_{n\to\infty} \mu_n(A)$$

exists for each $A \in \sigma(\bigcup\limits_n F_n) = F_\infty$.
Furthermore, using uniform integrability, μ is countably additive. Also μ is of bounded variation, using the $\|\cdot\|_1$-boundedness of $(X_n)_{n\in\mathbb{N}}$ and finally, $\mu \ll P$. By (1), there exists $X_\infty \in L^1_E(\Omega, F_\infty, P)$ such that

$$\mu(A) = \int_A X_\infty \, dP$$

for each $A \in F_\infty$. By the martingale property, for each $n \in \mathbb{N}$ and each $A \in F_n$:

$$\int_A X_n = \mu_n(A) = \mu(A) = \int_A X_\infty$$

So, $X_n = E^{F_n} X_\infty$. From theorem II.1.3 it follows now that $(X_n)_{n \in \mathbb{N}}$ is $\|\cdot\|_1$-convergent to X_∞.

(ii) ⇒ (iii) is obvious.

(iii) ⇒ (i)

Let us first note that (iii) implies the $\|\cdot\|_1$-convergence of martingales indexed by a directed set I : this means, a net $(X_i, F_i)_{i \in I}$ such that each $X_i \in L^1_E(\Omega, F_i, P)$ and such that $E^{F_i} X_j = X_i$, for each $i,j \in I$, $i \leq j$. Indeed, suppose $(X_i)_{i \in I}$ is not $\|\cdot\|_1$-convergent. Then it is not $\|\cdot\|_1$-Cauchy. So there is an $\varepsilon > 0$ such that for each $i \in I$ there exist $i', i'' \geq i$, $i', i'' \in I$ such that $\|X_{i'} - X_{i''}\|_1 \geq 2\varepsilon$. So $\|X_{i'} - X_i\|_1$ or $\|X_{i''} - X_i\|_1$ is greater than ε. So there exists $i''' \geq i$, $i''' \in I$ such that $\|X_{i'''} - X_i\|_1 \geq \varepsilon$. So by induction we get a sequence $(i_n)_{n \in \mathbb{N}}$ for which $(X_{i_n}, F_{i_n})_{n \in \mathbb{N}}$ contradicts (iii).

To prove (i), suppose that $\mu : F \to E$ is a vectormeasure such that there exists a $K \in \mathbb{R}^+$ such that $\|\mu(A)\| \leq K\,P(A)$ for each $A \in F$.

It suffices to show that μ has a Radon-Nikodym-density. Indeed, suppose this has been proved and let ν be a general vectormeasure of bounded variation $|\nu|$ such that $\nu \ll P$. Since (Ω, F, P) is a complete measure space, it follows that $|\nu| \ll P$. So, by the classical Radon-Nikodym theorem there exists a function $\varphi \in L^1(\Omega, F, P)$ such that for every $A \in F$

$$|\nu|(A) = \int_A \varphi \, dP$$

Since $\|\mu(A)\| \leq |\mu|(A)$ for every $A \in F$, by what we supposed, we have the existence of a function $f \in L^1_E(\Omega, F, P)$ such that

$$\nu(A) = \int_A f \, d|\nu|$$

for every $A \in F$. Finally,

$$\nu(A) = \int_A f\varphi \, dP$$

for every $A \in F$.

So it suffices to prove that μ as above has a Radon-Nikodym density w.r.t. P. To show this, let π be an arbitrary partition of Ω into elements of F and define

$$X_\pi = \sum_{A\in\pi} \frac{\mu(A)}{P(A)} \chi_A$$

Then $(X_\pi, \sigma(\pi))_{\pi\in\Pi}$ is a uniformly bounded martingale w.r.t. the index set

$$\Pi = \{\text{all finite partitions of } \Omega \text{ into elements of } F\}$$

ordered by refinement. By what we have proved already : there is a $X \in L^1_E$ such that $\int_\Omega \|X_\pi - X\| \to 0$ for $\pi \in \Pi$. Fix $A \in F$. We have :

$$\lim_{\pi\in\Pi} \int_A X_\pi = \int_A X$$

Define $\pi_0 = \{A, \Omega\setminus A\} \in \Pi$. Since $\Pi(\pi_0) = \{\pi\in\Pi | \pi \geqslant \pi_0\}$ is cofinal in Π, it also follows that

$$\lim_{\pi\in\Pi(\pi_0)} \int_A X_\pi = \int_A X$$

Hence, from the definition of X_π :

$$\mu(A) = \int_A X$$

So E has (RNP). □

With the help of theorem II.2.2.1, we can give a shorter proof of the fact that c_0 lacks (RNP). Let $(e_n)_{n\in\mathbb{N}}$ denote the canonical basis of c_0. Define :

$$X_1 = e_1 \chi_{[0,1)}$$

$$X_2 = (e_1+e_2)\chi_{[0,\frac{1}{2})} + (e_1-e_2)\chi_{[\frac{1}{2},1)}$$

$$X_3 = (e_1+e_2+e_3)\chi_{[0,\frac{1}{4})} + (e_1+e_2-e_3)\chi_{[\frac{1}{4},\frac{1}{2})} + (e_1-e_2+e_3)\chi_{[\frac{1}{2},\frac{3}{4})}$$

$$+ (e_1-e_2-e_3)\, \chi_{[\frac{3}{4},1)}$$

and so on.

Let $\Omega = [0,1)$, $F = B[0,1)$, the Borelsets in $[0,1)$ and $P = \lambda$, the Lebesguemeasure on $[0,1)$. For every $n \in \mathbb{N}$, put $F_n = \sigma(X_1,\dots,X_n)$. Then $(X_n,F_n)_{n\in\mathbb{N}}$ is obviously a uniformly bounded martingale, diverging in the $L^1_{c_o}$-norm. Hence (II.2.2.1), c_0 lacks (RNP). In fact, $(X_n,F_n)_{n\in\mathbb{N}}$ also diverges everywhere. From section II.2.4 it will follow that c_0 is even "far away" from a (RNP) space. Using the same theorem II.2.2.1 we can also prove easily that $L^1(\Omega,F,P) = L^1(P)$ lacks (RNP) if $L^1(P)$ is not an $\ell^1(\Gamma)$-space. Indeed, take the measure

$$\mu : F \to L^1(P)$$

$$A \to \chi_A$$

Hence, obviously $|\mu|(\Omega) < \infty$ and $\mu \ll P$. If $L^1(P)$ had (RNP), then, by theorem II.2.2.1, using the same notation as in the proof of it, the martingale

$$(X_\pi = \sum_{A\in\pi} \frac{\mu(A)}{P(A)}\chi_A,\ \sigma(\pi))_{\pi\in\Pi}$$

is convergent in the L^1-norm. However for every $\pi,\pi' \in \Pi$, where $\pi' \geqslant \pi$ with

$$\pi = \{A_1,\dots,A_n\}$$

$$\pi' = \{A_{11},A_{12},\dots,A_{1q_1};\ \dots\ ;\ A_{n1},A_{n2},\dots,A_{nq_n}\}$$

and $A_i = \bigcup_{j=1}^{q_i} A_{iq_j}$, for every $i = 1,\dots,n$, we have

$$\|X_\pi - X_{\pi'}\|_1 = \int_\Omega \Big\| \sum_{i=1}^{n} \sum_{j=1}^{q_i} \Big(\frac{\chi_{A_i}}{P(A_i)} - \frac{\chi_{A_{ij}}}{P(A_{ij})}\Big)\chi_{A_{ij}}(\omega)\Big\|_1 \, dP(\omega)$$

$$= \int_\Omega \sum_{i=1}^{n} \sum_{j=1}^{q_i} \Big\|\frac{\chi_{A_i}}{P(A_i)} - \frac{\chi_{A_{ij}}}{P(A_{ij})}\Big\|_1 \chi_{A_{ij}}(\omega)\, dP(\omega)$$

But, for every $j = 1,\dots,q_i$ and $i = 1,\dots,n$

$$\left\| \frac{\chi_{A_i}}{P(A_i)} - \frac{\chi_{A_{ij}}}{P(A_{ij})} \right\|_1 = \int_\Omega \left\| \frac{\chi_{A_i}}{P(A_i)} - \frac{\chi_{A_{ij}}}{P(A_{ij})} \right\| dP$$

$$= \int_{A_{ij}} + \int_{A_i \setminus A_{ij}} + \int_{\Omega \setminus A_i} = 2 - 2 \frac{P(A_{ij})}{P(A_i)}$$

Hence

$$\|X_\pi - X_{\pi'}\|_1 = 2 - 2 \sum_{i=1}^{n} \sum_{j=1}^{q_i} \frac{P(A_{ij})^2}{P(A_i)} .$$

Hence, $\forall \pi \in \Pi$, $\exists \pi' \geq \pi$ such that $\|X_\pi - X_{\pi'}\|_1 > 1$, a contradiction. □

Theorem II.2.2.2 : Let $1 < p < \infty$. For a Banach space E the following assertions are equivalent :

(i) E has (RNP)

(ii) Every $\|\cdot\|_p$-bounded martingale is $\|\cdot\|_p$-convergent.

Proof : (ii) ⇒ (i)
This follows from theorem II.2.2.1, (iii) ⇒ (i).

(i) ⇒ (ii)
Since $p > 1$, $\|\cdot\|_p$-boundedness implies uniform integrability. Then from theorem II.2.2.1 it follows that there is a $X_\infty \in L^1_E$ such that $(X_n)_{n \in \mathbb{N}}$ is $\|\cdot\|_1$-convergent to X_∞. So $X_n = E^{F_n} X$ for each $n \in \mathbb{N}$. In view of theorem II.1.3 we only have to show that $X_\infty \in L^p_E$. Let $(X_{n_k})_{k \in \mathbb{N}}$ be a pointwise a.e. convergent (to X_∞) subsequence. By Fatou's lemma,

$$\int_\Omega \|X_\infty\|^p \leq \liminf_{k \to \infty} \int_\Omega \|X_{n_k}\|^p$$

$$\leq \sup_{n \in \mathbb{N}} \int_\Omega \|X_n\|^p < \infty$$

So $X_\infty \in L^p_E$. □

II.2.3. Mean convergence of martingales and geometry of Banach spaces

In section II.2.2 we showed that mean convergence of uniformly bounded martingales is equivalent to (RNP). We now show that the same martingale property is equivalent to a geometric property of the Banach space, called dentability, hence obtaining also a geometric characterization of (RNP). So the link has a probabilistic nature, namely martingale convergence.

Definition II.2.3.1 : Let $D \subset E$ be a bounded set in the Banach space E. We say that D is dentable if for each $\varepsilon > 0$ there is a $x \in D$ such that

$$x \notin \overline{Co}\ (D \setminus B_\varepsilon(x))$$

where $\overline{Co}$ denotes the closed convex hull and

$$B_\varepsilon(x) = \{x \in E \mid \|x\| \leq 1\} .$$

If every bounded subset of E is dentable, then E is called dentable.

Theorem II.2.3.2 (Huff [1974] and Rieffel [1967]) :
The following assertions are equivalent :

(i) E is dentable.

(ii) Every uniformly bounded martingale is mean convergent.

Proof : (i) ⇒ (ii) (Rieffel)
By the proof of theorem II.2.2.1, (iii) ⇒ (i), we see that it suffices to prove that every martingale of the form

$$(X_\pi, \sigma(\pi))_{\pi \in \Pi} = \Big(\sum_{A \in \pi} \frac{\mu(A)}{P(A)} \chi_A, \sigma(\pi)\Big)_{\pi \in \Pi}$$

is mean convergent where $\mu : F \to E$ is a vectormeasure such that there is a $K > 0$ with $\|\mu(A)\| \leq K\, P(A)$ for each $A \in F$, and

$$\Pi = \{\text{all finite partitions of } \Pi \text{ into elements of } F\}.$$

So let $\mu : \mathcal{F} \to E$ be as above. We have to show that dentability implies that $(X_\pi)_{\pi \in \Pi}$ is $\|\cdot\|_1$-convergent. This will be done after a rather intricate argument of Rieffel which follows now.

I. First we prove :
For every $A \in \mathcal{F}^+ = \{A \in \mathcal{F} | P(A) > 0\}$ and every $\varepsilon > 0$, $\exists A' \in \mathcal{F}^+$, $A' \subset A$ and $x \in E$ such that

$$\mathcal{A}_{A'}(\mu) \subset B_\varepsilon(x)$$

Here we denote $\mathcal{A}_{A'}(\mu)$ (<u>the average range of μ over A'</u>) for

$$\mathcal{A}_{A'}(\mu) = \{\frac{\mu(B)}{P(B)} \mid B \in \mathcal{F}^+,\ B \subset A'\}$$

Since $\mathcal{A}_A(\mu)$ is dentable, there is an $x \in \mathcal{A}_A(\mu)$ such that

$$x \notin \overline{\text{Co}}\ (\mathcal{A}_A(\mu) \setminus B_\varepsilon(x)) \ . \tag{1}$$

Since $x \in \mathcal{A}_A(\mu)$ it is of the form :

$$x = \frac{\mu(B_o)}{P(B_o)}$$

for some $B_o \in \mathcal{F}^+$, $B_o \subset A$. If $\mathcal{A}_{B_o}(\mu) \subset B_\varepsilon(x)$ the proof of I is finished. In the other case, there exists $\frac{\mu(A^*)}{P(A^*)} \in \mathcal{A}_{B_o}(\mu) \setminus B_\varepsilon(x)$. So

$$\frac{\mu(A^*)}{P(A^*)} \in \overline{\text{Co}}\ (\mathcal{A}_A(\mu) \setminus B_\varepsilon(x)) \tag{2}$$

So there exists a smallest positive integer k_1 such that there exists $A^* \in \mathcal{F}$, $A^* \subset B_o \subset A$, $P(A^*) \geq \frac{1}{k_1}$ and

$$\frac{\mu(A^*)}{P(A^*)} \in \overline{\text{Co}}\ (\mathcal{A}_A(\mu) \setminus B_\varepsilon(x))$$

Call A_1 one of the A^*'s above. Put $B_1 = B_o \setminus A_1$. If $P(B_1) = 0$ then $P(B_o) = P(A_1)$ and hence also $\mu(B_o) = \mu(A_1)$. Thus :

$$\frac{\mu(B_o)}{P(B_o)} = \frac{\mu(A_1)}{P(A_1)}$$

which is impossible, due to (1) and (2). So $P(B_1) > 0$.
If $A_{B_1}(\mu) \subset B_\varepsilon(x)$, the proof of I is finished. If $A_{B_1}(\mu) \not\subset B_\varepsilon(x)$, the same reasoning as before yields the existence of a smallest positive integer $k_2 \geq k_1$ for which there exist $A_2 \subset B_1$, $A_2 \in F$, $P(A_2) \geq \frac{1}{k_2}$ and

$$\frac{\mu(A_2)}{P(A_2)} \in \overline{Co}\ (A_A(\mu) \setminus B_\varepsilon(x))$$

Now put $B_2 = B_1 \setminus A_2$. Again $P(B_2) > 0$. If $A_{B_2}(\mu) \subset B_\varepsilon(x)$, then we are done. If not, there exists (as above) a smallest positive integer $k_3 \geq k_2$ such that there exist $A_3 \subset B_2$, $A_3 \in F$, $P(A_3) \geq \frac{1}{k_3}$ and

$$\frac{\mu(A_3)}{P(A_3)} \in \overline{Co}\ (A_A(\mu) \setminus B_\varepsilon(x))$$

Ifnone of the B_i's is good by continuing in the above way we obtain a sequence $(A_n)_{n\in\mathbb{N}}$ of pairwise disjoint sets in F and a non decreasing sequence $(k_n)_{n\in\mathbb{N}}$ in $\mathbb{N}$ such that, for every $n \in \mathbb{N}$

$$P(A_n) \geq \frac{1}{k_n}$$

and

$$\frac{\mu(A_n)}{P(A_n)} \in \overline{Co}\ (A_A(\mu) \setminus B_\varepsilon(x))$$

Now if $C \in F$ such that $C \subset B_o \setminus \bigcup_{k=1}^{n} A_k$ and such that

$$\frac{\mu(C)}{P(C)} \in \overline{Co}\ (A_A(\mu) \setminus B_\varepsilon(x))$$

then, by the definition of k_{n+1} :

$$P(C) \leq \frac{1}{k_{n+1} - 1} \tag{3}$$

Since $\lim_{n\to\infty} P(A_n) = 0$, it follows that $\lim_{n\to\infty} k_n = \infty$. Put $A' = B_o \setminus \bigcup_n A_n$. Then $P(A') > 0$. Indeed, suppose $P(A') = 0$ then, as before, $P(B_o) = P(\bigcup_n A_n)$ and $\mu(B_o) = \mu(\bigcup_n A_n)$. So :

$$\frac{\mu(B_o)}{P(B_o)} = \frac{\sum_n \mu(A_n)}{P(\bigcup_n A_n)}$$

$$= \sum_n \frac{P(A_n)}{P(\bigcup_n A_n)} \frac{\mu(A_n)}{P(A_n)} \in \overline{\text{Co}}\ (A_A(\mu) \setminus B_\varepsilon(x))$$

contradicting (1). So $P(A') > 0$.
Furthermore

$$A_{A'}(\mu) \subset B_\varepsilon(x)$$

Indeed, for any $A^* \subset A'$, $A^* \in F^+$ we have that $A^* \subset B_o \setminus \bigcup_{k=1}^{n} A_k$ for each $n \in \mathbb{N}$. So, if

$$\frac{\mu(A^*)}{P(A^*)} \notin B_\varepsilon(x)$$

then

$$\frac{\mu(A^*)}{P(A^*)} \in \overline{\text{Co}}\ (A_A(\mu) \setminus B_\varepsilon(x))$$

and hence, from (3), $P(A^*) \leqslant \frac{1}{k_{n+1} - 1}$, for each $n \in \mathbb{N}$. So $P(A^*) = 0$ which is not true. This finishes the proof of I.

II. From I it now follows that, given $\varepsilon > 0$ there is a smallest positive integer k_1 for which there is a $x_1 \in E$ and $A_1 \in F$ such that

$$P(A_1) \geqslant \frac{1}{k_1}$$

and

$$A_{A_1}(\mu) \subset B(x_1, \varepsilon)$$

Also there is a smallest $k_2 \in \mathbb{N}$, $k_2 \geqslant k_1$ such that there exist $x_2 \in E$ and $A_2 \in F$ with $A_2 \subset \Omega \setminus A_1$ such that

$$P(A_2) \geqslant \frac{1}{k_2}$$

and

$$A_{A_2}(\mu) \subset B(x_2,\varepsilon)$$

Thus we generate inductively a sequence $(A_n)_{n\in\mathbb{N}}$ of pairwise disjoint members of F^+ and a non-decreasing sequence $(k_n)_{n\in\mathbb{N}}$ in $\mathbb{N}$ and a sequence $(x_n)_{n\in\mathbb{N}}$ in E such that

$$P(A_n) \geqslant \frac{1}{k_n}$$

and

$$A_{A_n}(\mu) \subset B_\varepsilon(x_n) \quad .$$

Again as in I we see that $\lim_{n\to\infty} k_n = \infty$ and so $P(\Omega \setminus \bigcup_n A_n) = 0$. For $\pi \in \Pi$, put

$$X_\pi = \sum_{A\in\pi} \frac{\mu(A)}{P(A)} \chi_A \; .$$

We have only to show that $(X_\pi)_{\pi\in\Pi}$ is $\|\cdot\|_1$-Cauchy. Let $\varepsilon > 0$. Put $\delta = \frac{\varepsilon}{K}$. Let $n_\varepsilon \in \mathbb{N}$ be such that

$$\sum_{k=n_\varepsilon+1}^{\infty} P(A_k) \leqslant \frac{\varepsilon}{K} \; .$$

Put $A_0 = \sum_{k=n_\varepsilon+1}^{\infty} A_k$. Also put $\pi_\varepsilon = \{A_0, A_1, \ldots, A_{n_\varepsilon}\}$ and let $\pi \in \Pi(\pi_\varepsilon)$. So π is of the form

$$\pi = \{A_{0,1},\ldots,A_{0,k(0)},A_{1,1},\ldots,A_{1,k(1)},\ldots,A_{n_\varepsilon,1},\ldots,A_{n_\varepsilon,k(n_\varepsilon)}\}$$

where

$$P(A_j \,\Delta \sum_{i=1}^{k(j)} A_{j,i}) = 0$$

(Δ = symmetric difference). Hence :

$$\|X_\pi - X_{\pi_\varepsilon}\|_1 = \int \Big\| \sum_{\pi} \frac{\mu(A_{ji})}{P(A_{ji})} \chi_{A_{ji}} - \sum_{\pi_\varepsilon} \frac{\mu(A_j)}{P(A_j)} \chi_{A_j} \Big\|$$

$$= \int \Big\| \sum_{j=0}^{n_\varepsilon} \Big(\sum_{i=1}^{k(j)} \frac{\mu(A_{ji})}{P(A_{ji})} \chi_{A_{ji}} - \frac{\mu(A_j)}{P(A_j)} \chi_{A_j} \Big) \Big\|$$

$$\leqslant \sum_{i=1}^{k(0)} \int \Big\| \frac{\mu(A_{0i})}{P(A_{0i})} \chi_{A_{0i}} \Big\| + \int \Big\| \frac{\mu(A_0)}{P(A_0)} \chi_{A_0} \Big\|$$

$$+ \sum_{j=1}^{n_\varepsilon} \int \Big\| \sum_{i=1}^{k(j)} \frac{\mu(A_{ji})}{P(A_{ji})} \chi_{A_{ji}} - \frac{\mu(A_j)}{P(A_j)} \chi_{A_j} \Big\|$$

$$\leqslant 2\varepsilon + \sum_{j=1}^{n_\varepsilon} \Big[\sum_{i=1}^{k(j)} \Big\| \frac{\mu(A_{ji})}{P(A_{ji})} - \frac{\mu(A_j)}{P(A_j)} \Big\| P(A_{ji}) \Big]$$

$$\leqslant 2\varepsilon + \sum_{j=1}^{n_\varepsilon} \sum_{i=1}^{k(j)} \Big\| \frac{\mu(A_{ji})}{P(A_{ji})} - x_j \Big\| P(A_{ji})$$

$$+ \sum_{j=1}^{n_\varepsilon} \sum_{i=1}^{k(j)} \Big\| x_j - \frac{\mu(A_j)}{P(A_j)} \Big\| P(A_{ji})$$

$\leqslant 4\varepsilon$, finishing (i) $\Rightarrow$ (ii)

(ii) $\Rightarrow$ (i) (Huff)

Suppose that E is nondentable and let D be a bounded subset which is not dentable. Let $\varepsilon > 0$ be such that $x \in D$ implies $x \in \overline{\text{Co}}\,(D \setminus B_\varepsilon(x))$. Put $\Omega = [0,1)$, P the Lebesgue measure on the σ-algebra $\mathcal{F}$ of all Borel subsets of $[0,1)$. We shall define by induction a sequence of partitions $\pi_n = \{A_{n,1},\dots,A_{n,k(n)}\}$ of Ω into half-open intervals of the same form as $[0,1)$, and a sequence

$$X_n = \sum_{i=1}^{k(n)} x_{n,i}\chi_{A_{n,i}}$$

with range in D such that

(i) $\pi_{n+1} \geq \pi_n$ for each $n \in \mathbb{N} \cup \{0\}$

(ii) $F = \sigma(\bigcup_n \pi_n)$

(iii) $\|X_n(\omega) - X_{n+1}(\omega)\| \geq \varepsilon$, for each $n \in \mathbb{N} \cup \{0\}$ and $\omega \in \Omega$

(iv) $\left\| \int_{A_{n,i}} (X_n - X_{n+k}) \right\| \leq \frac{1}{2^n} P(A_{n,i})$ for each $n,k \in \mathbb{N} \cup \{0\}$, and each $i \in \{1,\ldots,k(n)\}$

(v) $\mu(A) = \lim_{n\to\infty} \int_A X_n$ exists for each $A \in F$.

Suppose momentarily that this is done. Then put

$$Y_n = \sum_{i=1}^{k(n)} \frac{\mu(A_{n,i})}{P(A_{n,i})} \chi_{A_{n,i}}$$

We have

$$\|X_n - Y_n\|_1 = \sum_{i=1}^{k(n)} \|x_{n,i} P(A_{n,i}) - \mu(A_{n,i})\|$$

$$= \lim_{j\to\infty} \sum_{i=1}^{k(n)} \left\| \int_{A_{n,i}} (X_n - X_{n+j}) \right\|$$

$$\leq \frac{1}{2^n} \tag{1}$$

Since $(Y_n, \sigma(\pi_n))_{n\in\mathbb{N}}$ is a uniformly bounded martingale, (ii) implies the mean convergence of $(Y_n)_{n\in\mathbb{N}}$. Hence, by (1) $(X_n)_{n\in\mathbb{N}}$ is mean convergent, a contradiction with (iii). So it suffices to construct $(X_n)_{n\in\mathbb{N}}$ as above with the five indicated properties.
Start with $\pi_0 = \{\Omega\}$ and $x_{0,1} \in D$ arbitrarily. Suppose inductively that

$$\pi_n = \{A_{n,1},\ldots,A_{n,k(n)}\} \quad \text{and} \quad X_n = \sum_{i=1}^{k(n)} x_{n,i}\, \chi_{A_{n,i}}$$

have been defined, where $A_{n,i}$ $(i = 1,\ldots,k(n))$ are half-open disjoint intervals and where $x_{n,i} \in D$. For each $i = 1,\ldots,k(n)$, choose $y_1^{(i)},\ldots,y_{m(i)}^{(i)} \in D$ and $\alpha_1^{(i)},\ldots,\alpha_{m(i)}^{(i)} \in \mathbb{R}^+$ such that $\sum_{j=1}^{m(i)} \alpha_j^{(i)} = 1$, $\|y_j^{(i)} - x_{n,i}\| \geq \varepsilon$ and

$$\|x_{n,i} - \sum_{j=1}^{m(i)} \alpha_j^{(i)}\, y_j^{(i)}\| < \frac{1}{2^{n+1}}$$

We do not require the $y_j^{(i)}$ to be distinct, and so we can arrange for $\alpha_j^{(i)} < \frac{1}{n+1}$ for each $j = 1,\ldots,m(i)$. Partition $A_{n,i}$ into half-open intervals $B_1^{(i)},\ldots,B_{m(i)}^{(i)}$ such that $P(B_j^{(i)}) = \alpha_j^{(i)}\, P(A_{n,i})$ for $j = 1,\ldots,m(i)$.

Let $\pi_{n+1} = \{B_j^{(i)} \mid j = 1,\ldots,m(i);\ i = 1,\ldots,k(n)\}$ and let

$$X_{n+1} = \sum_{i=1}^{k(n)} \sum_{j=1}^{m(i)} y_j^{(i)}\, \chi_{B_j^{(i)}}$$

Now (i), (ii) and (iii) follow immediately ((ii) since $\lim_{n\to\infty} \max\{P(A_{n,i}) \mid i = 1,\ldots,k(n)\} = 0$).

For (iv) we remark that

$$\|\int_{A_{n,i}} (X_n - X_{n+1})\| = \|x_{n,i}\, P(A_{n,i}) - \sum_{j=1}^{m(i)} y_j^{(i)}\, P(B_j^{(i)})\|$$

$$= \|x_{n,i} - \sum_{j=1}^{m(i)} \alpha_j^{(i)}\, y_j^{(i)}\| P(A_{n,i})$$

$$\leq \frac{1}{2^{n+1}}\, P(A_{n,i})$$

Telescoping $(X_n - X_{n+k})$ now yields (iv). From this and since $(X_n)_{n\in\mathbb{N}}$ is uniformly bounded, (v) follows easily. □

So, deleting the martingale convergence property as the intermediary property in II.2.2.1 and II.2.3.2, we have the well-known theorem :

Theorem II.2.3.3 : (Davis-Huff-Maynard-Phelps-Rieffel) :
The following assertions are equivalent :

(i) E is dentable

(ii) E has (RNP)

A carefull look at the proof of Rieffel and an easy modification of the proof of Huff yields also :

Theorem II.2.3.4 : The following assertions are equivalent :

(i) E is σ-dentable

(ii) E has (RNP)

Here σ-dentability is defined as follows :

Definition II.2.3.5 : Let $D \subset E$ be a bounded set in E. We say that D is σ-dentable if for each $\varepsilon > 0$ there is a $x \in D$ such that

$$x \notin \sigma(D \setminus B_\varepsilon(x))$$

where for $A \subset E$

$$\sigma(A) = \{ \sum_{n=1}^{\infty} \lambda_n x_n \mid x_n \in A,\ \lambda_n \geqslant 0 \text{ for each } n \in \mathbb{N},\ \sum_{n=1}^{\infty} \lambda_n = 1\}$$

There cannot be any confusion with the notation $\sigma(\mathcal{A})$ where $\mathcal{A}$ is a set of elements of $\mathcal{F}$, for obvious reasons.
We say that E is σ-dentable if every bounded subset of E is σ-dentable.

For other properties of dentability and σ-dentability we refer the reader to Diestel and Uhl [1977] . In this work we only wish to indicate the relation of martingale convergence with these notions.

We close this section by introducing the notions of bush and tree.

Definition II.2.3.6 : Let A be the set

$$A = \{(i,j) \in \mathbb{N} \times \mathbb{N} \| 1 \leq j \leq n_i\}$$

where $(n_i)_{i \in \mathbb{N}}$ is increasing in $\mathbb{N}$ such that $n_1 = 1$ and $n_i \geq 2n_{i-1}$ for each $i \in \mathbb{N}$. In the terminology of Mc Cartney [1980] this is called a bush domain. For each $j \in \mathbb{N}$, let $C_i = \{1,2,\ldots,n_i\}$ and let $S_{1,1} = \{1\}$ and for $i \geq 2$

$$\{S_{i,j} \| 1 \leq j \leq n_{i-1}\}$$

be a partition of C_i into pairwise disjoint sets of consecutive integers. A bush or ε-bush ($\varepsilon > 0$) is a subset $\{x_{i,j} \| (i,j) \in A\}$ of $B_E = \{x \in E \| \|x\| \leq 1\}$ such that for every $i \geq 2$ and $j \in C_{i-1}$

(i) $\# S_{i,j} \geq 2$ where $\#$ denotes the cardinality

(ii) $$x_{i-1,j} = \frac{1}{\# S_{i,j}} \cdot \sum_{k \in S_{i,j}} x_{i,k}$$

(iii) For each $(i,j) \in A$ and $k \in S_{i+1,j}$

$$\|x_{i,j} - x_{i+1,k}\| > \varepsilon$$

ε is called the separation constant of the bush. We have :

Theorem II.2.3.7 : For a Banach space E, the following assertions are equivalent :

(i) E has (RNP)

(ii) E does not contain an ε-bush for every $\varepsilon > 0$.

Proof : (i) ⇒ (ii)

Suppose E does contain an ε-bush. This gives rise in a trivial way to a uniformly bounded finitely generated martingale on $[0,1)$ such that $\|X_n(\omega) - X_{n+1}(\omega)\| > \varepsilon$ for each $\omega \in [0,1)$, contradicting the fact that E has (RNP), using theorem II.2.2.1.

(ii) ⇒ (i)

Suppose E does not have (RNP). By theorem II.2.3.3, E is not dentable. The construction of the martingale (Y_n, F_n) in the proof of (ii) ⇒ (i) in theorem II.2.3.2 yields an ε-bush for a certain $\varepsilon > 0$. Indeed using properties (iii) and (iv) in this proof we find a $n_0 \in \mathbb{N}$ such that if $n \in \mathbb{N}(n_0)$,

$$\|Y_n(\omega) - Y_{n+1}(\omega)\| > \frac{\varepsilon}{2} \qquad (1)$$

for all $\omega \in \Omega = [0,1)$. Now take any set $A_{n_0,i}$. Then

$$(Y_n|_{A_{n_0,i}}, F_n|_{A_{n_0,i}})_{n \in \mathbb{N}(n_0)}$$

is a martingale in B_E such that (1) is valid everywhere. The corresponding vectors

$$\{Y_n(\omega) \| \omega \in A_{n_0,i};\ n \in \mathbb{N}(n_0)\}$$

yield the $\frac{\varepsilon}{2}$-bush. □

Definition II.2.3.8 : A bush (or ε-bush) where $n_i = 2^i$ $(i \geq 2)$ is called a tree (or ε-tree). So we already have from theorem II.2.3.7 that if E has an ε-tree then E cannot have (RNP). The converse was open for a long time. Recently Bourgain and Rosenthal [1980] settled this in the negative. It was shown however to be true in Banach lattices by Talagrand [1981].

II.2.4. Almost everywhere convergence of martingales and (RNP)

We have reached one of the basic theorems of this book : the martingale a.e. convergence theorem. This states that in a Banach space with (RNP), every $\|\cdot\|_1$-bounded martingale converges a.e.. The classical reference for this theorem is Chatterji [1968]. Chatterji is indeed the first to state the theorem in this explicit way, in 1968. It must however be emphasized that the first proof of this fundamental theorem has been given by A. and C. Ionescu-Tulcea [1963], five years earlier.

The only thing that is not done in Ionescu-Tulcea [1963] is the invention of the name (RNP); however they work with the property, and they prove the martingale convergence theorem in complete detail and generality - see Ionescu-Tulcea [1963], theorem 4 and the remark preceding it.

However, Chatterji's proof is shorter and is still direct in the sense that it does not use Doob's theorem, the real version of the martingale convergence theorem. So, we present this proof.

Also two other proofs are given, since their methods will be very important throughout the book. The second is based upon Doob's theorem and the Kadec-renorming theorem in Banach spaces; the third proof is given by Chacon and Sucheston [1975], and consists of an important argument for reducing the problem to the a.e. convergence of martingales of the form $(E^{F_n}X, F_n)_{n\in\mathbb{N}}$, where $X \in L^1_E$, which is easy (theorem II.1.6).

For the first proof we start with an introductory result.

Theorem II.2.4.1 : Suppose that the martingale $(X_n, F_n)_{n\in\mathbb{N}}$ in L^1_E is $\|\cdot\|_1$-convergent to X_∞. Then it is pointwise a.e. convergent to X_∞.

Proof : For each $\varepsilon > 0$ and $\delta > 0$, choose $n_0 \in \mathbb{N}$ such that if $m,n \in \mathbb{N}(n_0)$ then $\|X_n - X_m\|_1 < \varepsilon\delta$. Fix $m \in \mathbb{N}(n_0)$. Since $(X_n - X_m, F_n)_{n\in\mathbb{N}(m)}$ is a martingale, the maximal lemma II.1.5 implies :

$$P(\sup_{n\in\mathbb{N}(m)} \|X_n - X_m\| > \varepsilon) \leq \frac{1}{\varepsilon} \sup_{n\in\mathbb{N}(m)} \|X_n - X_m\|_1 \leq \delta .$$

So $(X_n)_{n\in\mathbb{N}}$ is almost uniformly Cauchy, hence pointwise convergent and hence convergent to X_∞. □

We come now to one of the most important results of this book.

Remark II.2.4.2 : Only strong convergence a.e. is studied since for martingales weak convergence a.e. implies strong convergence a.e. - see remark II.2.4.7(2). For a proof of an even more general result, see II.3.1.6 ((iii) ⇒ (i)).

Theorem II.2.4.3 : (A. and C. Ionescu-Tulcea) : Let E have (RNP). Then every $\|\cdot\|_1$-bounded martingale $(X_n,F_n)_{n\in\mathbb{N}}$ converges pointwise a.e. to to an integrable function, and conversely.

Proof (of Chatterji) : For each $A \in \cup_n F_n$, define

$$\mu(A) = \lim_{n\to\infty} \int_A X_n$$

Apply the Lebesgue decomposition theorem (Diestel and Uhl [1977] , p.31) to μ. This gives :

$$\mu = \nu + \kappa$$

where ν and κ are additive and of bounded variation and $|\nu|$ and $|\kappa|$ are mutually singular with $|\nu| \ll P$ and κ P-singular. So ν is extendable to F_∞ into a σ-additive measure, of course still of bounded variation. We still denote this measure by ν. Using (RNP) of E, let $X_\infty \in L^1_E$ be such that, for each $A \in F_\infty$:

$$\nu(A) = \int_A X_\infty \, dP$$

Put $Y_n = E^{F_n} X_\infty$. Then from section II.1, $(Y_n,F_n)_{n\in\mathbb{N}}$ is an L^1_E-bounded martingale which converges to X_∞ both in the $\|\cdot\|_1$-sense and in the pointwise a.e.-sense. Define $Z_n = X_n - Y_n$ for each $n \in \mathbb{N}$. $(Z_n,F_n)_{n\in\mathbb{N}}$ is also an L^1_E-bounded martingale. We only have to show that $\lim_{n\to\infty} Z_n = 0$, a.e..

We have that $\kappa(A) = \int_A Z_n$ for each $A \in F_n$. Since $|\kappa|$ is P-singular on $\cup_n F_n$, for every $\delta > 0$ and $\varepsilon < 1$ there is a set $A \in \cup_n F_n$ such that

$$P(\Omega \setminus A) + |\kappa|(A) < \frac{\varepsilon\delta}{2}$$

So there is a $n_o \in \mathbb{N}$ such that $A \in F_{n_o}$. Using the maximal lemma II.1.5 yields :

$$P(\sup_{n\in \mathbb{N}(n_o)} \|Z_n\| > \varepsilon)$$

$$= P(\{\sup_{n\in \mathbb{N}(n_o)} \|Z_n\| > \varepsilon\} \setminus A) + P(\{\sup_{n\in \mathbb{N}(n_o)} \|Z_n\| > \varepsilon\} \cap A)$$

$$\leq \frac{\varepsilon\delta}{2} + \frac{1}{\varepsilon} \sup_{n\in \mathbb{N}(n_o)} \int_A \|Z_n\|$$

$$\leq \frac{\varepsilon\delta}{2} + \frac{1}{\varepsilon} |\kappa|(A)$$

$$\leq \frac{\varepsilon\delta}{2} + \frac{\delta}{2} < \delta$$

So $\lim_{n\to\infty} Z_n = 0$, a.e..
The converse follows immediately from theorem II.2.2.1, finishing the proof. □

We now give a second proof of theorem II.2.4.3. For this, we need two preliminary results. The first is the classical renorming theorem of Kadec-Klee, which is very important for the rest of our book. For this reason we present the proof, although it is beyond the scope of this book. See also Bessaga and Pełczynski [1970], theorem 3.1, p.177-178.

Theorem II.2.4.4 (Kadec-Klee) : Let E be a separable Banach space. Then on E there exists an equivalent norm $|\!|\!| \cdot |\!|\!|$ and a countable norming set $D \subset E'$ such that if $x'(x_n) \to x'(x)$ for each $x' \in D$ and if $|\!|\!| x_n |\!|\!| \to |\!|\!| x |\!|\!|$ then $x_n \to x$ in E.

Proof : Since E is separable, the set

$$B_{E'} = \{x' \in E' \| \|x'\| \leq 1\}$$

is w^*-metrizable, for instance - as is well-known - by the metric d :

$$d(x',y') = \sum_{n=1}^{\infty} \frac{1}{2^n} |x'(z_n) - y'(z_n)|$$

where $\{z_n \| n \in \mathbb{N}\}$ is an arbitrary dense subset of $B_E = \{x \in E \| \|x\| \leq 1\}$. Since $B_{E'}$ is w*-compact and w*-metrizable, it is w*-separable.

Therefore let $\{x'_n \| n \in \mathbb{N}\}$ be w*-dense in $B_{E'}$. Put, for every $x \in E$:

$$w_o(x) = \|x\| = \sup_{n \in \mathbb{N}} |x'_n(x)| \tag{1}$$

$$w_k(x) = \sup \{|x'(x) - y'(x)| \; x', y' \in B_{E'} \text{ and } d\,(x', y') \leq \frac{1}{k}\} \tag{2}$$

for $k \in \mathbb{N}$.
Then for every $x \in E$ put

$$|||x||| = \sum_{k=0}^{\infty} \frac{1}{2^k} w_k(x) \tag{3}$$

Evidently, $|||\cdot|||$ is equivalent to $\|\cdot\|$, since

$$\|x\| \leq |||x||| \leq 3\|x\|$$

for every $x \in E$. Since $\{x'_n \| n \in \mathbb{N}\}$ is dense in $B_{E'}$, we have that

$$w_k(x) = \sup \{|x'_i(x) - x'_j(x)| \| d\,(x'_i, x'_j) \leq \frac{1}{k}\} \tag{4}$$

for every $x \in E$ and every $k \in \mathbb{N}$. If for a sequence $(x_m)_{m \in \mathbb{N}}$ we have that

$$\lim_{m \to \infty} x'_n(x_m) = x'_n(x_o) \tag{5}$$

for every $n \in \mathbb{N}$, then

$$\liminf_{m \in \mathbb{N}} w_k(x_m) \geq w_k(x_o) \tag{6}$$

for every $k \in \{0\} \cup \mathbb{N}$, as is trivial to see. Hence also

$$\liminf_{m \in \mathbb{N}} |||x_m||| \geq |||x_o||| \, . \tag{7}$$

Suppose, in addition, that

$$\lim_{m\to\infty} \|x_m\| = \|x_o\| , \tag{8}$$

then of course (6) is an equality also :

$$\lim_{m\to\infty} w_k(x_m) = w_k(x_o) \tag{9}$$

for every $k \in \{0\} \cup \mathbb{N}$, as is seen by a small calculation using (6) and (8). Define the sequence $(f_{x_n})_{n\in\{0\}\cup\mathbb{N}}$ by

$$f_{x_n}(x') = x'(x_n) ,$$

for $x' \in B_{E'}$ and $n \in \{0\} \cup \mathbb{N}$. Then $(f_{x_n})_{n\in\{0\}\cup\mathbb{N}}$ is an equicontinuous sequence of bounded functions on the w^*-metric space $B_{E'}$, converging on the dense set $\{x'_m \| m \in \mathbb{N}\}$ to f_{x_o}. The equicontinuity of $(f_{x_n})_{n\in\mathbb{N}\cup\{0\}}$ is seen by (9) and the fact that $(w_k(\cdot))_{k\in\mathbb{N}}$ decreases. Hence by the classical Arzela theorem, $(f_{x_n})_{n\in\mathbb{N}}$ converges to f_{x_o} uniformly on $B_{E'}$; i.e.

$$\lim_{n\to\infty} \|x_n - x_o\| = 0 .$$

Hence also

$$\lim_{n\to\infty} \|\!|x_n - x_o\|\!| = 0 . \qquad \square$$

At the same time we have proved that for every w^*-dense countable subset of $B_{E'}$ we have an equivalent Kadec-renorming of E.

The following result is a very important but easy result on sequences of submartingales. This result will be used several times throughout the book. While using it we shall also extend it in chapter VIII. For the moment only the martingale case is considered.

<u>Theorem II.2.4.5 (Neveu [1975])</u> : Let $\{(X^i_n, F_n)_{n\in\mathbb{N}} \| i \in I\}$ be a countable family of real submartingales such that :

$$\sup_{n\in\mathbb{N}} \int_\Omega \sup_{i\in I} (X^i_n)^+ < \infty .$$

Then each of the submartingales $(X_n^i, F_n)_{n\in\mathbb{N}}$ converges a.e. to a function $X^i \in L^1$ and the submartingale $(\sup_{i\in I} X_n^i, F_n)_{n\in\mathbb{N}}$ converges a.e. to $\sup_{i\in I} X^i$.

Proof : From the boundedness assumption and the real submartingale convergence theorem (for an independent self-contained proof of this, see chapter VIII, but for our application further on we may suppose it are martingales), it follows that, for each $i \in I$, $(X_n^i)_{n\in\mathbb{N}}$ converges to an integrable function, say X^i and also that $(\sup_{i\in I} X_n^i)_{n\in\mathbb{N}}$ converges, say to $X_\infty \in L^1$. Obviously $X_\infty \geq \sup_{i\in I} X^i$. The proof will be finished if we can establish the opposite inequality. To obtain this it suffices to show that

$$\int_\Omega X_\infty = \int_\Omega \sup_{i\in I} X_\infty^i$$

Let $\{I_p | p \in \mathbb{N}\}$ be a sequence of finite subsets of I, increasing to I with p. Then $\int_\Omega \sup_{i\in I_p} X_n^i$ increases with p and also with n since for each $p \in \mathbb{N}$, $(\sup_{i\in I_p} X_n^i, F_n)_{n\in\mathbb{N}}$ is a submartingale. Also

$$S = \sup_{\substack{p\in\mathbb{N}\\ n\in\mathbb{N}}} \int_\Omega \sup_{i\in I_p} X_n^i = \sup_{n\in\mathbb{N}} \int(\sup_{i\in I} X_n^i) \leq \sup_{n\in\mathbb{N}} \int \sup_{i\in I} (X_n^i)^+ < \infty .$$

Hence, for every $\varepsilon > 0$ there exists p_ε, $n_\varepsilon \in \mathbb{N}$ such that

$$\int \sup_{i\in I_p} X_n^i \geq S - \varepsilon$$

for every $p \in \mathbb{N}(p_\varepsilon)$ and $n \in \mathbb{N}(n_\varepsilon)$ (since $(\int_\Omega \sup_{i\in I_p} X_n^i)$ increases with p and with n). Since $X_\infty - \sup_{i\in I_p} X_\infty^i = \lim_{n\to\infty} (\sup_{i\in I} X_n^i - \sup_{i\in I_p} X_n^i)$, Fatou's lemma implies

$$\int (X_\infty - \sup_{i\in I_p} X_\infty^i) \leq \liminf_{n\in\mathbb{N}} \int (\sup_{i\in I} X_n^i - \sup_{i\in I_p} X_n^i)$$

$$\leq S - (S - \varepsilon) = \varepsilon$$

whenever $p \in \mathbb{N}(n_\varepsilon)$. So $\int(X_\infty - \sup_{i\in I} X_\infty^i) \leq \varepsilon$ for every $\varepsilon > 0$, finishing the proof. □

Finally we can give a second proof of theorem II.2.4.3 :

II.2.4.6 : Second proof of theorem II.2.4.3 :
We may and do suppose that E is separable. Let $(X_n, F_n)_{n\in\mathbb{N}}$ be an L_E^1-bounded martingale. For each $A \in \bigcup_n F_n$, $\lim_{n\to\infty} \int_A X_n =: \mu(A)$ exists. Since $(X_n)_{n\in\mathbb{N}}$ is L_E^1-bounded, μ is of bounded variation. Hence μ is strongly additive. Now apply the Carathéodory-Hahn-Kluvanek extension theorem - see Diestel and Uhl [1977] , p.27 - to see that μ can be extended to a countably additive vectormeasure of bounded variation on $F_\infty = \sigma(\bigcup_n F_n)$ such that $\mu \ll P$. Using (RNP) we obtain a function $X_\infty \in L_E^1$ such that

$$\mu(A) = \int_A X_\infty \qquad (1)$$

for each $A \in F_\infty$. Also, for each $x' \in D$ where D is as in theorem II.2.4.4, $(x'(X_n), F_n)_{n\in\mathbb{N}}$ is a real martingale which is L^1-bounded. Hence it converges a.e. and its limit is $x'(X_\infty)$ by (1). Since D is countable we can apply theorem II.2.4.5. So

$$\sup_{x'\in D} x'(X_n) \to \sup_{x'\in D} x'(X_\infty), \text{ a.e..}$$

This means :

$$|\|X_n\|| \to |\|X_\infty\||, \text{ a.e.}$$

Consequently by II.2.4.4 it follows that $X_n \to X_\infty$, a.e., since D is countable. □

Remarks II.2.4.7 : (1) The first proof is independent of the real martingale convergence theorem of Doob while the second is not. So also in this sense, the first proof is more elementary. However the second method is powerful for obtaining results quickly as we shall also see several times later on.
(2) The above proof also shows that every a.e. weakly convergent L_E^1-bounded

martingale is strongly convergent a.e..

Finally, we present a third proof of the theorem of A. and C. Ionescu-Tulcea. We give it in complete detail since the method is basic and will be of use on several other occasions further on in this book. The method is due to Chacon and Sucheston.

II.2.4.8 : Third proof of theorem II.2.4.3

1° Reduction to the case $\sup_{n\in\mathbb{N}} \|X_n\| \in L^1$

This method goes back to a method of proof of the real martingale convergence theorem; see Lamb [1973] and Meyer [1972]. Fix $\lambda > 0$. Define a stopping time σ as ∞ if $\sup_{n\in\mathbb{N}} \|X_n\| \leq \lambda$. Otherwise, σ is the first $n \in \mathbb{N}$ such that $\|X_n\| > \lambda$. Then $\int_\Omega \sup_{n\in\mathbb{N}} \|X_{n\wedge\sigma}\| < \infty$. Indeed, if $\omega \in \{\sigma = \infty\}$ then $\sup_{n\in\mathbb{N}} \|X_{n\wedge\sigma}(\omega)\| \leq \lambda$. On $\{\sigma < \infty\}$, $\lim_{n\to\infty} \|X_{n\wedge\sigma}\| = \|X_\sigma\|$. So using Fatou's lemma

$$\int_{\{\sigma<\infty\}} \|X_\sigma\| \leq \liminf_{n\in\mathbb{N}} \int_{\{\sigma<\infty\}} \|X_{n\wedge\sigma}\| \leq \liminf_{n\in\mathbb{N}} \int_\Omega \|X_{n\wedge\sigma}\|$$

$$\leq \liminf_{n\in\mathbb{N}} \int_\Omega \|X_n\| \qquad (*)$$

$$\leq \sup_{n\in\mathbb{N}} \int_\Omega \|X_n\|$$

Since $\|X_{n\wedge\sigma}\| \leq \|X_\sigma\|$ on $\{\sigma < \infty\}$ it now follows that

$$\int_\Omega \sup_{n\in\mathbb{N}} \|X_{n\wedge\sigma}\| = \int_{\{\sigma=\infty\}} \sup_{n\in\mathbb{N}} \|X_{n\wedge\sigma}\| + \int_{\{\sigma<\infty\}} \sup_{n\in\mathbb{N}} \|X_{n\wedge\sigma}\|$$

$$\leq \lambda + \sup_{n\in\mathbb{N}} \int_\Omega \|X_n\| < \infty$$

Furthermore, $(X_{n\wedge\sigma}, F_{n\wedge\sigma})_{n\in\mathbb{N}}$ is a martingale.

By the maximal lemma II.1.5.(ii), $(X_{n\wedge\sigma})_{n\in\mathbb{N}}$ coincides with $(X_n)_{n\in\mathbb{N}}$ except on a set of small measure if λ is large.

2° We can suppose now that $\sup_{n\in\mathbb{N}} \|X_n\| \in L^1$

In this case, $(X_n)_{n\in\mathbb{N}}$ is uniformly integrable and hence by theorem II.2.2.1 it converges in the $\|\cdot\|_1$-sense to a function $X_\infty \in L^1_E$. Finally, theorem II.2.4.1 finishes the proof. □

II.3. Convergence of martingales in general Banach spaces

In this section we shall present some convergence results of martingales with values in general Banach spaces.

II.3.1. A theorem of Korzeniowski

Here the limit is found as an application of a measurable selection theorem for multifunctions, which makes the proof worth giving in detail. First some generalities concerning probability measures associated with a measurable function.

Theorem II.3.1.1 : Let X be an E-valued measurable function on the probability space $(\Omega,\mathcal{F},P)$. Then X induces a probability space $(E,\mathcal{B}(E),\mu)$ where $\mathcal{B}(E)$ denotes the σ-algebra of the Borelsets in E, in the following way : For each $B \in \mathcal{B}(E)$: $\mu(B) = P(X^{-1}(B)) = P(\{X\in B\})$. μ is called the probability distribution measure of X or p.m. of X. We have :

Lemma II.3.1.2 : Let X and μ be as above. Then for each Borel measurable function f on E :

$$\int_\Omega f(X(\omega))dP(\omega) = \int_E f(x)d\mu(x)$$

Proof : For $B \in \mathcal{B}(E)$ and $f = \chi_B$ we have equality by definition of μ. By linearity of integrals, the equality is therefore true for step-functions f on $\mathcal{B}(E)$. Now let f be Borel measurable and positive. Then there is an increasing sequence of stepfunctions $(f_m)_{m \in \mathbb{N}}$ converging to f everywhere. So the monotone convergence theorem establishes the desired equality for positive functions. Hence it is proved for any Borel measurable function on E. □

The key to the proof of the result of Korzeniowski is the following lemma in which a measurable selection theorem is used : the selection theorem of Kuratowski and Ryll-Nardzewski - see Parthasarathy [1972], theorem 5.1, p.50-52 for a proof. We only state the result :

Theorem II.3.1.3 (Kuratowski and Ryll-Nardzewski) : Let X be a set and S a countably additive family of subsets of X. Let Y be a complete separable metric space and let $F : X \to 2^Y$ be such that

$$\{x \in X \| F(x) \cap G \neq \phi\} \in S$$

for every open $G \subset Y$. Then there is a measurable selection $f : X \to Y$, in the sense that $f^{-1}(G) \in S$ for every open $G \subset Y$.

Lemma II.3.1.4 : Let F be a Polish space. Consider the following measurable maps :

$$\varphi : (\Omega, F, P) \to (F, \mathcal{B}(F))$$

and

$$f : (E, \mathcal{B}(E)) \to (F, \mathcal{B}(F))$$

such that $f^{-1}(x)$ is closed for each $x \in F$. If

$$P(\{\omega | \varphi(\omega) \in f(E)\}) = 1$$

then there exists a measurable function $X : \Omega \to E$ such that

$$\varphi(\omega) = f(X(\omega)) \text{ a.e.}.$$

Proof : Define the multifunction $T : F \to \mathcal{B}(E)$ as follows :

$$T(x) = \begin{cases} f^{-1}(x) & \text{if } x \in f(E) \\ 0 & \text{otherwise} \end{cases}$$

Suppose for the moment that T has a measurable selection, i.e. there exists w.r.t. certain σ-algebras- see further on - a measurable function t such that $t(x) \in T(x)$ for each $x \in F$ and such that $X = t \circ \varphi$ is measurable. Then X is the function we are looking for. Indeed,

$$f \circ X = f \circ t \circ \varphi$$

So for each $\omega \in \Omega$

$$f(X(\omega)) = f(t(\varphi(\omega))) \in f(T(\varphi(\omega)))$$

Since $P(\{\omega|\varphi(\omega) \in f(E)\}) = 1$ we see that

$$f(X(\omega)) \in f(f^{-1}(\varphi(\omega))) = \{\varphi(\omega)\} \text{ , a.e..}$$

So :

$$f(X(\omega)) = \varphi(\omega), \text{ a.e..}$$

Thus we only have to prove the existence of a selection of T in such a way that $X = t \circ \varphi$ is measurable. By theorem II.3.1.3, we have a function $t : F \to E$, measurable w.r.t. the completion $\overline{\mathcal{B}}(F)$ of $\mathcal{B}(F)$ w.r.t. $\varphi(P)$ and $\mathcal{B}(E)$, provided that we can show that $\{x \in F \| T(x) \cap G \neq \phi\} \in \overline{\mathcal{B}}(F)$ for every open G. This can be seen as follows. We have

$$\{x \in F | T(x) \cap G \neq \phi\} = \text{proj}_F(\text{graph}(T) \cap (F \times G))$$

where proj_F denotes the projection on F.
Now $f^{-1}(x)$ is closed for each $x \in f(E)$, and $f(E) \in \overline{\mathcal{B}}(F)$. This is seen by using theorems 3.1 and 3.4 in Parthasarathy [1967], p.16 to show that f(E) is analytic. Apply then Sion [1960], theorem 2.5 to see that f(E) is Suslin; hence, as is well-known, $f(E) \in \overline{\mathcal{B}}(F)$. Now it follows that $\text{graph}(T) \in \overline{\mathcal{B}}(F) \times \mathcal{B}(E)$. Thus $\{x \in F | T(x) \cap G \neq \phi\}$ is Suslin since Borel subsets of Suslin sets (hence of Polish spaces) are Suslin and continuous

images of Suslin sets are Suslin. So

$$\{x \in F \mid T(x) \cap G \neq \phi\} \in \overline{\mathcal{B}}(F) .$$

The above mentioned selection theorem now yields a measurable function $t : (F,\overline{\mathcal{B}}(F)) \to (E,\mathcal{B}(E))$ such that $t(x) \in T(x)$, for each $x \in F$. Since φ is $(\mathcal{F},\mathcal{B}(F))$-measurable we see that $X = t \circ \varphi$ is $(\mathcal{F},\mathcal{B}(E))$-measurable. □

In order to have the full generality of Korzeniowski's theorem we must introduce a topology $\mathcal{T}$ on E. Here $\mathcal{T}$ is any locally convex Hausdorff topology which is weaker than the norm topology. It follows from Schwartz [1973], p.122 that $(E,\mathcal{T})$ is a Radon space since it is a Suslin space. This means that every Borel probability measure μ is a Radon measure, i.e. : relative to the class $\mathcal{K}_{\mathcal{T}}$ of $\mathcal{T}$-compact sets in E it satisfies the inner regularity property

$$\mu(B) = \sup \{\mu(K) \mid K \subset B,\ K \in \mathcal{K}_{\mathcal{T}}\}$$

for each $B \in \mathcal{B}(E,\mathcal{T})$, the class of $\mathcal{T}$-Borel sets in E.

We need some more definitions. Let T be a completely regular Hausdorff space. Denote by $C_b(T)$ the space of all real valued bounded continuous functions on T.

Definition II.3.1.5 : Let $(\mu_i)_{i\in I}$ be a net of Radon probabilities on T and μ a Radon probability on T also. We say that $(\mu_i)_{i\in I}$ converges weakly to μ if

$$\lim_{i\in I} \int_T f \, d\mu_i = \int_T f \, d\mu$$

for every $f \in C_b(T)$. We denote this by $\mu_i \xrightarrow{w} \mu$.
In our context T will be $(E,\mathcal{T})$ where $\mathcal{T}$ is a topology on E as above. We have to following result :

Theorem II.3.1.6 (Korzeniowski [1978a]) : Let $(X_n,\mathcal{F}_n)$ be an L^1_E-bounded martingale and let μ_n be the p.m. of X_n for each $n \in \mathbb{N}$. Let $\mathcal{T}$ be a locally convex Hausdorff topology on E, weaker than the norm topology. The following assertions are equivalent :

(i) $(X_n)_{n \in \mathbb{N}}$ converges strongly a.e.;

(ii) $(X_n)_{n \in \mathbb{N}}$ converges in probability.

(iii) $\{X_n(\omega) \mid n \in \mathbb{N}\}$ is $\mathcal{T}$-relatively compact, a.e..

(iv) $(\mu_n)_{n \in \mathbb{N}}$ converges weakly.

Proof : (i) ⇒ (ii) ⇒ (iii) is trivial

(iii) ⇒ (i)

Note that this generalizes remark II.2.4.2. Let $|||\cdot|||$ denote an equivalent Kadec renorming of E w.r.t. the countable set $D \subset E'$ (see II.2.4.4). Since $(X_n, F_n)_{n \in \mathbb{N}}$ is an L^1_E-bounded martingale, for every $x' \in D$, there exists $X^\infty_{x'} \in L^1$ such that

$$\lim_{n\to\infty} x'(X_n) = X^\infty_{x'} \text{ , a.e.}$$

and by theorem II.2.4.5 also

$$\lim_{n\to\infty} |||X_n||| = \lim_{n\to\infty} \sup_{x' \in D} |x'(X_n)| = \sup_{x' \in D} |X^\infty_{x'}|, \text{ a.e. .}$$

By (iii), for every $\omega \in \Omega \setminus N$ (with $P(N) = 0$), the $\mathcal{T}$-closure of $(X_n(\omega))_{n \in \mathbb{N}}$ is $\mathcal{T}$-compact, hence $\mathcal{T}$-sequentially compact. Let $X_\infty(\omega)$ be its $\mathcal{T}$-clusterpoint. It now follows that

$$x'(X_\infty(\omega)) = X^\infty_{x'}(\omega), \text{ a.e.}$$

for every $x' \in D$, since D is countable. By II.2.4.4 it now follows that $(X_n)_{n \in \mathbb{N}}$ converges to X_∞, strongly a.e..

(i) ⇒ (iv)

In view of lemma II.3.1.2 it suffices to show that

$$\int_\Omega f(X_n(\omega))\, dP(\omega) \to \int_\Omega f(X(\omega))\, dP(\omega) \qquad (1)$$

where $X = \lim_{n\to\infty} X_n$ (using (i)). This X is in L^1_E since $(X_n)_{n \in \mathbb{N}}$ is L^1_E-

bounded, using Fatou's lemma. But since $f \in C_b(E,\mathcal{T})$, $f(X_n) \to f(X)$, a.e. . Since f is bounded, the dominated convergence theorem implies (1). So it remains to show (iv) ⇒ (i).

(iv) ⇒ (i)

We can and do suppose that E is separable. Let $|||\cdot|||$ denote an equivalent Kadec renorming of E w.r.t. the countable set $D \subset E'$ (see II.2.4.4). Let μ be the weak limit of $(\mu_n)_{n\in\mathbb{N}}$. For each $x' \in D$, $(x'(X_n))_{n\in\mathbb{N}}$ converges to a function $X_{x'} \in L^1$ on $\Omega \setminus N$ where N is a null-set, because $(x'(X_n),\mathcal{F}_n)_{n\in\mathbb{N}}$ is an L^1-bounded (scalar) martingale. Define :

$$\varphi(\omega) = (X_{x'})_{x'\in D} \quad \text{and} \quad f(x) = (x'(x))_{x'\in D}$$

Thus $\varphi : (\Omega,\mathcal{F},P) \to (\mathbb{R}^\infty, \mathcal{B}(\mathbb{R}^\infty))$ and $f : (E,\mathcal{B}(E),\mu) \to (\mathbb{R}^\infty,\mathcal{B}(\mathbb{R}^\infty))$. Now φ and f have the same distribution law. Indeed, for each $t_1,\ldots,t_m \in \mathbb{R}$ and $x'_1,\ldots,x'_m \in D$ we have :

$$\int_\Omega e^{i\sum_{j=1}^m t_j X_{x'_j}}\, dP = \lim_{n\to\infty} \int_\Omega e^{i\sum_{j=1}^m t_j x'_j(X_n)}\, dP$$

$$= \lim_{n\to\infty} \int_E e^{i\sum_{j=1}^m t_j x'_j(x)}\, d\mu_n(x) \qquad \text{(see lemma II.3.1.2)}$$

$$= \int_E e^{i\sum_{j=1}^m t_j x'_j(x)}\, d\mu$$

Therefor $P(\{\omega\in\Omega \mid \varphi(\omega) \in f(E)\}) = 1$. Now apply lemma II.3.1.4. So there is a measurable function X such that $X_{x'} = x'(X)$ a.e. for each $x' \in D$. This means that for each $x' \in D$: $\lim_{n\to\infty} x'(X_n) = x'(X)$, a.e.. Since $\sup_{n\in\mathbb{N}} \int \sup_{x'\in D} |x'(X_n)| = \sup_{n\in\mathbb{N}} |||X_n|||_1 < \infty$ it follows from theorem II.2.4.5 that $|||X_n||| \to |||X|||$, a.e.. Hence $X_n \to X$, a.e. . □

Corollary II.3.1.7 (Chatterji [1976]) : Let $(X_n,F_n)_{n\in\mathbb{N}}$ be an L^1_E-bounded martingale such that a.e., the set $\{X_n(\omega)\|n\in\mathbb{N}\}$ is weakly relatively compact in E. Then $(X_n)_{n\in\mathbb{N}}$ converges strongly a.e..

In Chatterji [1976], Chatterji proves this result in a more elementary way than in theorem II.3.1.6. For this reason and due to the importance of his method, we repeat his argument, yielding a second proof of corollary II.3.1.7. A number of preliminary results are needed.

Lemma II.3.1.8 : Let H be a total subset of E' and let $(X_n,F_n)_{n\in\mathbb{N}}$ be a martingale with values in E, such that :

(i) $\{x'(X_n)\,|\,n\in\mathbb{N}\}$ is uniformly integrable for each $x'\in H$.

(ii) there exists $X\in L^1_E$ such that $x'(X_n(\omega))\to x'(X(\omega))$, a.e. for each $x'\in H$.

Then $X_n\to X$, a.e..

Proof : Both assumptions imply that $x'(X_n) = E^{F_n}x'(X)$ for each $x'\in H$ and each $n\in\mathbb{N}$. Then

$$\int_A x'(X_n) = \int_A x'(X)$$

for each $A\in F_n$ and for each $n\in\mathbb{N}$. By the totality of H :

$$\int_A X_n = \int_A X$$

for each $A\in F_n$ and for each $n\in\mathbb{N}$. So $X_n = E^{F_n}X$. It follows now from theorem II.1.6 that $X_n\to X$, a.e.. □

The following result is a lemma for theorem II.3.1.7 but has also some importance on its own. We therefore state it as a proposition :

Proposition II.3.1.9 : If $(X_n,F_n)_{n\in\mathbb{N}}$ is an L^1_E-bounded martingale such that $x'(X_n)(\omega)\to x'(X)(\omega)$, a.e. for each $x'\in H$, a total subset of E' and if $X\in L^1_E$, then $X_n\to X$, a.e..

Proof : Indeed, if $(X_n, F_n)_{n\in\mathbb{N}}$ is uniformly integrable, this follows from lemma II.3.1.8. If not the result follows from the following lemma, proved exactly as in II.2.4.8(1°).

Lemma II.3.1.10 : If P is a property which is such that if the martingale $(X_n, F_n)_{n\in\mathbb{N}}$ has P, then every stopped martingale $(X_{n\wedge\sigma}, F_n)_{n\in\mathbb{N}}$ has P where σ is a stopping time. Then if

$$[P \text{ and } \sup_{n\in\mathbb{N}} \|X_n\| \in L^1_E] \Rightarrow (X_n)_{n\in\mathbb{N}} \text{ converges a.e. },$$

then

$$[P \text{ and } \sup_{n\in\mathbb{N}} \int_\Omega \|X_n\| < \infty] \Rightarrow (X_n)_{n\in\mathbb{N}} \text{ converges a.e..}$$

We take for P here (on $(X_n, F_n)_{n\in\mathbb{N}}$):

$$x'(X_n)(\omega) \rightarrow x'(X)(\omega), \text{ a.e.}$$

for each $x' \in H$, a total subset of E', where $X \in L^1_E$. Obviously $(X_{n\wedge\sigma}, F_n)_{n\in\mathbb{N}}$ has property P since

$$x'(X_{n\wedge\sigma})(\omega) \rightarrow x'(Y)(\omega), \text{ a.e.}$$

where

$$Y \begin{cases} = X & \text{on } \{\sigma = \infty\} \\ = X_\sigma & \text{on } \{\sigma < \infty\} \end{cases}$$

so $Y \in L^1_E$. □

II.3.1.11 : Second proof of corollary II.3.1.7.

Let a.e., $X(\omega)$ be a weak accumulation point of $\{X_n(\omega) | n \in \mathbb{N}\}$. Certainly for every $x' \in E'$, $(x'(X_n), F_n)_{n\in\mathbb{N}}$ converges a.e.. Since we may and do suppose that E is separable there is a countable norming total subset H of E'. We have $x'(X_n) \rightarrow x'(X)$, a.e. for all $x' \in E'$. So, since H is countable, there is a null set N such that on $\Omega \setminus N$, $x'(X_n) \rightarrow x'(X)$, for each $x' \in H$. X is scalarly measurable, hence measurable since E is separable. Now, for each $x' \in H$

$$|x'(X)| = \lim_{n\to\infty} |x'(X_n)|, \text{ a.e..}$$

So :

$$|x'(X)| \leq \lim_{n\to\infty} \|X_n\|, \text{ a.e. .}$$

This last limit exists since $(\|X_n\|, F_n)_{n\in\mathbb{N}}$ is an L^1-bounded submartingale in $\mathbb{R}$ (see also chapter III). Hence

$$\|X\| \leq \lim_{n\to\infty} \|X_n\|, \text{ a.e.}$$

and consequently

$$\int_\Omega \|X\| \leq \liminf_{n\to\infty} \int_\Omega \|X_n\| < \infty \quad .$$

An appeal to proposition II.3.1.9 finishes the proof. □

For a special case of this corollary we refer to Chatterji [1973].

II.3.2. The theorem of Burkholder and Shintani

In case a Banach space has (RNP) we have already proved theorem II.4.2.3 stating that all mean bounded martingales converge a.e.. Suppose now that E is a general Banach space. The problem is now to identify the class of L^1_E-bounded martingales that converge a.e.. This is done in the following theorem . First we need some introductory notions. Let $M^1_E = M^1_E(\Omega,F,P)$ be the space of all L^1_E-bounded martingales $X = (X_n,F_n)_{n\in\mathbb{N}}$ where $(F_n)_{n\in\mathbb{N}}$ is a fixed increasing sequence of sub-σ-algebras of F. We equip M^1_E with the norm

$$\|X\|_1 = \sup_{n\in\mathbb{N}} \|X_n\|_1$$

It follows that $M^1_{E,\|\cdot\|_1}$ is a Banach space.

Definition II.3.2.1 : Let $(X_n,F_n)_{n\in\mathbb{N}}$ be a martingale. Denote $d_n = X_n - X_{n-1}$ for each $n \in \mathbb{N}$, where $X_0 = 0$. $(d_n)_{n\in\mathbb{N}}$ is called the difference sequence of $(X_n,F_n)_{n\in\mathbb{N}}$. So $X_n = \sum_{i=1}^{n} d_n$. We say that $(X_n,F_n)_{n\in\mathbb{N}}$ is of bounded variation if $\sum_{i=1}^{\infty} \|d_i\| < \infty$, a.e. .Here $\|d_i\|$ is the function $\omega \to \|d_i(\omega)\|$. Denote

$$BV = \{X \in M^1_E | X \text{ is of bounded variation}\}$$

$$AE = \{X \in M^1_E | X \text{ converges a.e.}\}$$

It follows from Fatou's lemma that the limit of every martingale in AE is in L^1_E.
Clearly $BV \subset AE$. However, more is true :

Theorem II.3.2.2 (Burkholder-Shintani [1978]) : AE is closed in M^1_E and BV is dense in AE.

Proof : (i) AE is closed in M^1_E (and hence $\overline{BV} \subset AE$) :
Let $X \in \overline{AE}$ and $\varepsilon > 0$. Choose $Y \in AE$ such that $\|X - Y\|_1 < \varepsilon^2$.
Using the maximal lemma II.1.5 we find :

$$P(\sup_{n\in\mathbb{N}} \|X_n - Y_n\| > \varepsilon) \leq \frac{1}{\varepsilon} \|X - Y\|_1 < \varepsilon$$

where $X = (X_n,F_n)_{n\in\mathbb{N}}$ and $Y = (Y_n,G_n)_{n\in\mathbb{N}}$. Now Y converges a.e. .
Hence so does X. So $X \in AE$.

(ii) $AE \subset \overline{BV}$:
Let $X = (X_n,F_n)_{n\in\mathbb{N}} \in AE$. Then there is $X_\infty \in L^1_E$ such that $X_n \to X_\infty$, a.e..
Put $Y = (Y_n,F_n)_{n\in\mathbb{N}}$, $Z = (Z_n,F_n)_{n\in\mathbb{N}}$ where $Y_n = E^{F_n}X_\infty$ and $Z_n = X_n - Y_n$ for each $n \in \mathbb{N}$. From theorem II.1.3 and theorem II.1.6 it follows that $Y_n \to X_\infty$, a.e. and in the L^1_E-sense. Let now $\varepsilon > 0$. We shall construct an a.e. finite stopping time τ such that the stopped martingale $X^\tau = (X_{\tau\wedge n},F_n)_{n\in\mathbb{N}}$ satisfies $\|X - X^\tau\|_1 < \varepsilon$. Indeed this will finish the

proof since $X^\tau \in BV$. The rest of the proof is now devoted to the construction of τ.

Define $Z'_n = \sup_{k\in\mathbb{N}(n)} E^{F_n}\|Z_k\|$ and put $Z' = (Z'_n, F_n)_{n\in\mathbb{N}}$. Then $Z' \in M^1_{\mathbb{R}}$, $Z'_n \geq \|Z_n\|$, for each $n \in \mathbb{N}$ and $\|Z'\|_1 = \|Z\|_1$ using I.3.5.7 and II.1.4. Furthermore, since $\lim_n Z_n = 0$ and using Fatou :

$$\int_\Omega \liminf_{n\in\mathbb{N}} Z'_n \leq \liminf_{n\in\mathbb{N}} \int_\Omega (Z'_n - \|Z_n\|) = \|Z'\|_1 - \|Z\|_1 = 0$$

So $\liminf_{n\in\mathbb{N}} Z'_n = 0$. Define the stopping time τ_j by :

$$\tau_j(\omega) = \inf\{n\in\mathbb{N} \mid \text{exactly } j \text{ of } Z'_1(\omega),\ldots,Z'_n(\omega) \text{ are less than } 2^{-j}\}$$

Then τ_j is finite a.e.. Put $\tau = \tau_j$ where $j \in \mathbb{N}$ is such that $2^{-j} < \frac{\varepsilon}{4}$ and $\|Y_n - X_\infty\|_1 < \frac{\varepsilon}{8}$ for all $n \in \mathbb{N}(j)$.

I. $\|Y - Y^\tau\|_1 \leq \frac{\varepsilon}{2}$

Indeed : $\tau \in T(j)$. Hence, if $n \in \mathbb{N}(j)$, then $\tau \wedge n \in T(j)$ and so, by lemma II.1.4 :

$$\int_\Omega \|Y_{\tau\wedge n} - Y_j\| \leq \int_\Omega \|Y_n - Y_j\| < \frac{\varepsilon}{4}$$

So :

$$\|Y - Y^\tau\|_1 = \sup_{n\in\mathbb{N}} \int_\Omega \|Y_n - Y_{n\wedge\tau}\|$$

$$= \lim_{n\to\infty} \int_\Omega \|Y_n - Y_{n\wedge\tau}\| \qquad \text{(also by lemma II.1.4)}$$

$$\leq \lim_{n\to\infty} \int_\Omega \|Y_n - Y_j\| + \lim_{n\to\infty} \int_\Omega \|Y_j - Y_{n\wedge\tau}\|$$

$$\leq \frac{\varepsilon}{2} \ .$$

II. $\|Z - Z^{\tau}\|_1 \leq \frac{\varepsilon}{2}$

Indeed : $\|Z_n - Z_{\tau\wedge n}\| \leq Z'_n - Z'_{\tau\wedge n} + 2^{-j+1}$ (1) because if $n \leq \tau$, then (1) reduces to $0 \leq 2^{-j+1}$, and if $\tau < n$, then

$$\begin{aligned}\|Z_n - Z_\tau\| &\leq \|Z_n\| + \|Z_\tau\| \\ &\leq Z'_n + Z'_\tau \\ &= Z'_n - Z'_\tau + 2\, Z'_\tau \\ &\leq Z'_n - Z'_\tau + 2^{-j+1} \quad .\end{aligned}$$

So :

$$\int_\Omega \|Z_n - Z_{\tau\wedge n}\| \leq \int_\Omega Z'_n - \int_\Omega Z'_{\tau\wedge n} + 2^{-j+1} = 2^{-j+1}$$

Consequently

$$\|Z - Z^{\tau}\|_1 \leq 2^{-j+1} < \frac{\varepsilon}{2} \ .$$

Finally :

$$\|X - X^{\tau}\|_1 \leq \|Y - Y^{\tau}\|_1 + \|Z - Z^{\tau}\|_1 < \varepsilon \ . \qquad \square$$

Combining this result with theorem II.2.4.3 we get :

Corollary II.3.2.3 : The following assertions for a Banach space E are equivalent :

(i) $\overline{BV} = M^1_E$ for all probability spaces (Ω, F, P) and all increasing sequences $(F_n)_{n\in\mathbb{N}}$ of sub-σ-algebras.

(ii) $AE = M^1_E$ for all probability spaces (Ω, F, P) and all increasing sequences $(F_n)_{n\in\mathbb{N}}$ of sub-σ-algebras of F.

(iii) E has (RNP).

II.4. Notes and Remarks

II.4.1. A maximal lemma involving lim sup instead of sup can be found in Millet and Sucheston [1980b] . See also Bellow and Egghe [1981] and Egghe [1980d] . In fact it can be derived very easily from the proof of corollary I.3.5.3.

II.4.2. For earlier work on vector-valued martingales, see Chatterji [1960] , [1964] and Scalora [1961].
We refer also to an extension of theorem II.2.4.3 with a new proof which can be found in Uhl [1972] . The proof is in the context of martingales consisting of strongly measurable Pettis-integrable functions and is even new in case $E = \mathbb{R}$ (i.e. : a new proof of Doob's martingale convergence theorem).

II.4.3. For convergence of martingales $(X_n, \mathcal{F}_n)_{n \in \mathbb{N}}$ for which the functions X_n belong to some Orlicz spaces, see Uhl [1969b] , Uhl [1969c] . In the last article the underlying measure space may be only finitely additive.

II.4.4. Extensions to locally convex spaces of theorems II.2.2.1 and II.2.3.2 have been studied in Chi [1975] , for Fréchet spaces, in Saab [1978] for locally convex spaces in which every bounded set is metrizable and in Egghe [1978a] and [1980a] for sequentially complete locally convex spaces. See also Blondia [1981a] and [1981b] where a weaker type of measurable function is studied (still strongly measurable in the Banach space case). Finally see also Gilliam [1976] and Rodriguez and Salinas [1980] . For martingale convergence in locally convex spaces, see Pestman [1981] .

II.4.5. The proofs in section II.2.3 are not the original ones. For those, see Maynard [1973] , Davis and Phelps [1974] . The intimate relation between the convergence of adapted sequences and the geometry of Banach spaces will be shown for a second time in chapter IV where we shall discuss results of P. Mc Cartney and R. O'Brien and of G. Edgar.

II.4.6. Some results on the lines of section II.3.3 can also be found in Métivier [1967], but - as Chatterji remarks - the proof of lemma 2 (p.193-194) is wrong. See also Uhl [1969b] and [1969c]. Some applications of theorem II.3.1.7 appear also in Chatterji [1976]. They are applications of Gaussian processes and multiparameter processes.

II.4.7. In this remark we return to theorem II.2.3.3 (dentability is equivalent with (RNP)). Theorems of this kind - though very interesting and important - are not part of our purpose to prove; in this book it is the convergence of adapted sequences we are interested in. But of course, theorem II.2.3.3 follows immediately from our previous martingale study. The original proof of theorem II.2.3.3 was much more complicated : the simplification is due to the fact that we have let martingales enter the scene. Another example of this is the article of Kunen and Rosenthal [1982]. Using the martingale a.e. convergence theorem, or rather the quasi-martingale a.e. convergence theorem (cf section V.1), they succeed in giving a probabilistic proof of the next theorem. The fact that they are better suited than martingales in connection with "denting" problems is obvious from the argument in (ii) ⇒ (i) of theorem II.2.3.2 : $(X_n, F_n)_{n \in \mathbb{N}}$ is a quasi-martingale constructed naturally from the assumption that E is nondentable. See also V.1.14.

<u>Theorem II.4.7.1</u> : The following assertions are equivalent :

(i) E has (RNP).

(ii) Every closed bounded convex subset of E is the closed convex hull of its denting points.

(iii) Every closed bounded convex subset of E is the closed convex hull of its strongly exposed points.

For terminology, see Diestel and Uhl [1977].

We refer to Kunen and Rosenthal [1982] for the proof. The probabilistic argument in their proof is mainly the same as in the proof of (ii) ⇒ (i) in theorem II.2.3.2, but is more complicated; see also V.1.14.

Another example of such a "probabilistic intervention" is given by Van Dulst in an unpublished note. For this reason and also since the proof is simple, we present it in full detail. The argument of Van Dulst reproves the theorem of James, stating that if E has the asymptotic norming property then E has (RNP).

Definitions II.4.7.2 :

1. A subset $\Phi \subset B_{E'}$ is called norming if for every $x \in E$:

$$x = \sup_{x' \in \Phi} |x'(x)|$$

2. A sequence $(x_n)_{n \in \mathbb{N}}$ in E with $\|x_n\| = 1$ for every $n \in \mathbb{N}$ is said to be asymptotically normed by a set $\Phi \subset B_{E'}$, if for every $\varepsilon > 0$ there is $x' \in \Phi$ and $N \in \mathbb{N}$ such that

$$x'(x_n) > 1 - \varepsilon$$

for $n \in \mathbb{N}(N)$.

3. E is said to have the asymptotic norming property (ANP) if E has an equivalent norm for which there exists a norming set $\Phi \subset B_{E'}$ with the property that every sequence $(x_n)_{n \in \mathbb{N}}$ in E with $\|x_n\| = 1$ for every $n \in \mathbb{N}$, which is asymptotically normed by Φ, has a convergent subsequence.

Lemma II.4.7.3 : In the third definition above, we may assume Φ countable, if E is separable.

Proof :Let $(x_n)_{n \in \mathbb{N}}$ be dense in $S_E = \{x \in E \mid \|x\| = 1\}$. For every $n \in \mathbb{N}$, choose a sequence $(x'_{n,m})_{m \in \mathbb{N}}$ in Φ such that

$$x_n = \sup_{m \in \mathbb{N}} |x'_{n,m}(x_n)| \qquad (1)$$

Then $\Phi' = \{x'_{n,m} \| n,m \in \mathbb{N}\} \subset \Phi$ is a norming set as well and if a sequence $(x_n)_{n\in\mathbb{N}} \subset S_E$ is asymptotically normed by Φ, then it is also by Φ' as is easily seen from (1). □

<u>Theorem II.4.7.4</u> : Let E be a separable Banach space. If E has (ANP), then E has (RNP).

<u>Proof</u> : Since (RNP) is an isomorphism invariant, we can suppose that $\|\cdot\|$ on E is the norm satisfying definition II.4.7.2 (3).
By lemma II.4.7.3 we can assume Φ to be countable. Put on span (Φ), the linear hull of Φ, the following norm : for $y \in$ span (Φ)

$$|||y||| = \inf \{ \sum_{i=1}^{n} |\lambda_i| \, | \, y = \sum_{i=1}^{n} \lambda_i y_i ; \ y_i \in \Phi; \ 1 \leq i \leq n; \ n \in \mathbb{N}\}$$

Let Y be the completion of (span $(\Phi), |||\cdot|||$). Since E is separable and since Φ is norming it is easy to see that E embeds isometrically into Y'. We now show that we can assume that $\Phi = B_Y$. Indeed : suppose that $(x_n)_{n\in\mathbb{N}}$ is asymptotically normed (A.N.) by B_Y. We have to show that there is a convergent subsequence of $(x_n)_{n\in\mathbb{N}}$. Since $(x_n)_{n\in\mathbb{N}}$ is A.N. by B_Y we have that for every $\varepsilon > 0$, there exists $y \in B_Y$ and $N \in \mathbb{N}$ such that

$$y(x_n) > 1 - \varepsilon$$

for every $n \geq N$. Since $Y = \overline{\text{span}}\ (\Phi)$ we can assume that $y \in$ span (Φ). So y is of the form

$$y = \sum_{i=1}^{p} \lambda_i y_i$$

where $y_i \in \Phi$ for every $i \in \{1,\dots,p\}$ and where we can assume that $\sum_{i=1}^{p} |\lambda_i| \leq 1$ (since $y \in B_Y$ and due to the definition of $|||y|||$). Since y is independent of $(x_n)_{n\in\mathbb{N}}$, so is p. Now from

$$\sum_{i=1}^{p} \lambda_i y_i(x_n) > 1 - \varepsilon$$

it follows that for every $n \in \mathbb{N}$, there is an $i \in \{1,\dots,p\}$ such that $y_i(x_n) > 1 - \varepsilon$ (since $\sum_{i=1}^{p} |\lambda_i| \leq 1$). Since p is independent of n, there is thus an $i \in \{1,\dots,p\}$ for which there is a subsequence $(n_k)_{k \in \mathbb{N}}$ in $\mathbb{N}$ such that

$$y_i(x_{n_k}) > 1 - \varepsilon$$

Thus $(x_{n_k})_{n \in \mathbb{N}}$ is A.N. by Φ and so there is a convergent subsequence. This sequence is of course also a convergent subsequence of $(x_n)_{n \in \mathbb{N}}$.
So we can assume that $\Phi = B_Y$.
Now let $(X_n, F_n)_{n \in \mathbb{N}}$ be a uniformly bounded martingale in $E \subset Y$, and let $\varepsilon > 0$.
Obviously there is a weak*-measurable function X such that

(i) $\lim_{n\to\infty} y(X_n) = y(X)$, a.e.

(ii) $\lim_{n\to\infty} \|X_n\| = \|X\|$, a.e.

((i) follows from the real martingale convergence theorem together with Alaoglu's theorem stating that bounded sets in a dual are weak*-relatively compact; (ii) follows from theorem II.2.4.5, using the fact that B_Y contains a countable norming subset).
Let $(y_n)_{n \in \mathbb{N}}$ be a countable dense subset in B_Y. Let N be a nullset such that on $\Omega \setminus N$, (i) and (ii) are valid everywhere, for every $y = y_n$, $n \in \mathbb{N}$. Fix $\omega \in \Omega \setminus N$. If $\|X(\omega)\| = 0$, then $\lim_{n\to\infty} X_n(\omega) = X(\omega)$. If $\|X(\omega)\| \neq 0$, then choose y_k such that $y_k\left(\frac{X(\omega)}{\|X(\omega)\|}\right) > 1 - \varepsilon$. Then there exists $n_0 \in \mathbb{N}$ such that $y_k\left(\frac{X_n(\omega)}{\|X(\omega)\|}\right) > 1 - \varepsilon$ for every $n \in \mathbb{N}(n_0)$.
So $\left(\frac{X_n(\omega)}{\|X(\omega)\|}\right)_{n \in \mathbb{N}}$ is A.N. by B_Y. So $\left(\frac{X_n(\omega)}{\|X(\omega)\|}\right)_{n \in \mathbb{N}}$ and hence $(X_n(\omega))_{n \in \mathbb{N}}$ has a convergent subsequence. This can only be $X(\omega)$. Since $(X_n)_{n \in \mathbb{N}}$ is uniformly bounded it now follows that $\lim_{n\to\infty} X_n = X$, a.e. Then theorem II.2.2.1 implies that E has (RNP). □

These examples of intervention of martingale properties into characterizations of Banach properties are not the only ones, as we mentioned before. Other examples are also found in the work of Pisier [1975], Burkholder [1981a] , [1983] and [1981b], Schwartz [1981], Edgar [1976], Woyczynski [1975] and [1978], Maurey and Pisier [1976], and others.

Chapter III : SUB- AND SUPERMARTINGALE CONVERGENCE THEOREMS

In this chapter we deal with the problem of sub- and supermartingale convergence in Banach lattices.

We start with the positive submartingale convergence theorem of Heinich [1978a]. We give Heinich's proof but we apply also the renorming theorem of Davis-Ghoussoub-Lindenstrauss [1981]. This method - based on a lattice-version of the Kadec-renorming theorem - will also prove to be useful in chapter VIII.

In the next section it is furthermore shown that for general submartingales $(X_n, F_n)_{n \in \mathbb{N}}$, the condition $\sup_{n \in \mathbb{N}} \int_\Omega \|X_n^+\| < \infty$ is not sufficient to have a.e. convergence in a Banach lattice with (RNP) (as it is in the real case). Additional conditions are given in order to get positive results.

The next section describes convergence of supermartingales. It shows that supermartingales do not behave as well as submartingales in Banach lattices.

The chapter closes with some submartingale convergence results in Banach lattices which do not necessarily have (RNP).

In this chapter and also in chapter VIII we use several classical results from Banach lattice theory. Since perhaps not everybody is familiar with Banach lattice theory, we mention the results to be used in a preliminary section, stating them and giving references for the reader interested in a proof.

III.1. Preliminary results

Definition III.1.1 : Let E be a Banach lattice. We say that E is ordercontinuous if order convergence implies norm convergence.

The Banach lattice c_0 is an obvious example of an order-continuous Banach lattice, which we shall use later on.

Theorem III.1.2 (Lindenstrauss and Tzafriri [1979], p.28; Schaefer [1974], p.94) : Let E be a Banach lattice. The following assertions are equivalent:

(i) E is ordercontinuous.

(ii) Every orderinterval $[x,y] = \{z \in E | x \leqslant z \leqslant y\}$ is weakly compact.

(iii) ℓ^∞ is not lattice isomorphic with a sublattice of E.

Theorem III.1.3 (Lindenstrauss and Tzafriri [1979], p.34) : Let E be a Banach lattice. The following assertions are equivalent :

(i) E is weakly sequentially complete.

(ii) $c_0 \not\hookrightarrow E$.

(iii) Every norm bounded increasing sequence in E is norm convergent.

Definition III.1.4 : A Banach lattice E is called an AL-space if for every $x,y \in E^+$ (the positive cone of E) one has

$$\|x + y\| = \|x\| + \|y\|$$

Theorem III.1.5 (Schaefer [1974], p.242) (Schlotterbeck) : Let E be a Banach lattice. The following assertions are equivalent :

(i) E is lattice isomorphic to an AL-space.

(ii) For every sequence $(x_n)_{n \in \mathbb{N}}$ in E^+ such that Σx_n converges unconditionally, one has that $\Sigma \|x_n\| < \infty$.

From the theorem of Dvoretzky-Rogers we know that if in (ii) of the above theorem we take $(x_n)_{n \in \mathbb{N}}$ arbitrarily in E, then E is finite dimensional. What kind of spaces do we have here now? We have the following representation theorem of Kakutani.

Theorem III.1.6 (Schaefer [1974], p.114) (Kakutani) : For every AL-space E, there exists a locally compact space X and a strictly positive Radon-measure μ on X such that E is lattice isomorphic with $L^1(\mu)$ (and, of course every such space is an AL-space).
Another characterization of AL-spaces goes as follows :

Theorem III.1.7 (Schaefer [1974], p.113) : For E to be isomorphic with an AL-space it is necessary and sufficient that each directed, norm-bounded family has a supremum.
The last result we mention is a lemma to theorem III.2.2 of Heinich on the a.e. convergence of submartingales.

Theorem III.1.8 (Heinich [1978a], p.279) : Let $(x_n)_{n\in\mathbb{N}}$ and $(y_n)_{n\in\mathbb{N}}$ be two sequences in E^+ such that

(i) $x_n \leqslant y_n$ for every $n \in \mathbb{N}$

(ii) There exists an $x_0 \in E^+$ such that $y_n \to x_0$ in E and $x_n \to x_0$, weakly.

If the orderinterval $[0,x_0]$ is weakly compact, then $x_n \to x_0$ in E.

III.2. Heinich's theorem on the convergence of positive submartingales

Definition III.2.1 : Let E be a Banach lattice and $(X_n,F_n)_{n\in\mathbb{N}}$ be an adapted sequence. We say that $(X_n,F_n)_{n\in\mathbb{N}}$ is a submartingale if $X_n \leqslant E^{F_n}X_{n+1}$, a.e., for each $n \in \mathbb{N}$. We say that $(X_n,F_n)_{n\in\mathbb{N}}$ is a supermartingale if $X_n \geqslant E^{F_n}X_{n+1}$, a.e., for each $n \in \mathbb{N}$. As in lemma II.1.4 we can prove that $(X_n,F_n)_{n\in\mathbb{N}}$ is a sub-(super-)martingale if and only if for every $\sigma \in T$ and $\tau \in T(\sigma)$

$$E^{F_\sigma}X_\tau \geqslant X_\sigma \text{ (resp. } \leqslant\text{), a.e. .}$$

In this section we shall prove that positive L^1_E-bounded submartingales converge strongly a.e. in Banach lattices with (RNP). We present first the original proof of Heinich (Heinich [1978a]), and then, as in the second proof of the martingale convergence theorem II.2.4.6, we indicate a second proof, now based on the lattice renorming theorem of Davis-Ghoussoub-Lindenstrauss. This renorming theorem we do not present here since its proof is very long and since it is out of the scope of the book; we refer the reader to Davis-Ghoussoub-Lindenstrauss [1981].

Theorem III.2.2 (Heinich) : Let $(X_n, F_n)_{n\in\mathbb{N}}$ be an L^1_E-bounded positive submartingale. Suppose that the Banach lattice E has (RNP). Then $(X_n)_{n\in\mathbb{N}}$ converges strongly a.e..

Proof : We can of course suppose E to be separable. The sequence

$$(E^{F_m} X_n)_{n\in\mathbb{N}(m)}$$

increases in E. Also it is norm bounded in E, since for every $\lambda > 0$ we have from lemma II.1.5

$$P\left(\sup_{n\in\mathbb{N}(m)} \| E^{F_m} X_n \| > \lambda\right) \leq \frac{1}{\lambda} \sup_{\tau\in T(F_m)} \int \| E^{F_m} X_\tau \| , \qquad (1)$$

where $T(F_m)$ denotes the set of all bounded stopping times w.r.t. the constant sequence $(F_m)_{n\in\mathbb{N}(m)}$. Hence $T(F_m) \subset T(m) \subset T$. So

$$(1) \leq \frac{1}{\lambda} \sup_{\tau\in T} \int \|X_\tau\| = \frac{1}{\lambda} \sup_{n\in\mathbb{N}} \int \|X_n\| < \infty \quad ,$$

since $(X_n, F_n)_{n\in\mathbb{N}}$ is an L^1_E-bounded submartingale and by the equivalent definition in III.2.1. From theorem III.1.3, using the fact that $c_o \not\subset E$, it now follows that

$$(E^{F_m} X_n)_{n \in \mathbb{N}(m)}$$

converges in norm to a function, say Y_m. Put $Z_m = Y_m - X_m$, for every $m \in \mathbb{N}$. By the construction of Y_m, via the increasing sequence $(E^{F_m} X_n)_{n\in\mathbb{N}(m)}$, we see that $(Y_m, F_m)_{m\in\mathbb{N}}$ is an L^1_E-bounded martingale and that $\lim_{m\to\infty} \int Z_m = 0$. Using theorem II.2.4.3, there exists a function

$X_\infty \in L^1_E$ such that

$$\lim_{m\to\infty} Y_m = X_\infty \quad , \tag{2}$$

strongly a.e.. We show that for every $x' \in E'$, $\lim_{n\to\infty} x'(X_n) = x'(X_\infty)$, a.e.. Of course it suffices to take $x' \in (E')^+$, in which case $(x'(X_n),F_n)_{n\in\mathbb{N}}$ is a scalar L^1-bounded submartingale, converging to say $X_{x'} \in L^1$, according to Doob's theorem. But since $\lim_{n\to\infty} \int x'(Z_n) = 0$, there exists a subsequence $(n_k)_{k\in\mathbb{N}}$ such that $\lim_{k\to\infty} x'(Z_{n_k}) = 0$, a.e.. By this and (2), it now follows that

$$\lim_{k\to\infty} x'(X_{n_k}) = x'(X_\infty) \; ,$$

a.e.. So $X_{x'} = x'(X_\infty)$, a.e.. Since $\lim_{n\to\infty} (Y_n \vee X_\infty) = X_\infty$, a.e. we see that $\lim_{n\to\infty} x'(X_n \vee X_\infty) = x'(X_\infty)$, a.e. and so $\lim_{n\to\infty} x'(X_n \wedge X_\infty) = x'(X_\infty)$, a.e., since $\lim_{n\to\infty} x'(X_n) = x'(X_\infty)$, a.e..
Let Ω_o be such that $P(\Omega_o) = 1$ and such that for every $\omega \in \Omega_o$, the sequence

$$(X_n(\omega) \wedge X_\infty(\omega))_{n\in\mathbb{N}}$$

belongs to $[0,X_\infty(\omega)]$. This orderinterval is weakly compact, by theorem III.1.2 and hence weakly sequentially compact by Eberlein's theorem. So let $X'_\infty(\omega)$ be a weak clusterpoint of the above sequence. Then X'_∞ is measurable and $x'(X_\infty) = x'(X'_\infty)$, a.e.. Hence $X_\infty = X'_\infty$, a.e.. By theorem III.1.8,

$$\lim_{n\to\infty} (X_n \wedge X_\infty) = X_\infty \; ,$$

strongly a.e.. So, from $X_n \wedge X_\infty \leq X_n \leq Y_n$ for every $n \in \mathbb{N}$, we now have that

$$\lim_{n\to\infty} X_n = X_\infty \; , \text{ a.e. .} \qquad \square$$

A second proof is easily given on the lines of II.2.4.6, if we have to our disposal a renorming theorem where the new norm has all the properties of the norm in the theorem II.2.4.4 of Kadec-Klee, and in addition the new norm is a lattice norm. This is needed to make sure that $(\|X_n\|, F_n)_{n \in \mathbb{N}}$ is still a submartingale. Now such a result exists : the renorming theorem of Davis-Ghoussoub-Lindenstrauss [1981].

Theorem III.2.3 (Davis-Ghoussoub-Lindenstrauss) : If E is a separable ordercontinuous Banach lattice then there exists an equivalent lattice norm $\|\cdot\|$ on E and a countable norming set $D \subset E'^+$, such that $x_n \to x$ in E whenever

$$x'(x_n) \to x'(x) \text{ for each } x' \in D$$

and

$$\|x_n\| \to \|x\| .$$

The proof of this is beyond the scope of this book, and can be found in Davis-Ghoussoub-Lindenstrauss [1981].

We remark that all (RNP) Banach lattices are ordercontinuous. Indeed, a Banach lattice E is ordercontinuous if ℓ^∞ is not a sublattice of E (theorem III.1.2). (RNP) spaces do have this property since we have already shown that c_o does not have (RNP) and since (RNP) is hereditary for closed subspaces.

Remarks III.2.4 :

1) We refer the reader interested in a new and selfcontained proof of the positive real submartingale convergence theorem to VIII.3.1; however it presupposes knowledge of superpramarts and amarts (see later on).
2) In theorem III.2.2 above, the (RNP) of E is implied by the convergence property. Indeed, in view of theorem II.2.2.1, suppose that $(X_n,F_n)_{n\in\mathbb{N}}$ is a uniformly bounded martingale. Then $(X_n^+,F_n)_{n\in\mathbb{N}}$ and $(X_n^-,F_n)_{n\in\mathbb{N}}$ are uniformly bounded positive submartingales. So they are strongly convergent a.e.. Hence also $(X_n)_{n\in\mathbb{N}}$ since $X_n = X_n^+ - X_n^-$ for every $n \in \mathbb{N}$. However, with a bit more work we have the following more

interesting result which is not so well-known. The proof goes back to the Krickeberg result showing the possibility of writing an L^1-bounded martingale as the difference of two positive martingales.

Theorem III.2.5 (Ghoussoub and Talagrand [1979b]): If E is a Banach lattice and if every L^1_E-bounded positive martingale is strongly convergent a.e., then E has (RNP). We may even suppose the martingales to be finitely generated, uniformly bounded and positive.

Proof : First we note that c_o cannot be a sublattice of E. It is indeed easy to construct a positive uniformly bounded martingale in c_o not converging a.e. : take $A_{n,i}$ in $[0,1)$, $n \in \mathbb{N}$, $i \in \{2^n,\dots,2^{n+1}-1\}$ such that $P(A_{n,i}) = 2^{-n}$, where P is the Lebesguemeasure on $[0,1)$, for every i,n and such that $A_{n,i}$ is the disjoint union of $A_{n+1,2i}$ and $A_{n+1,2i+1}$. Define

$$F : ([0,1),P) \to c_o$$

as

$$F(A) = (P(A \cap A_{n,i}))_{n,i}$$

ordered lexicographically. Hence $F(A) \in c_o$ for every measurable set. $|F|([0,1)) < \infty$, $F \ll P$ and it is trivial to see that there is no function $X \in L^1_{c_o}$ such that

$$F(A) = \int_A X$$

for every measurable A. Of course, $(X_n,F_n)_{n\in\mathbb{N}}$ where

$$X_n = \sum_{i=2^n}^{2^{n+1}-1} \frac{F(A_{n,i})}{2^{-n}} \chi_{A_{n,i}}$$

and $F_n = \sigma(\{A_{n,i} \| i = 2^n,\dots,2^{n+1}-1\})$ is a positive uniformly bounded finitely generated martingale not converging a.e.. So $c_o \not\subset E$.
Now let $(X_n,F_n)_{n\in\mathbb{N}}$ be an L^1_E-bounded martingale.

Then, adapting the proof of the real Krickeberg decomposition, (cf. also the proof of theorem III.2.2), the sequence

$$(E^{F_m} X_n^+)_{n \in \mathbb{N}(m)}$$

increases in E. Furthermore this sequence is norm bounded in E. Indeed, for every $\lambda > 0$, we have from lemma II.1.5

$$P(\sup_{n \in \mathbb{N}(m)} \| E^{F_m} X_n^+ \| > \lambda) \leq \frac{1}{\lambda} \sup_{\tau \in T(F_m)} \int \| E^{F_m} X_\tau^+ \| \ , \qquad (1)$$

where $T(F_m)$ denotes the set of all bounded stopping times w.r.t. the constant sequence $(F_m)_{n \in \mathbb{N}(m)}$. Hence $T(F_m) \subset T(m) \subset T$. So

$$(1) \leq \frac{1}{\lambda} \sup_{\tau \in T} \int \| X_\tau^+ \|$$

$$\leq \frac{1}{\lambda} \sup_{\tau \in T} \int \| X_\tau \| = \frac{1}{\lambda} \sup_{n \in \mathbb{N}} \int \| X_n \| < \infty \ ,$$

since $(X_n, F_n)_{n \in \mathbb{N}}$ is an L^1_E-bounded martingale. From theorem III.1.3 it now follows that $(E^{F_m} X_n^+)_{n \in \mathbb{N}(m)}$ converges in norm to a function, say Y_m. It is now clear that $(Y_m, F_m)_{m \in \mathbb{N}}$ is a positive martingale. The same is true for $(Z_m, F_m)_{m \in \mathbb{N}}$ where $Z_m = Y_m - X_m$ for every $m \in \mathbb{N}$. From the fact that positive L^1_E-bounded martingales converge a.e., it now follows that $(Y_n)_{n \in \mathbb{N}}$ and $(Z_n)_{n \in \mathbb{N}}$ converge a.e.. So also $(X_n)_{n \in \mathbb{N}}$ implying (RNP), due to theorem II.2.2.1. If we had taken $(X_n)_{n \in \mathbb{N}}$ to be uniformly bounded and consisting of stepfunctions, then $(Y_n)_{n \in \mathbb{N}}$ and $(Z_n)_{n \in \mathbb{N}}$ would obviously have the same property. □

III.3. Convergence of general submartingales

For real submartingales it is known (cf Neveu [1975]) that

if

$$\sup_{n\in\mathbb{N}} \int X_n^+ < \infty \qquad (1)$$

then $(X_n)_{n\in\mathbb{N}}$ converges a.e.. In this section it will be shown that this cannot be true in general Banach lattices with (RNP). Necessary and sufficient conditions on the Banach lattices will be given in order that this will be true. Also, in general Banach lattices with (RNP), we shall impose additional requirements on the submartingale, in order to obtain convergence.

Condition (1) in Banach lattices may take the forms

$$\sup_{n\in\mathbb{N}} \int \|X_n^+\| < \infty \qquad (2)$$

$$(\int_\Omega X_n^+)_{n\in\mathbb{N}} \text{ is order bounded} \qquad (3)$$

Each of the conditions (2) or (3) will be studied. We first need a definition and a lemma.

<u>Definition III.3.1</u> : Let $(X_n,\mathcal{F}_n)_{n\in\mathbb{N}}$ be an adapted sequence. It is called <u>predictable</u> if X_n is $\mathcal{F}_{n-1}$-measurable for each $n \in \mathbb{N}$.

<u>Lemma III.3.2</u> : Let $(X_n,\mathcal{F}_n)_{n\in\mathbb{N}}$ be a submartingale. Then $X_n = M_n + A_n$, for each $n \in \mathbb{N}$, where $(M_n,\mathcal{F}_n)_{n\in\mathbb{N}}$ is a martingale and where $(A_n,\mathcal{F}_n)_{n\in\mathbb{N}}$ is predictable such that $A_1 = 0$ and $(A_n)_{n\in\mathbb{N}}$ is increasing a.e.. This decomposition is unique. This decomposition is called <u>Doob's decomposition</u> (cf. also V.2.15).

<u>Proof</u> : Write $A_1 = 0$, $M_1 = X_1$ and for each $n \geq 2$:

$$M_n = X_1 + \sum_{i=2}^{n} (X_i - E^{\mathcal{F}_{i-1}} X_i)$$

$$A_n = \sum_{i=2}^{n} E^{\mathcal{F}_{i-1}}(X_i - X_{i-1}) \quad .$$

Everything is now obvious except uniqueness. Suppose $X_n = M_n + A_n = M'_n + A'_n$ for each $n \in \mathbb{N}$, where $(M'_n,\mathcal{F}_n)_{n\in\mathbb{N}}$ and $(A'_n,\mathcal{F}_n)_{n\in\mathbb{N}}$ have the same properties as $(M_n, F_n)_{n\in\mathbb{N}}$ resp. $(A_n, F_n)_{n\in\mathbb{N}}$. Then $M_n - M'_n = A'_n - A_n$ which is F_{n-1}-measurable. Since $(M_n - M'_n, F_n)_{n\in\mathbb{N}}$ is a martingale it now follows that for each $n,k \in \mathbb{N}$: $M_n - M'_n = M_k - M'_k$. But $M_1 = M'_1$ since $A_1 = A'_1 = 0$. So $M_n = M'_n$ for each $n \in \mathbb{N}$ and hence also $A_n = A'_n$ for each $n \in \mathbb{N}$. □

Theorem III.3.3 (Szulga and Woyczynski [1976]) : Let E be a separable Banach lattice. The following assertions are equivalent :

(i) E has (RNP).

(ii) Every submartingale $(X_n, F_n)_{n\in\mathbb{N}}$ such that $\sup_{n\in\mathbb{N}} \int_\Omega \|X_n^+\| < \infty$ and $\sup_{n\in\mathbb{N}} \int_\Omega \|M_n^-\| < \infty$ converges strongly a.e. to an integrable function (M_n as in lemma III.3.2).

(iii) Every submartingale $(X_n, F_n)_{n\in\mathbb{N}}$ such that $\sup_{n\in\mathbb{N}} \int_\Omega \|X_n^+\|^p < \infty$ and $\sup_{n\in\mathbb{N}} \int_\Omega \|M_n^-\|^p < \infty$ for some $p \in \,]1,+\infty[$ converges strongly a.e. to an integrable function.

Proof : (iii) ⇒ (i) and (ii) ⇒ (i) are easy if we take martingales $(X_n, F_n)_{n\in\mathbb{N}}$ since in this case $X_n = M_n$ for each $n \in \mathbb{N}$ and so the result follows from theorems II.2.2.1 and II.2.2.2.

(i) ⇒ (ii)

From $X_n = M_n + A_n$ as in the lemma we have $X_n \geq M_n$ a.e. and so $X_n^+ \geq M_n^+$, a.e. for each $n \in \mathbb{N}$. Hence

$$\sup_{n\in\mathbb{N}} \int_\Omega \|M_n\| \leq \sup_{n\in\mathbb{N}} \int_\Omega \|M_n^+\| + \sup_{n\in\mathbb{N}} \int_\Omega \|M_n^-\|$$

$$\leq \sup_{n\in\mathbb{N}} \int_\Omega \|X_n^+\| + \sup_{n\in\mathbb{N}} \int_\Omega \|M_n^-\|$$

$$< \infty$$

Using theorem II.2.4.3 it follows that there exists $M_\infty \in L_E^1$ such that

$(M_n)_{n\in\mathbb{N}}$ converges a.e. to M_∞. From $A_n = X_n - M_n$ it follows that $A_n \leqslant X_n^+ + M_n^-$. Hence $\sup_{n\in\mathbb{N}} \int_\Omega \|A_n\| < \infty$. But $(\|A_n\|)_{n\in\mathbb{N}}$ increases. Thus $\sup_{n\in\mathbb{N}} \|A_n\| \in L^1$, and so is finite a.e. .

Since (RNP) is hereditary for closed subspaces and since c_o does not have (RNP) (chapter II) we see that c_o is not isomorphic to a subspace of E. In such Banach lattices one has the property that norm bounded monotone sequences are convergent (theorem III.1.3). So there exists a measurable function A_∞ such that $(A_n)_{n\in\mathbb{N}}$ converges to A_∞, a.e.. Fatou's lemma now yields

$$\int_\Omega \|A_\infty\| \leqslant \liminf_{n\in\mathbb{N}} \int_\Omega \|A_n\| \leqslant \sup_{n\in\mathbb{N}} \int_\Omega \|A_n\| < \infty .$$

So $A_\infty \in L^1_E$. Consequently $X_\infty =: M_\infty + A_\infty \in L^1_E$ and $(X_n)_{n\in\mathbb{N}}$ converges to X_∞, a.e. .

<u>(i) ⇒ (iii)</u>

This proof is the same as the previous one, now using theorem II.2.2.2.

□

Another result, in the same direction as theorem III.3.3, but now supposing order boundedness of $(\int_\Omega X_n^+)_{n\in\mathbb{N}}$ is the following also due to J. Szulga and W. Woyczynski [1976]. We repeat that an operator is called <u>order bounded</u> if it transforms order bounded sets into order bounded sets.

<u>Theorem III.3.4 (Szulga-Woyczynski)</u> : Let E be a separable Banach lattice and F a separable Banach lattice with (RNP). Suppose $T : E \to F$ is a positive continuous linear operator such that its transpose $T' : F' \to E'$ is order bounded. If $(X_n, F_n)_{n\in\mathbb{N}}$ is a submartingale with values in E such that $(\int_\Omega X_n^+)_{n\in\mathbb{N}}$ is order bounded then there exists $Y_\infty \in L^1_F$ such that the submartingale $(TX_n, F_n)_{n\in\mathbb{N}}$ converges to Y_∞, a.e. .

<u>Proof</u> : If $X_n = M_n + A_n$ is the Doob decomposition of $(X_n, F_n)_{n\in\mathbb{N}}$ as in lemma III.3.2 then, by uniqueness, $TX_n = TM_n + TA_n$ is the Doob decomposition of the submartingale $(TX_n, F_n)_{n\in\mathbb{N}}$. So in view of theorem III.3.3 it suffices to prove that

$$\sup_{n\in\mathbb{N}} \int_\Omega \|(TX_n)^+\| < \infty \quad \text{and} \quad \sup_{n\in\mathbb{N}} \int_\Omega \|(TM_n)^-\| < \infty$$

Now,

$$\sup_{n\in\mathbb{N}} \int_\Omega \|(TX_n)^+\| \leq \sup_{n\in\mathbb{N}} \int_\Omega \|T(X_n^+)\|$$

$$= \sup_{n\in\mathbb{N}} \int_\Omega \sup_{\substack{\|y'\|\leq 1 \\ y'\in F'^+}} y'T(X_n^+)$$

$$= \sup_{n\in\mathbb{N}} \int_\Omega \sup_{\substack{\|y'\|\leq 1 \\ y'\in F'^+}} (T'y')(X_n^+)$$

So, since T' is order bounded, there exists $x_o' \in E'^+$ such that $|T'y'| \leq x_o'$ for every $y' \in F'^+$ with $\|y'\| \leq 1$. So :

$$\sup_{n\in\mathbb{N}} \int_\Omega \|(TX_n)^+\| \leq \sup_{n\in\mathbb{N}} \int_\Omega x_o'(X_n^+) \tag{1}$$

$$= \sup_{n\in\mathbb{N}} x_o' \left(\int_\Omega X_n^+\right)$$

Since $(\int_\Omega X_n^+)_{n\in\mathbb{N}}$ is order bounded, there exists $x_o \in E^+$ such that $\int_\Omega X_n^+ \leq x_o$ for each $n \in \mathbb{N}$. So :

$$\sup_{n\in\mathbb{N}} \int_\Omega \|(TX_n)^+\| \leq x_o'(x_o) < \infty$$

Now

$$\int_\Omega (TM_n)^- = \int_\Omega (TM_n)^+ - \int_\Omega TM_n = \int_\Omega (TM_n)^+ - \int_\Omega TM_o$$

$$\leq \int_\Omega (TX_n)^+ - \int_\Omega TM_o$$

The same method which yields (1) also yields :

$$\sup_{n\in\mathbb{N}} \int_\Omega \|(TM_n)^-\| \leq \sup_{n\in\mathbb{N}} x_o' \left(\int_\Omega (TM_n)^-\right)$$

So :

$$\sup_{n\in\mathbb{N}} \int_\Omega \|(TM_n)^-\| \leq \sup_{n\in\mathbb{N}} x_o' \left(\int_\Omega (TX_n)^+\right) + \left|x_o' \left(\int_\Omega TM_o\right)\right|$$

$$\leq x_o'(x_o) + \left|x_o' \left(\int_\Omega (TM_o)\right)\right| < \infty \quad ,$$

finishing the proof. □

Since $\ell^1(\Gamma)$ for any index set Γ has (RNP) and since obviously $(Id(\ell^1(\Gamma),\ell^1(\Gamma)))'$, the adjoint of the identity operator on $\ell^1(\Gamma)$, is order bounded since it is equal to $Id(\ell^\infty(\Gamma),\ell^\infty(\Gamma))$, we get

Corollary III.3.5 (Szulga and Woyczynski [1976]) : A submartingale $(X_n,F_n)_{n\in\mathbb{N}}$ with values in $\ell^1(\Gamma)$, such that $(\int_\Omega X_n^+)_{n\in\mathbb{N}}$ is order bounded, is convergent a.e. to an integrable function.

In view of the previous corollary one might ask in which Banach lattices all submartingales $(X_n,F_n)_{n\in\mathbb{N}}$ such that $(\int_\Omega X_n^+)_{n\in\mathbb{N}}$ is order bounded are convergent a.e.. Let us first remark that in the spaces $\ell^1(\Gamma)$, where Γ is a directed set, conditions (2) and (3) of the beginning of this section are equivalent. Indeed, the same argument as in the proof of theorem III.3.4 shows that there is $x_o' \in \ell^1(\Gamma)' = \ell^\infty(\Gamma)$ such that

$$\sup_{n\in\mathbb{N}} \int_\Omega \|X_n^+\| \leq \sup_{n\in\mathbb{N}} x_o' \left(\int_\Omega X_n^+\right) \leq \|x_o'\| \sup_{n\in\mathbb{N}} \int_\Omega \|X_n^+\| .$$

Here we apply again that the identity operator on $\ell^\infty(\Gamma)$ is order bounded.

In general Banach lattices, conditions (2) and (3) are not necessarily the same. So we might also ask in which Banach lattices all submartingales $(X_n,F_n)_{n\in\mathbb{N}}$ such that

$$\sup_{n\in\mathbb{N}} \int_\Omega \|X_n^+\| < \infty$$

converge a.e.. Incidentally the two classes are the same and reduce to $\ell^1(\Gamma)$, see further on. We first need some preliminary results.

Lemma III.3.6 : Let E be a Banach lattice. The following assertions are equivalent :

(i) E is an AL-space.

(ii) For every positive $X \in L^1_E$ and G sub-σ-algebra of $\sigma(X)$

$$\|E^G X\| = E^G\|X\| .$$

(iii) For every positive $X \in L^1_E$

$$\|\int_\Omega X\| = \int_\Omega \|X\| .$$

Proof : (i) ⇒ (ii)

For stepfunctions this is easily seen while for a general positive function $X \in L^1_E$ we take a sequence $(X_n)_{n\in\mathbb{N}}$ of positive stepfunctions such that

$$\lim_{n\to\infty} \|X_n - X\| = 0, \text{ a.e. and in the } L^1_E\text{-sense.}$$

This can easily be done.

(ii) ⇒ (iii) is trivial

(iii) ⇒ (i)

Let $x,y \in E^+$ and form

$$X = 2x\,\chi_{[0,\frac{1}{2})} + 2y\,\chi_{[\frac{1}{2},1)}$$

(iii) now implies $\|x+y\| = \|x\| + \|y\|$. □

Theorem III.3.7 (Szulga [1979]) : Let E be an order continuous Banach lattice. Then the following assertions are equivalent :

(i) E is isomorphic to an AL-space,

(ii) Every submartingale $(X_n, F_n)_{n\in \mathbb{N}}$ such that $\sup_{n\in\mathbb{N}} \int_\Omega \|X_n^+\| < \infty$ satisfies $\sup_{n\in\mathbb{N}} \int_\Omega \|X_n\| < \infty$.

(iii) Every martingale $(X_n, F_n)_{n\in \mathbb{N}}$ such that $\sup_{n\in\mathbb{N}} \int_\Omega \|X_n^+\| < \infty$ satisfies $\sup_{n\in\mathbb{N}} \int_\Omega \|X_n\| < \infty$.

Proof : (i) ⇒ (ii)

For $x \in E$ one has $|x| = 2x^+ - x$. Then

$$|X_n| = 2\, X_n^+ - X_n$$

$$\int_\Omega |X_n| = 2 \int_\Omega X_n^+ - \int_\Omega X_n$$

$$\leq 2 \int_\Omega X_n^+ - \int_\Omega X_1$$

since $(X_n, F_n)_{n\in\mathbb{N}}$ is a submartingale. So, using lemma III.3.6 :

$$\int_\Omega \|X_n\| \leq 2 \int_\Omega \|X_n^+\| + \int_\Omega \|X_1\|$$

Hence :

$$\sup_{n\in\mathbb{N}} \int_\Omega \|X_n\| \leq 2 \sup_{n\in\mathbb{N}} \int_\Omega \|X_n^+\| + \int_\Omega \|X_1\| < \infty$$

(ii) ⇒ (iii) is trivial.

(iii) ⇒ (i)

We follow the proof given in Egghe [1982c] . Suppose E is not an AL-space. Then by Schlotterbeck's theorem (theorem III.1.5), let $(x_n)_{n\in\mathbb{N}}$ be in E^+ such that $(x_n)_{n\in\mathbb{N}}$ is commutatively convergent and such that $\Sigma\|x_n\| = \infty$. Put

$$X_1 = x_1 \;\chi_{[0,1)}$$

$$X_2 = x_{11} \;\chi_{[0,\frac{1}{2})} + x_{12} \;\chi_{[\frac{1}{2},1)}$$

where

$$x_{11} = x_1 - 2\, x_2$$

$$x_{12} = x_1 + 2\, x_2$$

So :

$$x_{11}^{+} \leq x_1 \leq \sum_{j=1}^{\infty} x_j$$

Inductively, suppose $X_1,\ldots,X_{k-1}$ have been constructed. Put

$$X_k = \sum_{i=1}^{k-1} x_{i,1}\, \chi_{[1-\frac{1}{2^{i-1}},\, 1-\frac{1}{2^i})} + x_{k-1,2}\, \chi_{[1-\frac{1}{2^{k-1}},\, 1)} \tag{1}$$

where

$$x_{k-1,1} = \frac{1}{2^{k-2}}\, x_{k-2,2} - 2^{k-1}\, x_k \tag{2}$$

$$x_{k-1,2} = 2(1-\frac{1}{2^{k-1}})\, x_{k-2,2} + 2^{k-1}\, x_k \tag{3}$$

So, $(X_n,\sigma(X_1,\ldots,X_n))_{n\in\mathbb{N}}$ is a martingale. We have :

$$\begin{aligned} x_{k-1,2}^{+} = x_{k-1,2} &\leq 2\, x_{k-2,2} + 2^{k-1}\, x_k \\ &\leq 2(2x_{k-3,2} + 2^{k-2}\, x_{k-1}) + 2^{k-1}\, x_k \\ &\leq \ldots\ldots \\ &\leq 2^{k-1} \sum_{i=1}^{k} x_i\,. \end{aligned}$$

Furthermore

$$\begin{aligned} x_{i,1}^{+} &= (\frac{1}{2^{i-1}}\, x_{i-1,2} - 2^i\, x_{i+1}) \vee 0 \\ &\leq \frac{1}{2^{i-1}}\, x_{i-1,2} \end{aligned}$$

$$\leq \sum_{j=1}^{i} x_j$$

It follows that

$$\sup_{n\in\mathbb{N}} \int_\Omega \|X_k^+\| < \infty .$$

Also from $\|x_{i,1}\| \geq 2^i \|x_{i+1}\| - \frac{1}{2^{i-1}} \cdot \|x_{i-1,2}\|$

$$\geq 2^i \|x_{i+1}\| - \frac{1}{2^{i-1}} \|\sum_{j=1}^{\infty} x_j\|$$

we see that

$$\sup_{n\in\mathbb{N}} \int_\Omega \|X_k\| \geq \sum_{i=1}^{\infty} \|x_{i+1}\| - \sum_{i=1}^{\infty} \frac{1}{2^{i-1}} \|\sum_{j=1}^{\infty} x_j\| = \infty$$

finishing the proof. □

The next theorem now solves the problems posed after corollary III.3.5.

<u>Theorem III.3.8 (Szulga [1979])</u> : Let E be a Banach lattice. The following assertions are equivalent :

(i) E has (RNP) and is an AL-space.

(ii) E is isomorphic to a sublattice of $\ell^1(\Gamma)$, for a certain index set Γ.

(iii) Every submartingale $(X_n, F_n)_{n\in\mathbb{N}}$ for which

$$\sup_{n\in\mathbb{N}} \int_\Omega \|X_n^+\| < \infty$$

is convergent strongly a.e. to an integrable function.

(iv) Every submartingale $(X_n, F_n)_{n\in\mathbb{N}}$ such that

$$\left(\int_\Omega X_n^+\right)_{n\in\mathbb{N}}$$

is order bounded is convergent strongly a.e. to an integrable function.

Proof : (i) ⇒ (ii)

From Kakutani's representation theorem of AL-spaces (theorem III.1.6), E is isomorphic to an $L^1(\Phi,\Sigma,\mu)$ for some measure space (Φ,Σ,μ). Since $L^1(\Phi,\Sigma,\mu)$ must also have (RNP), it follows that μ is purely atomic (see Diestel and Uhl [1977] or chapter II).

(ii) ⇒ (iii) and (ii) ⇒ (iv)

This follows from corollary III.3.5 and the remarks following this corollary.

(iii) ⇒ (i)

(iii) implies that every uniformly bounded martingale converges in L^1_E. So, from theorem II.2.2.1, E must have (RNP). If E is not an AL-space, then the same construction as in the proof of (iii) ⇒ (i) in theorem III.3.7 yields a submartingale $(X_n,F_n)_{n\in\mathbb{N}}$ such that

$$\sup_{n\in\mathbb{N}} \int_\Omega \|X_n^+\| < \infty$$

but such that $(X_n)_{n\in\mathbb{N}}$ does not converge a.e. to an integrable function, contradicting (iii).

(iv) ⇒ (i)

If E is not an AL-space, then the same construction as in the proof of (iii) ⇒ (i) in theorem III.3.7 yields a submartingale $(X_n,F_n)_{n\in\mathbb{N}}$ such that

$$\int_\Omega X_n^+ = \sum_{i=1}^{n-1} x_{i,1}^+ \frac{1}{2^i} + x_{i-1,2}^+ \frac{1}{2^{n-1}}$$

$$\leq \sum_{i=1}^{n-1} \left(\sum_{j=1}^{i} x_j\right) \frac{1}{2^i} + \sum_{i=1}^{n} x_i$$

$$\leq 2 \sum_{i=1}^{\infty} x_i \ .$$

Thus the sequence $(\int_\Omega X_n^+)_{n\in\mathbb{N}}$ is order bounded and $(X_n)_{n\in\mathbb{N}}$ does not converge a.e. to an integrable function, a contradiction.
So E is a space $L^1(\Phi,\Sigma,\mu)$ for a certain measure space (Φ,Σ,μ) by the Kakutani representation theorem. Now (iv) implies (iii) since $Id(L^\infty,L^\infty)$ is order bounded (cfr. remarks after Corollary III.3.5). This finishes the proof. □

III.4. Convergence of supermartingales

This section is devoted to the study of convergence of supermartingales. As we have seen, submartingales behave very well in Banach lattices : the theorem of Heinich is indeed a very positive result since it is valid in Banach lattices with (RNP).

On the other hand, supermartingales do not converge so easily. The real theorem which states that positive supermartingales converge a.e. is only true in the $\ell^1(\Gamma)$-spaces as we shall show. Even some weaker convergence properties of supermartingales require E to be an $\ell^1(\Gamma)$. These results are contained in the next theorem. The main part of the proof is due to Benyamini and Ghoussoub [1978].

Theorem III.4.1 : Let E be a Banach lattice. The following assertions are equivalent :

(i) E is isomorphic to a sublattice of $\ell^1(\Gamma)$, for a certain set Γ.

(ii) Every uniformly integrable super-(or sub-) martingale is L_E^1-convergent.

(iii) Every positive uniformly bounded supermartingale is L_E^1-convergent.

(iv) Every positive L_E^1-bounded supermartingale converges weakly a.e..

(v) Every positive supermartingale converges strongly a.e..

Proof : The proof runs as follows :

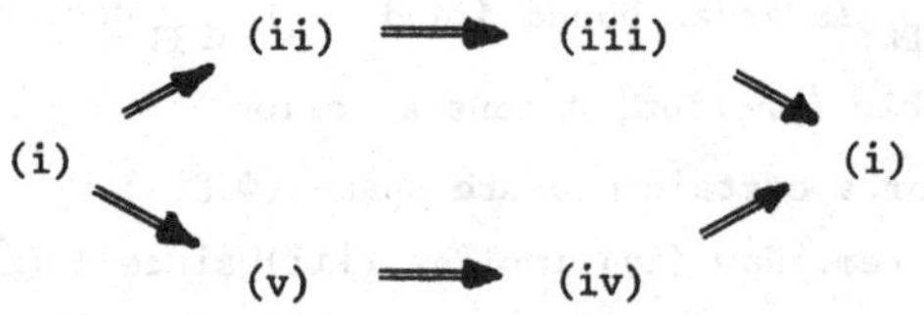

Amongst these implications, the following are completely trivial :
(ii) ⇒ (iii), (v) ⇒ (iv).

<u>(i) ⇒ (v)</u> :
Let $(X_n,\mathcal{F}_n)_{n\in\mathbb{N}}$ be a positive supermartingale. Then $(-X_n,\mathcal{F}_n)_{n\in\mathbb{N}}$ is a submartingale and obviously $\sup_{n\in\mathbb{N}} \int_\Omega \|(-X_n)^+\| = 0 < \infty$. So, from theorem III.3.8, $(X_n)_{n\in\mathbb{N}}$ converges strongly a.e..

<u>(i) ⇒ (ii)</u>
Follows immediately from theorem III.3.8.

<u>(iii) ⇒ (i)</u> and <u>(iv) ⇒ (i)</u>
We first show that E is an AL-space. Suppose not. Then by Schlotterbeck's theorem (theorem III.1.5) there exists a disjoint sequence $(x_n)_{n\in\mathbb{N}}$ in E^+ such that Σx_n is commutatively convergent and such that $\Sigma\|x_n\| = \infty$. Hence there exists a sequence $(m_k)_{k\in\mathbb{N}}$ such that

$$\sum_{i=m_k+1}^{m_{k+1}} \|x_i\| \geq 1$$

for each $k \in \mathbb{N}$. We can suppose

$$\sum_{i=m_k+1}^{m_{k+1}} \|x_i\| = 1$$

since we may multiply $(x_n)_{n\in\mathbb{N}}$ by a number in $[0,1]$; this does not disturb the commutative convergence of Σx_n.
For every $k \in \mathbb{N}$ we take $m_{k+1} - m_k$ disjoint subintervals $(A_{i,k})_{i=m_k+1}^{m_{k+1}}$ of

$[0,1]$ such that the length of each interval $A_{i,k}$ is $\frac{\|x_i\|}{\alpha_k}$ where the sequence $(\alpha_k)_{n\in\mathbb{N}}$ is chosen later on. Put $A_k = \bigcup_{i=m_k+1}^{m_{k+1}} A_{i,k}$. Put $\Omega_k = [0,1]$, λ_k the Lebesguemeasure on Ω_k and F_k the Borel-σ-algebra on Ω_k. Put also $\Omega = \prod_{k\in\mathbb{N}} \Omega_k$, $F = \prod_{k\in\mathbb{N}} F_k$, $P = \prod_{k\in\mathbb{N}} \lambda_k$. Define :

$$X_k(\omega) = \begin{cases} \alpha_k \dfrac{x_i}{\|x_i\|} + \sum\limits_{j>m_{k+1}} x_j & \text{for } \omega_k \in A_{i,k},\ m_k < i \leqslant m_{k+1} \\ \sum\limits_{j>m_{k+1}} x_j & \text{for } \omega_k \notin A_k \end{cases}$$

where ω_k = the kth-coordinate of $\omega \in \Omega = \prod_{k\in\mathbb{N}} \Omega_k$. Then for each $\omega \in \Omega$:

$$\|X_k(\omega)\| \leqslant \begin{cases} \alpha_k + \|x\| & \text{for } \omega_k \in A_k \\ \|x\| & \text{for } \omega_k \notin A_k \end{cases}$$

where $x = \sum_{j=1}^{\infty} x_j$, and hence $\int\|X_k\| \leqslant 1 + \|x\|$ since $P(A_k \times \prod_{j\neq k} \Omega_j) = \lambda_k(A_k) = \frac{1}{\alpha_k}$. So every $X_k \in L^1_E$ and $(X_k)_{k\in\mathbb{N}}$ is L^1_E-bounded. Obviously, $(X_k)_{k\in\mathbb{N}}$ is an independent sequence and also a supermartingale :

$$\int X_k = (1 - P(A_k \times \prod_{j\neq k} \Omega_j)) \sum_{j>m_{k+1}} x_j$$

$$+ \sum_{i=m_k+1}^{m_{k+1}} P(A_{i,k} \times \prod_{j\neq k} \Omega_j)\,[\alpha_k \frac{x_i}{\|x_i\|} + \sum_{j>m_{k+1}} x_j]$$

$$= \sum_{i=m_k+1}^{m_{k+1}} x_n + \sum_{j>m_{k+1}} x_j = \sum_{j>m_k} x_j \leqslant \int X_{k-1}$$

So, by the same calculation and by independence, for each $A \in \sigma(X_1,\dots,X_{k-1})$

$$\int_A X_{k-1} \geq \int_\Omega X_k . P(A) = \int_A X_k$$

Now we finish proving (iii) ⇒ (i). Choose $\alpha_k = 1$ for every $k \in \mathbb{N}$. So $(X_k)_{k\in\mathbb{N}}$ is uniformly bounded and not L^1_E-convergent since for each $\omega \in \Omega$ we have

$$\|X_k(\omega) - X_{k+1}(\omega)\| \geq \min_{m_k+1 \leq i \leq m_{k+1}} \left\|\frac{x_i}{\|x_i\|}\right\| = 1$$

Hence $(X_n)_{n\in\mathbb{N}}$ is not L^1_E-convergent, a contradiction.
Now we finish proving (iv) ⇒ (i). Choose $\alpha_k = k$ and put $B_k = \{\omega\in\Omega \| \omega_k \in A_k\}$. Then $(B_k)_{k\in\mathbb{N}}$ are independent and $P(B_k) = \frac{1}{k}$ so $\Sigma P(B_k) = \infty$. So by Borel-Cantelli, almost all $\omega \in \Omega$ are in an infinite number of B_k. Now, if $\omega \in B_k$, $\|X_k(\omega)\| \geq k$. Then for almost all ω, the sequence $(\|X_n(\omega)\|)_{n\in\mathbb{N}}$ is not bounded; so $(X_n)_{n\in\mathbb{N}}$ is not weakly convergent a.e., a contradiction.
We have now shown that E is an AL-space, i.e. $E = L^1(\mu)$ for a certain measure. Suppose now that μ is not purely atomic. So $L^1[0,1]$ is isomorphic to a subspace of $L^1(\mu)$. However a uniformly bounded positive martingale which is not weakly convergent a.e. is constructed as in the example after theorem II.2.2.1 :
Let $\Omega = [0,1]$ with Lebesguemeasure and let

$$F_k = \sigma(\{[\frac{m-1}{2^k},\frac{m}{2^k}] \mid m = 1,2,\dots,2^k\})$$

Define, for $\omega \in [\frac{m-1}{2^k},\frac{m}{2^k}]$

$$X_k(\omega) = 2^k \chi_{[\frac{m-1}{2^k},\frac{m}{2^k}]}$$

Then $(X_k,F_k)_{k\in\mathbb{N}}$ is a positive $L^1[0,1]$-valued uniformly bounded martingale which is not weakly convergent a.e. (cf. remark II.2.4.7(2)). So μ must be purely atomic. Hence $E = L^1(\mu) = \ell^1(\Gamma)$ for some Γ.

Another argument proving that $E = L^1(\mu)$ is in fact an $\ell^1(\Gamma)$ follows from the fact that (iii) and (iv) both imply that E must have (RNP), due to the proof of theorem III.2.5. So, since $E = L^1(\mu)$ and has (RNP), $E = \ell^1(\Gamma)$ for a certain set Γ. □

Remark III.4.2 : In chapter V we shall show :

Theorem : For a Banach lattice E the following assertions are equivalent :

(i) E has (RNP) and every separable sublattice F of E has a quasi-interior point in F'. (For terminology see Schaefer [1974] or chapter V).

(ii) Every supermartingale of class (B) in E converges weakly a.e..

(iii) Every positive supermartingale of class (B) in E converges weakly a.e..

This is a result of Ghoussoub and Talagrand [1979a]. The result belongs to this chapter but the proof uses some concepts developed in chapter V. So for the proof we refer to chapter V.

III.5. Submartingale convergence in Banach lattices without (RNP)

We remarked already that (RNP) Banach lattices are order-continuous (see section III.1). Obviously c_0 is an example of an order-continuous Banach lattice without (RNP).

It is for this strictly larger class of Banach lattices that we shall prove a convergence result for submartingales. The method of proof, given by Davis-Ghoussoub-Lindenstrauss [1981], uses also theorem III.2.3.

Theorem III.5.1 (Davis-Ghoussoub-Lindenstrauss) : Let E be an order-continuous Banach lattice and $(X_n, \mathcal{F}_n)_{n \in \mathbb{N}}$ an L^1_E-bounded submartingale such that

$$0 \leq X_n \leq Y_n, \text{ a.e.}$$

for each $n \in \mathbb{N}$, where $(Y_n)_{n\in\mathbb{N}}$ is a sequence of measurable functions converging a.e. to function Y. Then there exists $X_\infty \in L^1_E$ such that $\lim_{n\to\infty} X_\infty = X$, a.e. .

<u>Proof</u> : By theorem III.2.3, we may suppose that $\|\cdot\|$ is the equivalent renorming with the properties mentioned in theorem III.2.3, and that D be the countable norming subset of E'^+.
We have $Y \leq X_n \vee Y \leq Y_n \vee Y$, a.e.. Thus $X_n \vee Y \to Y$, a.e.. So, since

$$X_n \wedge Y + X_n \vee Y = X_n + Y$$

it follows that for every $x' \in D$

$$\lim_{n\to\infty} x'(X_n \wedge Y) = \lim_{n\to\infty} x'(X_n)$$

exists a.e. since $(x'(X_n), F_n)_{n\in\mathbb{N}}$ is an L^1_E-bounded submartingale.
By theorem III.1.2, the order interval $[0,Y(\omega)]$ is weakly compact. So the sequence $(X_n(\omega) \wedge Y(\omega))_{n\in\mathbb{N}}$ has a weakly convergent subsequence. Denote the limit by $X_\infty(\omega)$. Then

$$\lim_{n\to\infty} x'(X_n(\omega) \wedge Y(\omega)) = x'(X_\infty(\omega)) \text{ a.e. },$$

for each $x' \in D$. Hence X_∞ is scalarly measurable and hence by Pettis' theorem (theorem I.1.1.1) X_∞ is strongly measurable. $X_\infty \in L^1_E$ by Fatou's lemma. Furthermore

$$x'(X_n \wedge Y) \to x'(X_\infty) \text{ a.e.}$$

for each $x' \in D$, and hence, by theorem II.2.4.5

$$\|X_n \wedge Y\| = \sup_{x'\in D} x'(X_n \wedge Y) \to \sup_{x'\in D} x'(X_\infty) = \|X_\infty\| \text{ a.e.}$$

since D is countable. An application of theorem III.2.3 finishes the proof. □

Remark III.5.2 : The version of theorem III.5.1 is a slight modification of the theorem of Davis-Ghoussoub-Lindenstrauss, given by M. Słaby [1982], but the proof remains essentially the same. An extension of theorem III.5.1 to subpramarts is given in theorem VIII.3.3.1, due to Słaby [1982a]

Corollary III.5.3 (Ghoussoub [1977]) : If E is ordercontinuous then every orderbounded submartingale converges strongly a.e..

Proof : Let x_o and $y_o \in E$ be such that

$$x_o \leq X_n(\omega) \leq y_o$$

for each $n \in \mathbb{N}$ and $\omega \in \Omega$. Then

$$0 \leq X_n(\omega) - x_o \leq y_o - x_o$$

and we can apply theorem III.5.1. □

Of the same type is a result in Ghoussoub and Talagrand [1979b], lemma 4, p.222.

III.6. Notes and remarks

An extension of theorem III.3.4 by Szulga [1979] goes as follows :

Definition III.6.1 : Let E be a Banach lattice and F a Banach space. Let $T : E \to F$ be a continuous linear operator. T is called cone absolutely summing (c.a.s.) if it transforms unconditionally convergent positive series into absolutely convergent ones. For properties of c.a.s. operators we refer the reader to Schaefer [1974], IV.3; see also results of Jeurnink [1982]. Szulga proved :

Theorem III.6.2 (Szulga) : Let $T : E \to F$ be a c.a.s. operator (E Banach lattice, F Banach space). Assume that F has (RNP). If $(X_n, \mathcal{F}_n)_{n \in \mathbb{N}}$ is a submartingale such that

$$\sup_{n\in\mathbb{N}} \left\| \int_\Omega X_n^+ \right\| < \infty \quad .$$

Then there exists $Y \in L^1_F$ such that $(TX_n)_{n\in\mathbb{N}}$ converges to Y, a.e. .
This is indeed a generalization of theorem III.3.4 since, if T satisfies the requirements of theorem III.3.4, we have for each sequence $(x_n)_{n\in\mathbb{N}}$ in E^+ :

$$\sum_{n=1}^{\infty} \|Tx_n\| = \sum_{n=1}^{\infty} \sup_{\substack{y'\in F' \\ \|y'\|\leq 1}} y'(Tx_n)$$

$$= \sum_{n=1}^{\infty} \sup_{\substack{y'\in F' \\ \|y'\|\leq 1}} T'y'(x_n)$$

$$\leq \sum_{n=1}^{\infty} x_o'(x_n) \quad \text{(see the proof of III.3.4).}$$

Hence T is c.a.s. .

A property also equivalent to the equivalent properties in lemma III.3.6 is

<u>Lemma III.6.3</u> : Let E be a Banach lattice. The following assertions are equivalent :

(i) E is an AL-space.

(ii) Every positive supermartingale is of class (B).

<u>Proof</u> : <u>(i) ⇒ (ii)</u>
Since $(X_n, F_n)_{n\in\mathbb{N}}$ is a supermartingale we have

$$E^{F_\sigma} X_\tau \leq X_\sigma$$

for every $\sigma \in T$ and $\tau \in T(\sigma)$ (cf. definition III.2.1). Hence, due to positivity

$$\|E^{F_\sigma} X_\tau\| \leq \|X_\sigma\| \tag{1}$$

Combining lemma III.3.6 with (1) and integrating we find

$$\int_\Omega \|X_\tau\| \leq \int_\Omega \|X_\sigma\|$$

So :

$$\sup_{\tau \in T} \int_\Omega \|X_\tau\| \leq \int_\Omega \|X_1\|$$

(ii) ⇒ (i)

Suppose E is not an AL-space. Using the Schlotterbeck theorem III.1.5 and the construction in theorem III.4.1, (iv) ⇒ (i), we find a positive supermartingale with the property that

$$P(\sup_{n \in \mathbb{N}} \|X_n\| = \infty) = 1 .$$

Thus from lemma II.1.5, $(X_n, \mathcal{F}_n)_{n \in \mathbb{N}}$ is not of class (B). □

III.6.4. In theorem III.2.2 the following result has been implicitly proved - cf. also the proof of III.2.5 :

Theorem III.6.4.1 (Szulga) : Suppose that the Banach lattice does not contain c_o. Let $(X_n, \mathcal{F}_n)_{n \in \mathbb{N}}$ be a submartingale such that

$$\sup_{n \in \mathbb{N}} \int_\Omega \|X_n^+\| < \infty$$

Then $(X_n, \mathcal{F}_n)_{n \in \mathbb{N}}$ can be written as

$$X_n = Y_n - Z_n$$

where $(Y_n, \mathcal{F}_n)_{n \in \mathbb{N}}$ is a positive L_E^1-bounded martingale and $(Z_n, \mathcal{F}_n)_{n \in \mathbb{N}}$ is a positive L_E^1-bounded supermartingale.

We use here exactly the same proof as in a part of the proof of theorem III.2.5, together with the fact that if $(X_n, \mathcal{F}_n)_{n \in \mathbb{N}}$ is a submartingale, then, trivially $(X_n^+, \mathcal{F}_n)_{n \in \mathbb{N}}$ is also a submartingale.

This decomposition is called the Krickeberg decomposition of the submartingale $(X_n, F_n)_{n\in\mathbb{N}}$ - cf. also Neveu [1975] for the real case. This decomposition is minimal in the sense that if

$$X_n = Y'_n - Z'_n$$

where $(Y'_n, F_n)_{n\in\mathbb{N}}$ is a positive martingale and where $(Z'_n, F_n)_{n\in\mathbb{N}}$ is a positive supermartingale, then, with the notation of theorem III.6.4.1,

$$Y'_n \geqslant Y_n \quad \text{and} \quad Z'_n \geqslant Z_n .$$

This is seen as follows :

$$X_n = Y_n - Z_n = Y'_n - Z'_n \quad \text{and} \quad Y_n = \lim_{p\to\infty} \uparrow E^{F_n} X_p^+$$

$\forall n \in \mathbb{N}$. So $Y'_n = X_n + Z'_n \geqslant X_n$. Thus, since $Y'_n \geqslant 0$, $Y'_n \geqslant X_n^+$. Hence, for every $m \in \mathbb{N}$ such that $m \leqslant n$, one has

$$Y'_m = E^{F_m} Y'_n \geqslant E^{F_m} X_n^+$$

Hence, for every $n \in \mathbb{N}$ and $p \in \mathbb{N}(n)$ one has

$$Y'_n \geqslant E^{F_n} X_p^+$$

and so $Y'_n \geqslant Y_n$ for every $n \in \mathbb{N}$. Hence also $Z'_n \geqslant Z_n$, for every $n \in \mathbb{N}$.

Chapter IV : BASIC INEQUALITIES FOR ADAPTED SEQUENCES

In this chapter we shall prove some inequalities - all of the same type - for adapted sequences with values in a Banach space, which have important applications to the convergence of adapted sequences. These inequalities allow us to deduce convergence from the way the adapted sequence ressembles a martingale. For example an inequality of the type

$$\limsup_{m,n \in \mathbb{N}} \|X_n(\omega) - X_m(\omega)\| \leq C.\limsup_{\substack{m \in \mathbb{N} \\ n \in \mathbb{N}(m)}} \|E^{F_m} X_n(\omega) - X_m(\omega)\|$$

supposed to be valid a.e. implies convergence a.e. of $(X_n)_{n \in \mathbb{N}}$ as soon as

$$\limsup_{\substack{m \in \mathbb{N} \\ n \in \mathbb{N}(m)}} \|E^{F_m} X_n - X_m\| = 0,$$

a.e.. It is the purpose of this chapter to investigate which conditions on $(X_n, F_n)_{n \in \mathbb{N}}$ and which conditions on the Banach space E imply the validity of such inequalities and others. This is done in section 1.

We shall see that the results are rather sharp in the sense that :

1) The necessity of some conditions is proved by describing the geometric example of Mc Cartney-O'Brien : this is another example of the relation between adapted sequences and the geometry of Banach spaces (cf. also sections II.2.3 and II.4.7). This is done in section 2.
2) Applications of the inequalities yield all known Banach space valued convergence results in their full generality and even some new results

or extensions of old results : this is carried out in chapter V and chapter VII.

Most of the results in the first section are due to Bellow and Egghe [1981] and [1982].

IV.1. Basic inequalities

Let E be a separable Banach space. We are always in a position to suppose this since we are working with adapted sequences $(X_n, F_n)_{n\in\mathbb{N}}$ consisting of Bochner integrable functions. Let T be the norm topology on E, or the weak topology $\sigma(E,E')$, or in case E is a dual Banach space, $E = F'$, T can be the weak*-topology $\sigma(F',F)$. T can be more general but we only have applications for the above mentioned topologies. So we restrict ourselves to these topologies. We denote by Q_T the set of all mappings $q : E \to \mathbb{R}^+$ satisfying

(i) q is a continuous seminorm on $E, \|\cdot\|$

(ii) There is a countable set $D_q \subset (E,T)'$ such that $\|x'\| \leq 1$ for every $x' \in D_q$ and such that

$$q(x) = \sup_{x'\in D_q} |x'(x)|$$

for all $x \in E$.

Examples of seminorms q belonging to Q_T are $q(\cdot) = |x'(\cdot)|$ with $x' \in (E,T)'$ and, if $T = \|\cdot\|$-topology, $q(\cdot) = \|\cdot\|$.

Before we can prove our first inequalities, we need some lemmas :

Lemma IV.1.1 : Let $(X_n, F_n)_{n\in\mathbb{N}}$ be an adapted sequence and let $\sigma \in T$ and $\gamma_j \in T(\sigma)$ for each $j \in \mathbb{N}$. Let q be an arbitrary continuous seminorm on $E, \|\cdot\|$. Then there is a sequence $(\delta_n)_{n\in\mathbb{N}}$ in $T(\sigma)$ such that

$$q(E^{F_\sigma} X_{\delta_n}(\omega) - X_\sigma(\omega)) \uparrow \sup_{j\in\mathbb{N}} q(E^{F_\sigma} X_{\gamma_j}(\omega) - X_\sigma(\omega))$$

Proof : For each $n \in \mathbb{N}$, let $\{A_1,\ldots,A_n\}$ be a partition of Ω with $A_i \in F_\sigma$ for each $1 \leq i \leq n$ and such that on A_i

$$q(E^{F_\sigma}X_{\gamma_i}(\omega) - X_\sigma(\omega)) = \sup_{1\leq j\leq n} q(E^{F_\sigma}X_{\gamma_j}(\omega) - X_\sigma(\omega))$$

Define δ_n by setting $\delta_n = \gamma_i$ on A_i $(1 \leq i \leq n)$. Then $\delta_n \in T(\sigma)$, by the localization property in T (see I.2.1) and

$$q(E^{F_\sigma}X_{\delta_n}(\omega) - X_\sigma(\omega)) = \sup_{1\leq j\leq n} q(E^{F_\sigma}X_{\gamma_j}(\omega) - X_\sigma(\omega))$$

This clearly increases to $\sup_{j\in\mathbb{N}} q(E^{F_\sigma}X_{\gamma_j}(\omega) - X_\sigma(\omega))$. □

Corollary IV.1.2 : Let $(X_n,F_n)_{n\in\mathbb{N}},\sigma,\gamma_j$ $(j\in\mathbb{N})$ and q be as in lemma IV.1.1. Suppose that Z is a finite positive measurable function such that a.e.

$$Z(\omega) \leq \sup_{j\in\mathbb{N}} q(E^{F_\sigma}X_{\gamma_j}(\omega) - X_\sigma(\omega))$$

Then there exists a sequence $(\tau_n)_{n\in\mathbb{N}}$ in $T(\sigma)$ such that for each $n \in \mathbb{N}$

$$P(\{\omega\in\Omega \,\|\, q(E^{F_\sigma}X_{\tau_n}(\omega) - X_\sigma(\omega)) \geq Z(\omega) - \frac{1}{2^n}\}) \geq 1 - \frac{1}{2^n}.$$

Proof : Apply lemma IV.1.1 to yield a sequence $(\delta_n)_{n\in\mathbb{N}}$ in $T(\sigma)$ such that

$$q(E^{F_\sigma}X_{\delta_n}(\omega) - X_\sigma(\omega)) \uparrow \sup_{j\in\mathbb{N}} q(E^{F_\sigma}X_{\gamma_j}(\omega) - X_\sigma(\omega))$$

Hence, for each $n \in \mathbb{N}$, there is a large enough $j_n \in \mathbb{N}$ such that

$$P(\{\omega\in\Omega \,\|\, q(E^{F_\sigma}X_{\delta_{j_n}}(\omega) - X_\sigma(\omega)) \geq \sup_{j\in\mathbb{N}} q(E^{F_\sigma}X_{\gamma_j}(\omega) - X_\sigma(\omega)) - \frac{1}{2^n}\})$$

$$\geq 1 - \frac{1}{2^n}.$$

Hence, also

$$P(\{\omega \in \Omega \| q(E^{F_\sigma} X_{\delta_{j_n}}(\omega) - X_\sigma(\omega)) \geq Z(\omega) - \frac{1}{2^n}\}) \geq 1 - \frac{1}{2^n}$$

Now take $\tau_n = \delta_{j_n}$ for each $n \in \mathbb{N}$. □

The following lemma is well-known and easy to prove and hence the proof is omitted. (cf. the argument of Pettis in Dunford and Schwartz [1957], p.318-319 works here too).

Lemma IV.1.3 : Let $(f_k)_{k \in \mathbb{N}}$ be an L^1_E-bounded sequence. Let G be a sub-σ-algebra of F. Assume that

$$\nu(A) = T - \lim_{k \to \infty} \int_A f_k$$

exists for each $A \in G$. Then :

(i) ν is strongly countably additive .

(ii) $|\nu|(A) \leq \sup_{k \in \mathbb{N}} \int_A \|f_k\|$, for each $A \in G$.

(iii) $\nu \ll P|G$.

We can now state and prove the main inequalities :

Theorem IV.1.4 (Bellow and Egghe) : Assume that E has (RNP) and that $(X_n, F_n)_{n \in \mathbb{N}}$ is an adapted sequence in L^1_E.

Case I

(A_T) Suppose that there is a subsequence $(X_{n_k})_{k \in \mathbb{N}}$ such that $(X_{n_k})_{k \in \mathbb{N}}$ is L^1_E-bounded and such that for each $A \in \bigcup_n F_n$

$$T - \lim_{k \to \infty} \int_A X_{n_k}$$

exists. Let q be any seminorm in Q_T. Then there exists an L^1_E-bounded martingale $(Y_n, F_n)_{n \in \mathbb{N}}$ converging a.e. to a function $Y \in L^1_E$ such that for every $q \in Q_T$ we have a.e.

$$\limsup_{m\in\mathbb{N}} q(Y_m(\omega) - X_m(\omega)) \tag{1}$$

$$\leq \limsup_{m\in\mathbb{N}} \big(\sup_{n\in\mathbb{N}(m)} q(E^{F_m} X_n(\omega) - X_m(\omega))\big)$$

Hence also a.e.

$$\limsup_{m,n\in\mathbb{N}} q(X_n(\omega) - X_m(\omega)) \tag{1'}$$

$$\leq 2 \limsup_{m\in\mathbb{N}} \big(\sup_{n\in\mathbb{N}(m)} q(E^{F_m} X_n(\omega) - X_m(\omega))\big)$$

<u>Case II</u>

(B_T) Suppose that there is a sequence $(\gamma_n)_{n\in\mathbb{N}}$ in $T(n)$ such that $\gamma_n \leq \gamma_{n+1}$ for each $n \in \mathbb{N}$, such that $(X_{\gamma_n})_{n\in\mathbb{N}}$ is L^1_E-bounded and such that for each $A \in \bigcup_n F_n$

$$T - \lim_{n\to\infty} \int_A X_{\gamma_n}$$

exists. Then there exists an L^1_E-bounded martingale $(Y_n, F_n)_{n\in\mathbb{N}}$ converging a.e. to a function $Y \in L^1_E$ such that for every $q \in Q_T$ there are increasing sequences $(\sigma_n)_{n\in\mathbb{N}}$ and $(\tau_n)_{n\in\mathbb{N}}$ in T with $\tau_n \geq \sigma_n \geq n$ for each $n \in \mathbb{N}$ and such that a.e.

$$\limsup_{m\in\mathbb{N}} q(Y_m(\omega) - X_m(\omega)) \leq \liminf_{n\in\mathbb{N}} q(E^{F_{\sigma_n}} X_{\tau_n}(\omega) - X_{\sigma_n}(\omega)) \tag{2}$$

Hence we have also a.e.

$$\limsup_{m,n\in\mathbb{N}} q(X_m(\omega) - X_n(\omega)) \leq 2 \liminf_{n\in\mathbb{N}} q(E^{F_{\sigma_n}} X_{\tau_n}(\omega) - X_{\sigma_n}(\omega)) \tag{2'}$$

<u>Proof</u> : <u>Case I</u>

Let $D_q \in (E,T)'$ be the countable (semi-)norming set mentioned above.

Define :

$$\mu_m(A) = T - \lim_{k\to\infty} \int_A X_{n_k}$$

for each $A \in F_m$ and each $m \in \mathbb{N}$. From lemma IV.1.3 it follows that μ_m is of bounded variation (on F_m), is countably additive and that $\mu_m \ll P|F_m$. So the Radon-Nikodym derivative $Y_m = \frac{d\mu_m}{d(P|F_m)}$ of μ_m w.r.t. $P|F_m$ exists in L^1_E and is F_m-measurable. Also lemma IV.1.3 shows that $(Y_m)_{m\in\mathbb{N}}$ is L^1_E-bounded and it is trivial that $(Y_m,F_m)_{n\in\mathbb{N}}$ is a martingale. Hence it converges a.e. (theorem II.2.4.3). For every $m \in \mathbb{N}$, $A \in F_m$ and $x' \in D_q$ we now have :

$$\left|\int_A x'(Y_m - X_m)\right| = \left|x'\left(\int_A (Y_m - X_m)\right)\right|$$

$$= \lim_{k\to\infty} \left|x'\left(\int_A X_{n_k} - X_m\right)\right|$$

$$= \lim_{k\to\infty} \left|x'\left(\int_A (E^{F_m}X_{n_k} - X_m)\right)\right|$$

$$\leq \sup_{k\in\mathbb{N}(m)} \int_A q(E^{F_m}X_{n_k} - X_m)$$

$$\leq \int_A^* \sup_{n\in\mathbb{N}(m)} q(E^{F_m}X_n - X_m) \qquad (1)$$

where $\int^*$ denotes the outer integral. If $\sup_{n\in\mathbb{N}(m)} q(E^{F_m}X_n - X_m)$ were integrable, we would have immediately that

$$|x'(Y_m(\omega) - X_m(\omega))| \leq \sup_{n\in\mathbb{N}(m)} q(E^{F_m}X_n(\omega) - X_m(\omega)) \text{ a.e.} \quad (2)$$

However we can still prove (2) but we have to be a bit careful. Denote by G_m the function $\sup_{n\in\mathbb{N}(m)} q(E^{F_m}X_n - X_m)$ and put

$$C_m = \{\omega\in\Omega \| G_m(\omega) = +\infty\} \in F_m$$

On C_m (2) is clear. On $\Omega \setminus C_m$ we have :

$$\Omega \setminus C_m = \bigcup_{j \in \mathbb{N}} \Omega_j \ , \text{ where } \Omega_j = \{G_m \leqslant j\} \quad .$$

Since (1) holds for every $A \in \mathcal{F}_m$ and since $\Omega_j \in \mathcal{F}_m$ for each $j \in \mathbb{N}$ it follows that (2) is valid a.e. on Ω_j, for each $x' \in D$. So (2) is valid on $\Omega \setminus C_m$, a.e. and hence on Ω, a.e.. From this and the fact that D is countable and semi-norming we have

$$q(Y_m(\omega) - X_m(\omega)) \leqslant G_m(\omega) \tag{3}$$

a.e.. Now, a.e.

$$\limsup_{n,m \in \mathbb{N}} q(X_n(\omega) - X_m(\omega))$$

$$\leqslant \limsup_{n \in \mathbb{N}} q(X_n(\omega) - Y_n(\omega)) + \limsup_{m,n \in \mathbb{N}} q(Y_n(\omega) - Y_m(\omega))$$

$$+ \limsup_{m \in \mathbb{N}} q(Y_m(\omega) - X_m(\omega))$$

$$\leqslant 2 \limsup_{n \in \mathbb{N}} q(X_n(\omega) - Y_n(\omega)) \quad ,$$

since $(Y_n)_{n \in \mathbb{N}}$ converges a.e. and since q is a continuous semi-norm on E. This, together with (3) finishes the proof of case I.

Case II

In exactly the same way as in case I and keeping the same notations we again have an L^1_E-bounded martingale $(Y_m, \mathcal{F}_m)_{m \in \mathbb{N}}$ and we also have for each $x' \in D_q$, now taking a stopping time $\sigma \in T$ instead of m and $(\gamma_n)_{n \in \mathbb{N}}$ instead of $(n_k)_{k \in \mathbb{N}}$:

$$\left| \int_A x'(Y_\sigma - X_\sigma) \right| \leqslant \int_A^* \sup_{n \in \mathbb{N}(\sigma)} q(E^{\mathcal{F}_\sigma} X_{\gamma_n} - X_\sigma)$$

and the same reasoning as in case I gives, a.e.

$$q(Y_\sigma(\omega) - X_\sigma(\omega)) \leqslant \sup_{p \in \mathbb{N}(\sigma)} q(E^{\mathcal{F}_\sigma} X_{\gamma_p}(\omega) - X_\sigma(\omega)) \tag{4}$$

We now apply corollary I.3.5.2 yielding a sequence $(\sigma_n)_{n\in\mathbb{N}}$ in T with $n\leqslant\sigma_n\leqslant\sigma_{n+1}$ for each $n\in\mathbb{N}$ such that, a.e.

$$\limsup_{n\in\mathbb{N}} q(Y_n(\omega)-X_n(\omega)) = \lim_{n\to\infty} q(Y_{\sigma_n}(\omega)-X_{\sigma_n}(\omega)) \qquad (5)$$

(4) and corollary IV.1.2 imply that for each $n\in\mathbb{N}$, there exists a stopping time $\tau_n\in T(\sigma_n)$ such that

$$P(\{\omega\in\Omega \| q(E^{F_{\sigma_n}}X_{\tau_n}(\omega)-X_{\sigma_n}(\omega)) \geqslant q(Y_{\sigma_n}(\omega)-X_{\sigma_n}(\omega)) - \frac{1}{2^n}\})$$

$$\geqslant 1-\frac{1}{2^n}$$

Put

$$B_n = \{\omega\in\Omega \| q(E^{F_{\sigma_n}}X_{\tau_n}(\omega)-X_{\sigma_n}(\omega)) < q(Y_{\sigma_n}(\omega)-X_{\sigma_n}(\omega)) - \frac{1}{2^n}\}$$

then $P(B_\infty)=0$ where $B_\infty = \limsup_{n\in\mathbb{N}} B_n$. On $\Omega\setminus B_\infty$ we have for n large enough

$$q(E^{F_{\sigma_n}}X_{\tau_n}(\omega)-X_{\sigma_n}(\omega)) \geqslant q(Y_{\sigma_n}(\omega)-X_{\sigma_n}(\omega)) - \frac{1}{2^n} \qquad (6)$$

$(\sigma_n)_{n\in\mathbb{N}}$ is increasing and by eventually passing to a subsequence if necessary we can suppose also that $(\tau_n)_{n\in\mathbb{N}}$ is increasing, since $\sigma_n\geqslant n$ for each $n\in\mathbb{N}$. (6) together with (5) now yields, a.e.

$$\limsup_{n\in\mathbb{N}} q(Y_n(\omega)-X_n(\omega) \leqslant \liminf_{n\in\mathbb{N}} q(E^{F_{\sigma_n}}X_{\tau_n}(\omega)-X_{\sigma_n}(\omega))$$

finishing the proof. □

We now prove a corollary to theorem IV.1.4, involving the Pettis norm $\|\cdot\|_{Pe}$. This corollary will be used in V.2.

<u>Corollary IV.1.5 (Bellow and Egghe)</u> : Suppose E has (RNP) and let $(X_n,F_n)_{n\in\mathbb{N}}$ be an adapted sequence. With the notations of theorem

IV.1.4, if condition (A_T) is satisfied for $T = \sigma(E,E')$, then for each $\sigma \in T$

$$\|Y_\sigma - Y_\sigma\|_{Pe} \leq 2 \liminf_{k \in \mathbb{N}(\sigma)} \|E^{F_\sigma} X_{n_k} - X_\sigma\|_{Pe} \qquad (1)$$

and if condition (B_T) is satisfied for $T = \sigma(E,E')$, then for each $\sigma \in T$

$$\|Y_\sigma - X_\sigma\|_{Pe} \leq 2 \liminf_{n \in \mathbb{N}(\sigma)} \|E^{F_\sigma} X_{\gamma_n} - X_\sigma\|_{Pe}$$

<u>Proof</u> : (1)

Fix $x' \in E'$ and $A \in F_\sigma$. From the proof of theorem IV.1.4 we have

$$x'(\int_A (Y_\sigma - X_\sigma)) = \lim_{k \in \mathbb{N}(\sigma)} \int_A x'(X_{n_k} - X_\sigma)$$

$$= \lim_{k \in \mathbb{N}(\sigma)} \int_A x' (E^{F_\sigma} X_{n_k} - X_\sigma)$$

$$\leq \liminf_{k \in \mathbb{N}(\sigma)} \int_A |x'(E^{F_\sigma} X_{n_k} - X_\sigma)|$$

$$\leq \liminf_{k \in \mathbb{N}(\sigma)} \|E^{F_\sigma} X_{n_k} - X_\sigma\|_{Pe} \quad .$$

This inequality holds for any $x' \in E'$. So, for every $A \in F_\sigma$

$$\|\int_A (Y_\sigma - X_\sigma)\| \leq \liminf_{k \in \mathbb{N}(\sigma)} \|E^{F_\sigma} X_{n_k} - X_\sigma\|_{Pe} \quad .$$

By theorem I.1.2.7, we now have (1). Inequality (2) is proved in exactly the same way. □

Later in this chapter we shall see that conditions (A_T) and (B_T) are indispensable. Before doing this, we shall indicate some cases in which (A_T) or (B_T) is satisfied. To do this we need some more preliminary lemmas.

Lemma IV.1.6 : Let $(X_n, F_n)_{n\in\mathbb{N}}$ be an adapted sequence of class (B). Denote $M = \sup_{\tau\in T} \|X_\tau\|_1$. Let q be any continuous seminorm on $E,\|\cdot\|$, and suppose that $c \in \mathbb{R}^+$ is such that for each $x \in E$, $q(x) \leq c\|x\|$. Fix $\sigma \in T$. Then we have

$$\int_\Omega \sup_{\tau\in T(\sigma)} q(E^{F_\sigma} X_\tau) = \sup_{\tau\in T(\sigma)} \int_\Omega q(E^{F_\sigma} X_\tau) \leq M.c < \infty$$

Proof : This is proved in exactly the same way as corollary I.3.5.7(ii).

□

The following lemma is easily seen and hence the proof is omitted.

Lemma IV.1.7 : Suppose $H \subset L^1_E(\Omega,F,P)$ is uniformly integrable. Then the set

$$\{E^G(X) \| X \in H, G \subset F \text{ is a sub-}\sigma\text{-algebra of } F\}$$

is also uniformly integrable.

The final lemma is a key result for the application of theorem IV.1.4.

Lemma IV.1.8 : Let $(f_n)_{n\in\mathbb{N}}$ be a sequence of functions in $L^1_E(\Omega,G,P|G)$ where G is a sub-σ-algebra of F. For each $h \in L^\infty_{\mathbb{R}}(\Omega,G,P|G)$, define

$$o(h) = \{\int_\Omega h\, f_n \| n\in\mathbb{N}\}$$

Suppose that

(a) For each $h \in L^\infty_{\mathbb{R}}(\Omega,G,P|G)$, the T-closure of the set o(h) is T-sequentially compact in E.

(b) $(f_n)_{n\in\mathbb{N}}$ is uniformly integrable.

Then there is a subsequence $(n_k)_{k\in\mathbb{N}}$ in $\mathbb{N}$ such that

$$T\text{-}\lim_{k\to\infty} \int_\Omega h\, f_{n_k}$$

exists for all $h \in L^\infty_{\mathbb{R}}(\Omega,G,P|G)$.

Proof : Let $G_o = \sigma(f_1,\dots,f_n,\dots)$. Since this σ-algebra is countably generated, let $(A_n)_{n\in\mathbb{N}}$ be a sequence in G_o such that $G_o = \sigma(A_1,\dots,A_n,\dots)$. Using (a) and a diagonal argument we find a subsequence $(f_{n_k})_{n\in\mathbb{N}}$ such that

$$(\int_{A_i} f_{n_k})_{k\in\mathbb{N}}$$

T-converges for every $i \in \mathbb{N}$. Let now $A \in G_o$ arbitrarily. Let $\varepsilon > 0$ and choose $i \in \mathbb{N}$ such that

$$\sup_{n\in\mathbb{N}} \int_{A_i \Delta A} \|f_n\| < \varepsilon$$

This is possible by (b) and the density of $(A_n)_{n\in\mathbb{N}}$ in G_o.
Hence

$$\|\int_A f_{n_k} - \int_{A_i} f_{n_k}\| < \varepsilon$$

for all $k \in \mathbb{N}$. From this it follows that $(\int_A f_{n_k})_{k\in\mathbb{N}}$ is T-Cauchy, hence T-convergent by (a). So the same is true for the sequence $(\int_\Omega g\, f_{n_k})_{k\in\mathbb{N}}$ where g is a G_o-stepfunction. Since every h as above is uniformly approximable by G_o-stepfunctions, the sequence $(\int_\Omega h\, f_{n_k})_{n\in\mathbb{N}}$ is T-Cauchy for every $h \in L^\infty_{\mathbb{R}}(\Omega,G_o,P|G_o)$, hence T-convergent, again using (a).
Now let $h \in L^\infty_{\mathbb{R}}(\Omega,G,P|G)$. Then $E^{G_o}h \in L^\infty_{\mathbb{R}}(\Omega,G_o,P|G_o)$ and so

$$\int_\Omega h\, f_{n_k} = \int_\Omega (E^{G_o}h)\, f_{n_k}$$

So $(\int_\Omega h\, f_{n_k})_{n\in\mathbb{N}}$ is T-convergent. □

We can now give very applicable criteria under which the condition (A_T) and (B_T) in theorem IV.1.4 are satisfied.

Theorem IV.1.9 : Let $(X_n,F_n)_{n\in\mathbb{N}}$ be an adapted sequence.

Case I

Suppose $(n_k)_{k\in\mathbb{N}}$ is a subsequence in $\mathbb{N}$ such that

(a) For each $m \in \mathbb{N}$ and $h \in L^\infty_{\mathbb{R}}(\Omega,F_m,P|F_m)$, the T-closure of the set

$$o(h) = \{\int_\Omega h\, X_{n_k} \| k \geq m\}$$

is T-sequentially compact and such that (b) or (b') or (b") below is satisfied

(b) $(X_{n_k})_{k\in\mathbb{N}}$ is uniformly integrable

(b') $(X_n,F_n)_{n\in\mathbb{N}}$ is of class (B)

(b") There is an $m_o \in \mathbb{N}$ such that for each $m \in \mathbb{N}(m_o)$, $(E^{F_m}X_{n_k})_{k\in\mathbb{N}(m)}$ is uniformly integrable.

Then condition (A_T) is satisfied.

Case II

Let $(\gamma_n)_{n\in\mathbb{N}}$ be a sequence in T such that $n \leq \gamma_n \leq \gamma_{n+1}$ for each $n \in \mathbb{N}$ such that

(a) For each $m \in \mathbb{N}$ and $h \in L^\infty_{\mathbb{R}}(\Omega,F_mP|F_m)$ the T-closure of the set

$$o(h) = \{\int_\Omega h\, X_{\gamma_n} \| n \geq m\}$$

is T-sequentially compact and such that (b) or (b') or (b") below is satisfied

(b) $(X_{\gamma_n})_{n\in\mathbb{N}}$ is uniformly integrable

(b') $(X_n,F_n)_{n\in\mathbb{N}}$ is of class (B)

(b") There is an $m_o \in \mathbb{N}$ such that for each $m \in \mathbb{N}(m_o)$, $(E^{F_m}X_{\sigma_n})_{n\in\mathbb{N}(m)}$ is uniformly integrable.

Then condition (B_T) is satisfied.

Proof : Case I

(b) ⇒ (b") by lemma IV.1.7.

(b') ⇒ (b") by lemma IV.1.6.

So if we have (b"), apply lemma IV.1.8 to

$$f_k = E^{\mathcal{F}_m} X_{n_k} \qquad (k \geq m)$$

for every fixed $m \in \mathbb{N}(m_o)$, together with the diagonalization procedure which produces the final subsequence appearing in condition (A_T).

Case II

The proof runs in exactly the same way. □

Theorem IV.1.9 as well as theorem IV.1.4 will be used in this chapter as well as in chapters V and VII. We conclude section IV.1 by giving some corrolaries.

Corollary IV.1.10 : Suppose that in theorem IV.1.4, condition (B_T) is satisfied. Then we have, with the same notations as in theorem IV.1.4 :

$(2'_D)$ ("Distributional form of (2')"). For every $\lambda > 0$

$$P(\limsup_{n,m \in \mathbb{N}} q(X_n(\omega) - X_m(\omega)) > \lambda)$$

$$\leq \liminf_{n \in \mathbb{N}} P(2q(E^{\mathcal{F}_{\sigma_n}} X_{\tau_n}(\omega) - X_{\sigma_n}(\omega)) > \lambda)$$

$(2'_I)$ ("Integral form of (2')")

$$\int_\Omega \limsup_{m,n \in \mathbb{N}} q(X_n - X_m) \leq 2 \liminf_{n \in \mathbb{N}} \int_\Omega q(E^{\mathcal{F}_{\sigma_n}} X_{\tau_n} - X_{\sigma_n})$$

Also the distributional and integral forms of inequality (2) in theorem IV.1.4 are valid.

Proof : For each $\lambda > 0$ we have

$$P(\limsup_{m,n \in \mathbb{N}} q(X_n(\omega) - X_m(\omega)) > \lambda)$$

$$\leq P(2 \liminf_{n \in \mathbb{N}} q(E^{\mathcal{F}_{\sigma_n}} X_{\tau_n}(\omega) - X_{\sigma_n}(\omega)) > \lambda)$$

$$\leq \liminf_{n \in \mathbb{N}} P(2q(E^{F_{\sigma_n}} X_{\tau_n}(\omega) - X_{\sigma_n}(\omega)) > \lambda)$$

proving $(2'_D)$. The proof of $(2'_I)$ follows easily from theorem IV.1.4 and the lemma of Fatou.

We have the same proof as for the distributional and integral forms of inequality (2) in theorem IV.1.4. □

The result of Edgar [1979] is even more a special case :

Corollary IV.1.11 (Edgar) : Let E be a separable dual Banach space and $(X_n, F_n)_{n \in \mathbb{N}}$ be an adapted sequence of class (B). Then

$$\limsup_{n,m \in \mathbb{N}} \|X_n(\omega) - X_m(\omega)\| \quad (1)$$

$$\leq 2 \limsup_{m \in \mathbb{N}} \sup_{n \in \mathbb{N}(m)} \|E^{F_m} X_n(\omega) - X_m(\omega)\|, \quad \text{a.e.}$$

$$\int_\Omega \limsup_{m,n \in \mathbb{N}} \|X_n - X_m\| \leq 2 \limsup_{\sigma \in T} \sup_{\tau \in T(\sigma)} \int_\Omega \|E^{F_\sigma} X_\tau - X_\sigma\| \quad (2)$$

Proof : We have

$$\limsup_{\sigma \in T} \sup_{\tau \in T(\sigma)} \int_\Omega \|E^{F_\sigma} X_\tau - X_\sigma\| < \infty$$

Furthermore condition (a) in theorem IV.1.9 (case I or case II) is satisfied if we take for T the w^*-topology (the Alaoglu-theorem together with the separability of E and hence of the predual of E is used here to obtain that the w^*-closure of bounded sets in E are w^*-sequentially compact). So (A_T) and (B_T) are valid according to theorem IV.1.9. Theorem IV.1.4 now yields (1) with $q(\cdot) = \|\cdot\|$ while corollary IV.1.10 yields with $q(\cdot) = \|\cdot\|$ here also :

$$\int_\Omega \limsup_{m,n\in\mathbb{N}} \|X_n - X_m\| \leq 2 \liminf_{n\in\mathbb{N}} \int_\Omega \|F^{\mathcal{F}_{\sigma_n}} X_{\tau_n} - X_{\sigma_n}\|$$

where $\tau_n \in T(\sigma_n)$ for each $n \in \mathbb{N}$. Hence

$$\int_\Omega \limsup_{m,n\in\mathbb{N}} \|X_n - X_m\| \leq 2 \limsup_{\sigma\in T} \sup_{\tau\in T(\sigma)} \int_\Omega \|E^{\mathcal{F}_\sigma} X_\tau - X_\sigma\| \qquad \square$$

IV.2. Failure of the inequalities

The inequalities thus proved so far are not valid without the additional $\mathcal{T}$-sequentially compactness conditions in theorem IV.1.4 and in theorem IV.1.9 (in theorem IV.1.4 expressed in terms of $\mathcal{T}$-convergence).

As a matter of fact, even the most restrictive case, corollary IV.1.11 of Edgar, is not valid if we replace the condition "E is a separable dual" by "E has (RNP)". This was remarked by Edgar [1979] in the same paper where he proves his inequality. Of course these (RNP) Banach spaces E where the inequality fails, cannot embed into a separable dual Banach space. Now it was for a long time an open problem, posed by J.J. Uhl Jr., if it were true that every (RNP) Banach space embeds in a separable dual. In 1979 this was proved in the negative by Mc Cartney and O'Brien [1980] based upon preliminary work of Mc Cartney [1980]. The method is purely geometrical. It was precisely this geometric example that led Edgar to the probabilistic interpretation of this example, being an example of an adapted sequence $(X_n, \mathcal{F}_n)_{n\in\mathbb{N}}$ for which the inequality (2) in IV.1.11 falls.

Some links between the geometry of Banach spaces and probability theory in Banach spaces were already described in chapter II, sections 2.3 and 4.7. This connection is illustrated here once more. Essentially the main thing is that in (RNP) spaces one can have "an approximate tree structure" (see later) without any tree while in dual spaces a w^*-argument proves that approximate trees are trees. Let us define what we mean by "approximate tree".

We have given already the definition of tree in II.2.3.8. These trees are in fact infinite sets. For the definition of "approximate tree" or "neighborly tree" in the terminology of Mc Cartney we need also the concept of finite tree.

Definition IV.2.1 : Let $A_n = \{(i,j) \| 1 \leq j \leq 2^{i-1};\ i = 1,\ldots,n\}$. A finite tree or finite ε-tree ($\varepsilon > 0$) is a set of the form

$$T = \{x_{i,j} \| (i,j) \in A_n\} \subset B_E$$

for a certain $n \in \mathbb{N}$, such that

(i) $$x_{i,j} = \frac{x_{i+1,2j-1} + x_{i+1,2j}}{2}$$

(ii) $$\|x_{i+1,2j-1} - x_{i+1,2j}\| > \varepsilon$$

for each $(i,j) \in A_n$. ε is called the separation constant of T.

Definition IV.2.2 (Mc Cartney [1980]) : Let $A = \{(i,j) \| 1 \leq j \leq 2^{i-1};\ i \in \mathbb{N}\}$. Let E be a Banach space. We say that E has the neighborly tree property, abbreviated (NTP) if there are numbers $\varepsilon > 0$ and $\delta > 0$ such that $\delta < \frac{\varepsilon}{4}$ and such that in B_E there are finite trees T_n such that

(i) each T_n is of the form $\{x^n_{i,j} \| (i,j) \in A_n\}$ as in definition IV.2.1.

(ii) each T_n has separation constant ε .

(iii) for each $(i,j) \in A$, there is a ball in E of radius δ which contains $\{x^n_{i,j} \| n \in \mathbb{N}(i)\}$.

It is intuitively clear that if the balls in E have some

compactness property such a neighborly tree structure is almost a tree structure. We indeed have

Theorem IV.2.3 (Mc Cartney) : Suppose that the dual E' of a Banach space has the NTP. Then E' contains a tree.

Proof : Let $(T_n)_{n\in\mathbb{N}}$ be the neighborly tree structure of definition IV.2.2. Let us call $B_{i,j}$ the ball mentioned in (iii) of this definition, and call the center of this ball $x_{i,j}$. If we endow each ball $B_{i,j}$, where $(i,j) \in A$, with the w^*-topology, we see from Tychonov's theorem that $\prod_{(i,j)\in A} B_{i,j}$ is compact. Define for each $n \in \mathbb{N}$ an element y_n of $\prod_{(i,j)\in A} B_{i,j}$ as follows :

$$y_n(i,j)\begin{cases} = x^n_{i,j} & \text{if } i \leq n \\ = x_{i,j} & \text{if } i > n \end{cases}$$

Since $\prod_{(i,j)\in A} B_{i,j}$ is also countably compact, the sequence $(y_n)_{n\in\mathbb{N}}$ has a cluster point y. y is a tree. Indeed, for each $(i,j) \in A$

$$\left\|\frac{y_{i+1,2j}+y_{i+1,2j-1}}{2} - y_{i,j}\right\|$$

$$= \sup_{\substack{\|x\|\leq 1\\ x\in E}} \left|x\left(\frac{y_{i+1,2j}+y_{i+1,2j-1}}{2} - y_{i,j}\right)\right|$$

$$= \sup_{\substack{\|x\|\leq 1\\ x\in E}} \lim_{k\to\infty} \left|x\left(\frac{x^{n^x_k}_{i+1,2j}+x^{n^x_k}_{i+1,2j-1}}{2} - x^{n^x_k}_{i,j}\right)\right|$$

$$= 0 \ ,$$

since $i+1 < n^x_k$ for k large enough and where $(n^x_k)_{k\in\mathbb{N}}$ is the subsequence of $\mathbb{N}$ corresponding to x in the definition of the cluster point y.
Clearly $y_{i,j} \in B_E$ for each $(i,j) \in A$.
Finally, the separation property is readily seen since $\delta < \frac{\varepsilon}{4}$. □

The crux of the matter is that in certain (RNP) spaces such a neighborly tree structure is possible.

Theorem IV.2.4 (Mc Cartney) : There exists a Banach space E such that E has the NTP and such that E contains no bush (hence E has (RNP) - see theorem II.2.3.7 - but has no tree).

Proof : Let $(e_n)_{n\in\mathbb{N}}$ be the canonical basis of c_o. Take the following tree $T = (t_{i,j})_{(i,j)\in A}$ in c_o : $t_{1,1} = e_1$, $t_{2,1} = e_1 + e_2$, $t_{2,2} = e_1 - e_2$, $t_{3,1} = e_1 + e_2 + e_3$, $t_{3,2} = e_1 + e_2 - e_3$, $t_{3,3} = e_1 - e_2 + e_3$, $t_{3,4} = e_1 - e_2 - e_3$, ... (cf. the remarks after II.2.2.1). Let $F = c_o \times_{\ell^1} (\Pi_{n\,\ell^1}\, \ell_n^\infty)$ where the ℓ^1-norms are used to define the productnorms and where ℓ_n^∞ is the space of all real functions on $\{1,2,\ldots,n\}$ with the sup norm. Truncate each element $t_{i,j}$ $(1\leq i\leq n,\ 1\leq j\leq 2^i)$ to get an element of ℓ_n^∞ by throwing away all zero components. The resulting finite tree is denoted $(t_{i,j}^n)_{(i,j)\in A_n}$.
Now fix $\alpha\in\,]\frac{3}{4},1[$. For each $n\in\mathbb{N}$ let $T_n = (x_{i,j}^n)_{(i,j)\in A_n}$ be the finite tree defined as follows : for each $(i,j)\in A$, $x_{i,j}^n$ has only two non-zero components : the c_o-component is $\alpha . t_{i,j}$ and the ℓ_n^∞-component is $(1-\alpha)t_{i,j}^n$. Thus trivially $\|x_{i,j}^n\| = 1$. Suppose now that j is odd. Then

$$\|x_{i,j}^n - x_{i,j+1}^n\| = \alpha\|t_{i,j} - t_{i,j+1}\| + (1-\alpha)\|t_{i,j}^n - t_{i,j+1}^n\|$$
$$= 2\ .$$

Also, if $i \geq 2$

$$\frac{x_{i,j}^n + x_{i,j+1}^n}{2} = x^n_{i-1,\frac{j+1}{2}}$$

So F has NTP since we can take $\varepsilon = 2$ and for each $(i,j)\in A$ and $n > m \geq i$ we have

$$\|x_{i,j}^m - x_{i,j}^n\| = 2 - 2\alpha < \frac{\varepsilon}{4}$$

if we take $\delta = 2 - 2\alpha$. Let E be the Banach space spanned by the set $\{x^n_{i,j} \| (i,j) \in A_n,\ n \in \mathbb{N}\}$. So E has NTP. We shall show now that E does not have a bush. It is enough to show that if E contained a bush, then $\prod_{n\,\ell^1} \ell^\infty_n$, being the separable dual of the Banach space $\prod_{n\,c_0} \ell^\infty_n$, would contain a bush (see theorem II.2.3.7 and the fact that separable dual spaces have (RNP)). This in turn is implied by an inequality of the form

$$(1-\alpha)\left\| \sum_{k=1}^{M} c_k\, x^{n_k}_{i_k,j_k} \right\| \leq \left\| \sum_{k=1}^{M} c_k\, \hat{x}^{n_k}_{i_k,j_k} \right\| \qquad (1)$$

where $c_k \in \mathbb{R}$ $(k = 1,\ldots,M)$ and where for each $n \in \mathbb{N}$ and $(i,j) \in A$, the vector $\hat{x}^n_{i,j}$ is defined as the element of $c_o \times_{\ell^1} (\prod_{n\,\ell^1} \ell^\infty_n)$ whose only non-zero component is the ℓ^∞_n-component and this is equal to $(1-\alpha)t^n_{i,j}$, being the ℓ^∞_n-component of $x^n_{i,j}$ itself. Why? Well, if E contained a bush then, by making the c_0-component of each bush element zero we would obtain a bush; indeed, the averaging property holds trivially, the new structure is obviously in the unit ball of $\prod_{n\,\ell^1} \ell^\infty_n$ and inequality (1) implies that the separation property is valid. Hence the new structure is a bush too. Thus we only have to prove (1). So let $M \in \mathbb{N}$, $c_k \in \mathbb{R}$ and i_k, j_k, n_k $(k \in \{1,\ldots,M\})$ in $\mathbb{N}$ be fixed. For purposes of calculation we denote

$$x^{n_k}_{i_k,j_k} = [\alpha t_{i_k,j_k}, (1-\alpha)t^{n_k}_{i_k,j_k}]$$

i.e. : we write down the c_0- and the $\ell^\infty_{n_k}$-component explicitly, the other components being zero. Denote in the same way

$$\hat{x}^{n_k}_{i_k,j_k} = [0, (1-\alpha)t^{n_k}_{i_k,j_k}]$$

Put $K = \max\{n_1,\ldots,n_M\}$. Hence $t^{n_k}_{i_k,j_k}$ is mapped in ℓ^∞_K in a natural way for each $k \in \{1,\ldots,M\}$. Denote this image by $\tilde{t}^{n_k}_{i_k,j_k}$. Now

$$\left\| \sum_{k=1}^{M} c_k\, x^{n_k}_{i_k,j_k} \right\| = \alpha \left\| \sum_{k=1}^{M} c_k\, [t_{i_k,j_k}, 0] \right\| + \left\| \sum_{k=1}^{M} c_k\, [0, (1-\alpha)t^{n_k}_{i_k,j_k}] \right\|$$

But

$$\left\| \sum_{k=1}^{M} c_k [0,(1-\alpha)t_{i_k,j_k}^{n_k}] \right\|$$

$$\geq \left\| \sum_{k=1}^{M} c_k [0,(1-\alpha)\tilde{t}_{i_k,j_k}^{n_k}] \right\|$$

$$= (1-\alpha) \left\| \sum_{k=1}^{M} c_k [0,\tilde{t}_{i_k,j_k}^{n_k}] \right\|$$

$$= (1-\alpha) \left\| \sum_{k=1}^{M} c_k [t_{i_k,j_k},0] \right\|$$

The first inequality is true since on the right hand side everything is "pushed" onto the ℓ_K^∞-components. So the inequality follows from the triangle inequality. The last equality follows from the definition of $t_{i_k,j_k}^{n_k}$. So

$$\left\| \sum_{k=1}^{M} c_k x_{i_k,j_k}^{n_k} \right\| \leq \left(\frac{\alpha}{1-\alpha} + 1\right) \left\| \sum_{k=1}^{M} c_k [0,(1-\alpha)t_{i_k,j_k}^{n_k}] \right\|$$

Hence (1) is proved. □

We are now ready to show that Edgar's inequality does not hold in certain (RNP) Banach spaces.

<u>Theorem IV.2.5 (Edgar [1979])</u> : There exists a Banach space E with (RNP) and an L_E^∞-bounded adapted sequence $(X_n,F_n)_{n\in\mathbb{N}}$ such that the inequality

$$\int_\Omega \limsup_{n,m\in\mathbb{N}} \|X_n - X_m\| \leq 2 \limsup_{\sigma\in T} \sup_{\tau\in T(\sigma)} \int_\Omega \|E^{F_\sigma} X_\tau - X_\sigma\| \quad (1)$$

<u>fails</u>.

<u>Proof</u> : In terms of adapted sequences an (ε,δ)-neighborly tree structure in a Banach space F is a sequence $(X_n,F_n)_{n\in\mathbb{N}}$ in L_F^∞, $\|\cdot\|_\infty$-bounded (by 1), adapted to the dyadic σ-algebras with $\|X_{n+1} - X_n\| > \varepsilon$

everywhere and $\|E^{F_\sigma} X_\tau - X_\sigma\| < \delta$ for every $\sigma \in T$ and $\tau \in T(\sigma)$. Now in theorem IV.2.4 a (RNP) Banach space E and such a neighborly tree structure is constructed for $\varepsilon = 2$ and δ fixed but arbitrarily small. So (1) cannot be true. □

So we have now reached our goal. But theorem IV.2.4 has a corollary, which solves the problem of J.J. Uhl Jr. in the negative, within such easy reach that we cannot resist finishing this part of Banach space theory with it.

Corollary IV.2.6 (Mc Cartney-O'Brien) : There is a separable Banach space E with (RNP) which does not embed in a separable dual space.

Proof : Let $n \geqslant 3$ and construct as in theorem IV.2.4 a Banach space E_n with (RNP) and with $(2,\frac{1}{n})$ - NTP. Put

$$E = \prod_{n\,\ell^1} E_n$$

the ℓ^1-product of the E_n. Then E has (RNP) (see Diestel and Uhl [1977]). Suppose now that $\Phi : E \to Y$ is an isomorphism of E into a dual space Y. Hence, there exists $c > 0$ such that for all $x \in E$,

$$\frac{1}{c}\,\|x\| \leqslant \|\Phi(x)\| \leqslant c\|x\|$$

Choose $n_o \in \mathbb{N}$ such that $n_o > 2c^2$. Let $i_{n_o} : E_{n_o} \to E$ be the canonical embedding of E_{n_o} into E. Let $(T_n^{n_o})_{n\in\mathbb{N}}$ be the neighborly tree structure in E_{n_o} with parameters $(2,\frac{1}{n_o})$. Put

$$\hat{T}_n = i_{n_o} \circ T_n^{n_o}$$

for each $n \in \mathbb{N}$. Then $(\Phi(\frac{1}{c}\hat{T}_n))_{n\in\mathbb{N}}$ is a neighborly tree structure in Y with separation constant $\varepsilon = \frac{2}{c^2}$ and with $\delta = \frac{1}{n_o} < \frac{1}{2c^2} = \frac{1}{4}\cdot\varepsilon$. Since Y is a dual space, theorem IV.2.3 shows that Y contains a tree. This is impossible since E has (RNP), being a separable dual space, and by theorem II.2.3.7. □

IV.3. Notes and remarks

IV.3.1. Another example of a (RNP) Banach space which does not embed in a separable dual space has been given by J. Bourgain and F. Delbaen [1980]. This example is however of a completely different nature and could not be used to disprove Edgar's inequality.

IV.3.2. The Banach space, constructed in Corollary IV.2.6 cannot be a Banach lattice : Indeed, due to a recent result of Talagrand [1981], every Banach lattice with (RNP) embeds in a separable dual space. This result is applied in chapter VIII; see from VIII.1.19 on.

IV.3.3. A problem which I have never seen formulated and which is related to the inequalities proved in section IV.1 is the following. First a definition

Definition IV.3.3.1 : Let $(X_n, \mathcal{F}_n)_{n\in\mathbb{N}}$ be an arbitrary adapted sequence with values in an arbitrary Banach space E. Let $A \in \mathcal{F} = \sigma(\bigcup_n \mathcal{F}_n)$ also arbitrary.

(a) We say that $(X_n, \mathcal{F}_n)_{n\in\mathbb{N}}$ is a martingale on A if

$$E^{\mathcal{F}_n} X_{n+1}(\omega) = X_n(\omega)$$

a.e. on A, for every $n \in \mathbb{N}$.

(b) Analogous definitions can be given for sub-(super-) martingales on A and for other types of adapted sequences which are to follow in the next chapters.

Problem IV.3.3.2 : Do we have the same convergence properties on A for these types of adapted sequences as we have on Ω for the corresponding class of adapted sequences, now defined on Ω? More concretely, since we know Doob's convergence theorem for martingales, is it true that real L^1-bounded martingales on A converge a.e. on A, where $A \in \mathcal{F}$?

If $A \in \bigcup_n F_n$, then of course there is no problem since, if $A \in F_{n_0}$, the sequence $(X_n|_A, F_n|_A)_{n\in\mathbb{N}(n_0)}$, is an L^1-bounded martingale on the finite measure space $(A, F|_A, P|_A)$, and hence converges a.e. on A. The problem arises when $A \in F \setminus \bigcup_n F_n$. Now it is not true that $(X_n|_A, F_n|_A)_{n\in\mathbb{N}}$ is a martingale on A, and the convergence on A is not clear. However, using some previous inequalities, we have quite trivially and even for adapted sequences with values in a separable dual Banach space :

Theorem IV.3.3.3 (Egghe) : Let $A \in F$ and let $(X_n, F_n)_{n\in\mathbb{N}}$ be a martingale on A with values in a separable dual Banach space. Suppose that

$$\sup_{\tau\in T} \int_\Omega \|X_\tau\| < \infty \quad .$$

Then there is a function $X_\infty \in L^1_E$ such that

$$\lim_{n\to\infty} X_n|_A = X_\infty|_A ,$$

a.e. on A.

Proof : Apply corollary IV.1.11 and the definition of "martingale on A" to see that $(X_n|_A)_{n\in\mathbb{N}}$ converges a.e. on A. Now the boundedness condition also implies that the limit X_∞ must be integrable by Fatou's lemma. □

Note that, even in the case $E = \mathbb{R}$, theorem IV.3.3.3 extends Doob's theorem considerably. Indeed, the L^1-boundedness condition for martingales implies that

$$\sup_{\tau\in T} \int |X_\tau| < \infty \quad ,$$

i.e. class (B).

Chapter V : CONVERGENCE OF GENERALIZED MARTINGALES IN BANACH SPACES - THE MEAN WAY

This chapter is devoted to the study of weak or strong a.e. convergence or $\|\cdot\|_1$- or $\|\cdot\|_{Pe}$-convergence of adapted sequences $(X_n,\mathcal{F}_n)_{n\in\mathbb{N}}$ in L^1_E such that the difference

$$(E^{\mathcal{F}_m}X_n - X_m)_{\substack{n\in\mathbb{N}(m)\\ m\in\mathbb{N}}}$$

goes to zero in a mean way. This can be in the Pettis way in the stopping time sense (called amarts) or in the L^1_E-way in the stopping time sense (called uniform amarts) or in a weak L^1-sense (called weak or weak sequential amarts) - see later on. These sequences all generalise martingales.

Each of the above mentioned adapted sequences will be studied in detail. For convergence the inequalities proved in the previous chapter are basic and will be often used.

Before starting the study of these adapted sequences, a word of motivation.

The first reason for studying convergence of these processes was (now about a decade ago) to have convergence theorems, with the same conditions as in Doob's theorem (for the real case) or as in the theorem of Ionescu-Tulcea (in the case of a Banach space with (RNP)), but for adapted sequences which are much more general than martingales, sub-martingales or supermartingales.

Of course, once we have such results, it is nice to have applications. They indeed exist, in various ways.

Uniform amarts arise naturally - in the form of quasi-martingales - in the geometric study of "denting"-problems in Banach spaces, such as in the theorem of Huff, and in the Bourgain-Phelps theorem. Both are of the same nature; we present once more Huff's theorem as an application. Uniform amarts are more general than quasi-martingales, however they are easier to study, especially when using the results of chapter IV.

Also amarts do have applications. We present them in the next chapter - we separate them from this chapter, since the applications are in the field of multiparameter martingale-convergence, which requires a good deal of introduction. It is also shown, by the theorem of Astbury (VI.2.3.1) that amart techniques are strongly linked with the Vitali-condition V, for adapted nets, which were studied almost 30 years ago by Krickeberg. The amart techniques in fact solve an old problem of Krickeberg. Also a didactical application of amarts is made in the notes and remarks of chapter VI.

Another type of application, is the possibility of giving probabilistic versions of topological or geometrical properties of Banach spaces. A lot of applications can be mentioned here - some of them quite deep. We mention the operator characterization of Egghe using amarts (V.2.27), the characterization of Asplund operators and Radon-Nikodym operators in terms of weak sequential amarts, by Edgar (V.3.14 and V.3.16), having as a corollary the Brunel-Sucheston characterization of the separability of a dual Banach space (V.3.15). It is also important to mention the Ghoussoub-Talagrand result, characterizing in Banach lattices, the existence of quasi-interior points in the dual, in terms of weak convergence a.e. of a class of weak sequential amarts (V.3.19 - see also V.3.20). The Brunel-Sucheston characterization of reflexivity in a Banach space, in terms of weak amarts, is worth mentioning (V.4.5). Finally, we remark that semiamarts, or in the case of an infinite dimensional Banach space uniform semi-amarts, have their natural roots and interest through the result of Edgar which we have just seen at the end of chapter IV, namely in IV.1.11 or, translated geometrically, through the results of Mc Cartney-O'Brien in section IV.2. These processes are studied in section V.5.

V.1. Uniform amarts

Historically, amarts (see section V.2) were introduced before uniform amarts but from some points of view it seems more natural to do it the other way round since in the definition of uniform amart, the L^1_E-norm is used while in the definition of amart, the Pettis norm is used (see the theory in V.2).

It turns out that in extending the notion of martingale to the notion of uniform amart we do not have to give up convergence properties. Most of the results in this section are due to A. Bellow [1977b] and [1978a].

Definition V.1.1 : Let $(X_n, F_n)_{n\in\mathbb{N}}$ be an adapted sequence. It is called a uniform asymptotic martingale (shortly uniform amart) if

$$\lim_{\sigma\in T} \sup_{\tau\in T(\sigma)} \int_\Omega \| E^{F_\sigma} X_\tau - X_\sigma \| = 0$$

Examples V.1.2

1. Every martingale is a uniform amart.
   Proof : It suffices to apply lemma II.1.4. □

2. Every a.e. convergent adapted sequence $(X_n, F_n)_{n\in\mathbb{N}}$ such that $(X_\sigma)_{\sigma\in T}$ is uniformly integrable, is a uniform amart.
   Proof : If $(X_n)_{n\in\mathbb{N}}$ is a.e. convergent we have that $(X_\sigma)_{\sigma\in T}$ is a.e. convergent, say to $X_\infty \in L^1_E$ (that $X_\infty \in L^1_E$ follows in fact from Fatou's lemma). Since $(X_\sigma)_{\sigma\in T}$ is uniformly integrable it follows that

$$\lim_{\sigma\in T} \|X_\sigma - X_\infty\|_1 = 0$$

So

$$\lim_{\sigma\in T} \sup_{\tau\in T(\sigma)} \|E^{F_\sigma} X_\tau - X_\sigma\|_1 = \lim_{\sigma\in T} \sup_{\tau\in T(\sigma)} \|E^{F_\sigma}(X_\tau - X_\sigma)\|_1$$

$$\leq \lim_{\sigma\in T} \sup_{\tau\in T(\sigma)} \|X_\tau - X_\sigma\|_1 = 0 \quad . \qquad \square$$

3. As a corollary of the previous example we have that every dominated a.e. convergent sequence $(X_n)_{n\in\mathbb{N}}$ is a uniform amart w.r.t. $\sigma(X_1,\ldots,X_n) = F_n$ and also w.r.t. $(F'_n)_{n\in\mathbb{N}}$ where F'_n is a σ-algebra such that $F'_n \supset F_n$ for each $n \in \mathbb{N}$.

4. A generalization of (1) is : every quasi-martingale is a uniform amart, where a quasi-martingale is defined as follows : an adapted sequence $(X_n, F_n)_{n\in\mathbb{N}}$ is called a quasi-martingale if

$$\sum_{n=1}^{\infty} \int_\Omega \| E^{F_n} X_{n+1} - X_n \| < \infty$$

Proof : For each $\varepsilon > 0$, choose $n_o \in \mathbb{N}$ such that

$$\sum_{n=n_o}^{\infty} \int_\Omega \| E^{F_n} X_{n+1} - X_n \| \leqslant \varepsilon$$

Now let $\sigma, \tau \in T(n_o)$ with $\tau \in T(\sigma)$. Then

$$\int_\Omega \| E^{F_\sigma} X_\tau - X_\sigma \|$$

$$\leqslant \sum_{k=\min\sigma}^{\max\sigma} \sum_{j=\min\tau}^{\max\tau} \int_{\{\tau=j\}\cap\{\sigma=k\}} \| E^{F_k} X_j - X_k \| \qquad (1)$$

Now $\{\tau = j\} \cap \{\sigma = k\} = \phi$ if $k > j$ and if $k = j$ then $\| E^{F_k} X_j - X_k \| = 0$. So

$$(1) = \sum_{\substack{k\in\{\min\sigma,\ldots,\max\sigma\} \\ j\in\{\min\tau,\ldots,\max\tau\} \\ k<j}} \int_{\{\tau=j\}\cap\{\sigma=k\}} \| E^{F_k} X_j - X_k \|$$

But :

$$E^{F_k} X_j - X_k$$

$$= E^{F_k}(X_j - X_k)$$

$$= E^{F_k}(X_j - X_{j-1} + X_{j-1} - X_{j-2} \dots - X_{k+1} + X_{k+1} - X_k)$$

$$= E^{F_k}(E^{F_{j-1}}X_j - X_{j-1} + E^{F_{j-2}}X_{j-1} - X_{j-2} \dots + E^{F_k}X_{k+1} - X_k)$$

since $k < j$. Since $\|E^{F_k}\| \leq 1$ we have

$$(1) \leq \sum_{\substack{k,j \\ k<j}} \int_{\{\tau=j\}\cap\{\sigma=k\}} \sum_{\ell=k+1}^{j} \|E^{F_{\ell-1}}X_\ell - X_{\ell-1}\|$$

$$\leq \sum_{\ell=\min\sigma+1}^{\max\tau} \int_\Omega \|E^{F_{\ell-1}}X_\ell - X_{\ell-1}\| \leq \varepsilon \quad \square$$

Due to our preparation in chapter IV, convergence of uniform amarts becomes trivial to prove.

<u>Theorem V.1.3 (A. Bellow)</u> : Let E have (RNP). Then every L^1_E-bounded uniform amart converges strongly a.e..

<u>Proof</u> : This follows readily from the definition of uniform amart and corollary IV.1.10, with $q(\cdot) = \|\ \|$. Here (B_T) is satisfied for T the norm topology since for each $A \in \bigcup_n F_n$, the net

$$\left(\int_A X_\tau\right)_{\tau\in T} \tag{1}$$

converges in norm. Indeed, let $A \in F_{n_o}$. Let $\tau \in T(\sigma)$, $\sigma \in T(n_o)$, then

$$\left\|\int_A X_\tau - \int_A X_\sigma\right\| = \left\|\int_A (E^{F_\sigma}X_\tau - X_\sigma)\right\| \leq \int_\Omega \|E^{F_\sigma}X_\tau - X_\sigma\|$$

So the convergence of (1) follows from the definition of uniform amart. In fact we proved that every uniform amart is an "amart" (see section V.2). $\square$

As we see from the above, it is much easier to study uniform amarts than to study the definition of quasi-martingale, although it is a special case of a uniform amart!

Uniform amarts are certainly generalizations of martingales. But how far away are uniform amarts from martingales. The next "Riesz-decomposition" theorem answers this question.

Theorem V.1.4 (A. Bellow) : Let E be any Banach space and let $(X_n,F_n)_{n\in\mathbb{N}}$ be a uniform amart. Then $(X_n,F_n)_{n\in\mathbb{N}}$ admits a unique decomposition

$$X_n = Y_n + Z_n$$

where $(Y_n,F_n)_{n\in\mathbb{N}}$ is a martingale and where $(Z_n,F_n)_{n\in\mathbb{N}}$ satisfies

$$\lim_{\tau\in T} \int_\Omega \|Z_\tau\| = 0 \tag{1}$$

as well as $\lim_{n\to\infty} Z_n = 0$, a.e..
An adapted sequence $(Z_n,F_n)_{n\in\mathbb{N}}$ satisfying (1) is called a uniform potential.

Proof : We could apply corollary IV.1.10 if we knew that E has (RNP). But in the proof of corollary IV.1.10 we only used (RNP) in applying theorem IV.1.4. There (RNP) was only used to define the martingale $(Y_n,F_n)_{n\in\mathbb{N}}$. This can be defined from the uniform amart property without the (RNP) property.
Indeed, fix $m\in\mathbb{N}$. Then $(E^{F_m}X_n)_{n\in\mathbb{N}(m)}$ is Cauchy in L^1_E : if $n_1,n_2\in\mathbb{N}(m)$, $n_1\leq n_2$

$$\|E^{F_m}X_{n_1} - E^{F_m}X_{n_2}\|_1 = \|E^{F_m}(X_{n_1}-X_{n_2})\|_1$$

$$= \|E^{F_m}(X_{n_1}-E^{F_{n_1}}X_{n_2})\|_1 \leq \|X_{n_1}-E^{F_{n_1}}X_{n_2}\|_1$$

Let $Y_m = \|\cdot\|_1 - \lim_{n\to\infty} E^{F_m}X_n$. This is precisely Y_m, as constructed in the proof of corollary IV.1.4 (which led to corollary IV.1.10). So the inequalities of corollary IV.1.10 apply. Thus we find, putting $Z_n = X_n - Y_n$ for each $n \in \mathbb{N}$:

$$\int \limsup_{n \in \mathbb{N}} \|Z_n\| \leq \limsup_{\sigma \in T} \sup_{\tau \in T(\sigma)} \int_\Omega \|E^{F_\sigma} X_\tau - X_\sigma\|$$

So $\lim_{n \to \infty} Z_n = 0$ a.e. Furthermore, since

$$Y_\sigma = \|\cdot\|_1 - \lim_{n \to \infty} E^{F_\sigma} X_n$$

for each $\sigma \in T$, we also have

$$\int_\Omega \|Y_\sigma - X_\sigma\| = \lim_{n \to \infty} \int_\Omega \|E^{F_\sigma} X_n - X_\sigma\|$$

So $\lim_{\sigma \in T} \|Z_\sigma\|_1 = 0$, using the uniform amart property.

Suppose now that for each $n \in \mathbb{N}$:

$$X_n = Y_n + Z_n = Y'_n + Z'_n$$

where also $(Y'_n, F_n)_{n \in \mathbb{N}}$ is a martingale and where also $\lim_{\sigma \in T} \|Z'_\sigma\|_1 = 0$. Then the martingale $(Z_n - Z'_n, F_n)_{n \in \mathbb{N}}$, being $(Y'_n - Y_n, F_n)_{n \in \mathbb{N}}$, converges to zero in the $\|\cdot\|_1$ sense and hence is constantly zero a.e.. So $Z_n = Z'_n$, a.e. and $Y_n = Y'_n$ a.e. for every $n \in \mathbb{N}$. □

Remarks V.1.5

1) A Riesz-decomposition theorem of another type, yielding a characterization of uniform amarts is contained in theorem V.5.13.
2) Theorem V.1.4 contains a second proof for theorem V.1.3. Indeed, if E has (RNP) and if $(X_n, F_n)_{n \in \mathbb{N}}$ is L^1_E-bounded, then it is easily seen that the martingale $(Y_n, F_n)_{n \in \mathbb{N}}$ is also L^1_E-bounded. An appeal to theorem II.2.4.3 of Ionescu-Tulcea and theorem V.1.4 finishes the proof.

We continue the theory of uniform amarts by proving two theorems concerning intrinsic properties of uniform amarts.

<u>Theorem V.1.6 (A. Bellow)</u> : Let $(X_n, F_n)_{n\in\mathbb{N}}$ be an L^1_E-bounded uniform amart. Then :

(i) It is of class (B).

(ii) $(\|X_n\|, F_n)_{n\in\mathbb{N}}$ is also a uniform amart.

(iii) There exists an $M \in \mathbb{R}^+$ such that for every $\lambda > 0$,
$P(\sup_{n\in\mathbb{N}} \|X_n\| > \lambda) \leqslant \frac{1}{\lambda}\cdot M.$
i.e. : the maximal inequality is valid for L^1_E-bounded uniform amarts.

<u>Proof</u> : (i) From theorem V.1.4 it follows that

$$\sup_{\tau\in T}\int_\Omega \|X_\tau\| \leqslant \sup_{\tau\in T}\int_\Omega \|Y_\tau\| + \sup_{\tau\in T}\int_\Omega \|Z_\tau\|$$

$$\leqslant \sup_{n\in\mathbb{N}}\int_\Omega \|Y_n\| + \sup_{\tau\in T}\int_\Omega \|Z_\tau\|$$

since $(Y_n, F_n)_{n\in\mathbb{N}}$ is a martingale. Now $\sup_{n\in\mathbb{N}}\int_\Omega \|Y_n\| < \infty$ from the L^1_E-boundedness of $(X_n)_{n\in\mathbb{N}}$ (cf. remark V.1.5.(2)). Choose $n_0 \in \mathbb{N}$ such that

$$\sup_{\tau\in T(n_0)}\int_\Omega \|Z_\tau\| < 1$$

(see theorem V.1.4). For every $\tau \in T$ we have

$$\int_\Omega \|Z_\tau\| = \int_{\{\tau\geqslant n_0\}} \|Z_\tau\| + \int_{\{\tau<n_0\}} \|Z_\tau\|$$

$$\leqslant \int_{\{\tau\geqslant n_0\}} \|Z_\tau\| + \sum_{k=1}^{n_0-1}\int_\Omega \|Z_k\|$$

Define

$$\tau' \begin{cases} = \tau & \text{if } \tau \geqslant n_0 \\ = n_0 & \text{if } \tau < n_0 \end{cases}$$

Then $\tau' \in T(n_o)$ and $\int_{\{\tau \geq n_o\}} \|Z_{\tau'}\| = \int_{\{\tau \geq n_o\}} \|Z_\tau\|$. Hence

$$\int_{\{\tau \geq n_o\}} \|Z_\tau\| \leq \int_\Omega \|Z_{\tau'}\| \leq \sup_{\gamma \in T(n_o)} \int_\Omega \|Z_\gamma\| < \infty$$

Therefore

$$\sup_{\tau \in T} \int_\Omega \|X_\tau\| < \sup_{n \in \mathbb{N}} \int_\Omega \|Y_n\| + 1 + \sum_{k=1}^{n_o - 1} \int_\Omega \|Z_k\| < \infty$$

(ii) Due to the fact that

$$\lim_{\sigma \in T} \int_\Omega \|X_\sigma - Y_\sigma\| = 0$$

the problem is reduced to $(Y_n, F_n)_{n \in \mathbb{N}}$. Indeed :

$$\int_\Omega \left| |E^{F_\sigma}\|Y_\tau\| - \|Y_\sigma\|| - |E^{F_\sigma}\|X_\tau\| - \|X_\sigma\|| \right|$$

$$\leq \int_\Omega |E^{F_\sigma}\|Y_\tau\| - E^{F_\sigma}\|X_\tau\| - (\|Y_\sigma\| - \|X_\sigma\|)|$$

$$\leq \int_\Omega |E^{F_\sigma}(\|Y_\tau\| - \|X_\tau\|)| + \int_\Omega |\|Y_\sigma\| - \|X_\sigma\||$$

$$\leq \int_\Omega |\|Y_\tau\| - \|X_\tau\|| + \int_\Omega |\|Y_\sigma\| - \|X_\sigma\||$$

$$\leq \int_\Omega \|Y_\tau - X_\tau\| + \int_\Omega \|Y_\sigma - X_\sigma\| \to 0$$

for σ, τ running through T. Now :

$$\int_\Omega |E^{F_\sigma}\|Y_\tau\| - \|Y_\sigma\|| = \int_\Omega (E^{F_\sigma}\|Y_\tau\| - \|Y_\sigma\|)$$

since $(\|Y_n\|, F_n)_{n \in \mathbb{N}}$ is a submartingale.

$$\int_\Omega (E^{F_\sigma}\|Y_\tau\| - \|Y_\sigma\|) = \int_\Omega \|Y_\tau\| - \int_\Omega \|Y_\sigma\| .$$

This goes to zero for σ,τ running through T since $(\int_\Omega \|Y_\gamma\|)_{\gamma\in T}$ is increasing and bounded by $\sup_{n\in\mathbb{N}} \int_\Omega \|Y_n\|$.

(iii) follows immediately from (i) and lemma II.1.5.(i). □

The next theorem is the "optional sampling theorem" for uniform amarts. For this, we need a lemma which has some interest in itself.

Lemma V.1.7 : Let $((X_n^m, F_n)_{n\in\mathbb{N}})_{m\in\mathbb{N}}$ be a sequence of uniform amarts such that for the adapted sequence $(X_n, F_n)_{n\in\mathbb{N}}$ we have that

$$\lim_{m\to\infty} \sup_{n\in\mathbb{N}} \int_\Omega \|X_n^m - X_n\| = 0$$

Then $(X_n, F_n)_{n\in\mathbb{N}}$ is a uniform amart.

Proof : This is trivial, using the fact that $\|E^{F_\sigma}\| \leq 1$ for every $\sigma \in T$ (see I.2.2.1). □

Theorem V.1.8 (A. Bellow) : Let $(X_n, F_n)_{n\in\mathbb{N}}$ be a uniform amart and let $(\tau_k)_{k\in\mathbb{N}}$ be an increasing sequence in T (not necessarily cofinal in T). Define for each $k \in \mathbb{N}$, $G_k = F_{\tau_k}$ and $X_k' = X_{\tau_k}$. Then $(X_k', G_k)_{k\in\mathbb{N}}$ is a uniform amart. If $(X_n)_{n\in\mathbb{N}}$ is L_E^1-bounded, then $(X_n')_{n\in\mathbb{N}}$ is L_E^1-bounded.

Proof : For each $k \in \mathbb{N}$, we have (from V.1.4)

$$X_{\tau_k} = Y_{\tau_k} + Z_{\tau_k}$$

where $(Y_n)_{n\in\mathbb{N}}$ and $(Z_n)_{n\in\mathbb{N}}$ are as in V.1.4. Hence for each $m \in \mathbb{N}$ we have

$$X_{\tau_k} = Y_{\tau_k} + Z_{\tau_k \vee m} + Z_{\tau_k \wedge m} - Z_m$$

From V.1.4 we see that

$$\lim_{m\to\infty} \sup_{k\in\mathbb{N}} \int_\Omega \|Z_{\tau_k \vee m} - Z_m\| = 0$$

Furthermore for each $m \in \mathbb{N}$ $(Z_{\tau_k \wedge m}, F_{\tau_k})_{k\in\mathbb{N}}$ is a uniform amart since it is a dominated (by $\sup_{k\in\{1,\ldots,m\}} \|Z_k\| \in L^1$), a.e.-convergent (to $Z_{\tau_\infty \wedge m}$ where $\tau_\infty = \lim_{k\to\infty} \tau_k$) sequence (apply V.1.2.(3)).
$(Y_{\tau_k}, F_{\tau_k})_{k\in\mathbb{N}}$ is trivially a martingale since $(\tau_k)_{k\in\mathbb{N}}$ increases, hence a uniform amart.
Apply lemma V.1.7 yielding that

$$(X'_k, G_k)_{k\in\mathbb{N}}$$

is a uniform amart.
If $(X_n)_{n\in\mathbb{N}}$ is L^1_E-bounded, then theorem V.1.6 implies that $(X_n, F_n)_{n\in\mathbb{N}}$ is of class (B). Hence $(X'_n)_{n\in\mathbb{N}}$ is L^1_E-bounded (and hence also of class (B) since we have already proved that it is a uniform amart). □

Remark V.1.9 : If we take in the above theorem $\tau_k = k \wedge \sigma$ where σ is a possibly infinite fixed stopping time, the above theorem is called the "optional stopping theorem".

The following easy result concerns a special stochastic basis $(F_n)_{n\in\mathbb{N}}$, namely the constant one.

Theorem V.1.10 (Edgar and Sucheston [1976a]) : Let E be any Banach space and $(X_n, F_n)_{n\in\mathbb{N}}$ be an adapted sequence of class (B) such that $F_n = F_m$ for all $n,m \in \mathbb{N}$. Then

$$\sup_{n\in\mathbb{N}} \|X_n\| \in L^1$$

Proof : Suppose $\int_\Omega \sup_{n\in\mathbb{N}} \|X_n\| = \infty$. Then for each $M \in \mathbb{N}$, choose $n_M \in \mathbb{N}$ such that

$$\int_\Omega \sup_{n\leqslant n_M} \|X_n\| \geqslant M$$

Now define

$$\tau_M = \inf\{k \leq n_M \mid \|X_k\| = \sup_{n \leq n_M} \|X_n\|\}$$

Precisely because $(F_n)_{n \in \mathbb{N}}$ is constant, $\tau_M \in T$ and $\int_\Omega \|X_{\tau_M}\| \to \infty$, contradicting the fact that $(X_n, F_n)_{n \in \mathbb{N}}$ is of class (B). □

Corollary V.1.11 : Let E be any Banach space and $(X_n, F_n)_{n \in \mathbb{N}}$ be any uniform amart such that $F_n = F_m$ for all $m, n \in \mathbb{N}$. Then

$$\sup_{n \in \mathbb{N}} \|X_n\| \in L^1$$

Proof : This follows trivially from V.1.10 or even from V.1.4. □

Remark V.1.12 : If in theorem V.1.10, $E = \mathbb{R}$, then the condition

$$\sup_{\tau \in T} \left|\int_\Omega X_\tau\right| < \infty$$

is sufficient instead of class (B).

Proof : If $\sup_{n \in \mathbb{N}} |X_n| \notin L^1$ then, since $X_1 \in L^1$, $\int_\Omega \sup_{n \in \mathbb{N}} X_n$ must be $+\infty$. Now do the same as was done on $\|X_n\|$ in the proof of theorem V.1.10. □

We mention the following strong convergence theorem for uniform amarts in order continuous Banach lattices.

Theorem V.1.13 (Słaby [1982]) : Let E be an order continuous Banach lattice and $(X_n, F_n)_{n \in \mathbb{N}}$ be an L^1_E-bounded uniform amart, such that

$$0 \leq X_n \leq X'_n, \text{ a.e.}$$

for each $n \in \mathbb{N}$, where $(X'_n)_{n \in \mathbb{N}}$ is a sequence of measurable functions converging a.e. to a function X'. Then there is a $X_\infty \in L^1_E$ such that

$$\lim_{n \to \infty} X_n = X_\infty \text{ , a.e..}$$

Proof : Apply theorem V.1.4. Then

$$X_n = Y_n + Z_n$$

for each $n \in \mathbb{N}$ where $(Y_n, F_n)_{n\in\mathbb{N}}$ is an L^1_E-bounded martingale and $(Z_n)_{n\in\mathbb{N}}$ converges strongly a.e. to zero. Now

$$0 \leq Y_n = X_n - Z_n \leq X'_n - Z_n$$

Since $\lim_{n\to\infty} (X'_n - Z_n) = X'$, a.e., theorem III.5.1 applies. Hence $(Y_n)_{n\in\mathbb{N}}$ converges strongly a.e. to an integrable function X_∞ (since $(Y_n)_{n\in\mathbb{N}}$ is L^1_E-bounded). So $\lim_{n\to\infty} X_n = X_\infty$, a.e.. □

Of course, corollary III.5.3 also applies now to uniform amarts.

As we have seen, quasi-martingales were easily studied through the notion of uniform amart. Now quasi-martingales arise naturally in the geometric theory of Banach spaces. The two best known examples are the theorem of Huff (II.2.3.2 (ii) ⇒ (i)) and the new proof of the Bourgain-Phelps theorem as given in Kunen and Rosenthal [1982]. They are both of the same nature and since Huff's proof is within easier reach, we adapt this one for our application (cf. also II.4.7).

V.1.14 : Uniform amart application in Huff's proof

We adapt the notation of II.2.3.2 (ii) ⇒ (i). As in Huff's proof, we suppose that (i) does not hold. Now we do not need to introduce the martingale $(Y_n, F_n)_{n\in\mathbb{N}}$. Indeed, $(X_n, F_n)_{n\in\mathbb{N}}$ is a uniformly bounded uniform amart which diverges everywhere; hence, by theorem V.1.3, E cannot have (RNP), contradicting (ii), by theorem II.2.2.1.

In fact, we prove here that $(X_n, F_n)_{n\in\mathbb{N}}$ is a quasi-martingale and apply V.1.2.4 : for every $n \in \mathbb{N}$,

$$\sum_{n=1}^{\infty} \int_\Omega \| E^{F_n} X_{n+1} - X_n \|$$

$$= \sum_{n=1}^{\infty} \left(\sum_{i=1}^{k(n)} \| x_{n,i} - \sum_{j=1}^{m(i)} \alpha_j^{(i)} y_j^{(i)} \| P(A_{n,i}) \right) < \sum_{n=1}^{\infty} \frac{1}{2^{n+1}} = \frac{1}{2} < \infty.$$

V.2. Amarts

We now generalise the concept of uniform amart to "amart" which is a property that is easier to check, when given an adapted sequence $(X_n, F_n)_{n\in\mathbb{N}}$.

Definition V.2.1 : Let $(X_n, F_n)_{n\in\mathbb{N}}$ be an adapted sequence. We say that it is an asymptotic martingale or shortly amart if

$$\limsup_{\sigma\in T} \sup_{\tau\in T(\sigma)} \| E^{F_\sigma} X_\tau - X_\sigma \|_{Pe} = 0$$

where $\|\cdot\|_{Pe}$ denotes the Pettis norm.

Obviously every uniform amart is an amart. But the definition as given above is not so easy to check. However there are two equivalent definitions which are much easier to deal with :

Theorem V.2.2 : The following assertions for an adapted sequence $(X_n, F_n)_{n\in\mathbb{N}}$ are equivalent :

(i) $(X_n, F_n)_{n\in\mathbb{N}}$ is an amart

(ii) There is a function $\mu_\infty : \bigcup_n F_n \to E$ such that

$$\lim_{\sigma\in T} \sup_{A\in F_\sigma} \| \int_A X_\sigma - \mu_\infty(A) \| = 0$$

(iii) The net

$$\left(\int_\Omega X_\tau\right)_{\tau\in T}$$

converges in E.

Proof : Since (i) ⇒ (ii) ⇒ (iii) is trivially proved we only have to carry out the proof of (iii) ⇒ (i)

$$\| E^{F_\sigma} X_\tau - X_\sigma \|_{Pe}$$

$$= \sup_{\|x'\|\leq 1} \int_\Omega | x'(E^{F_\sigma} X_\tau - X_\sigma) |$$

$$\leq 4 \sup_{\|x'\|\leq 1} \sup_{A\in \mathcal{F}_\sigma} \left| \int_A x'(E^{\mathcal{F}_\sigma} X_\tau - X_\sigma) \right|$$

(see Dunford and Schwartz [1957], p.97, lemma 5, cf. also I.1.2.7)

$$= 4 \sup_{A\in \mathcal{F}_\sigma} \left\| \int_A (E^{\mathcal{F}_\sigma} X_\tau - X_\sigma) \right\|$$

$$= 4 \sup_{A\in \mathcal{F}_\sigma} \left\| \int_A (X_\tau - X_\sigma) \right\|$$

Put

$$\sigma' \begin{cases} = \sigma & \text{on } A \\ = \max \tau & \text{on } \Omega \setminus A \end{cases}$$

$$\tau' \begin{cases} = \tau & \text{on } A \\ = \max \tau & \text{on } \Omega \setminus A \end{cases}$$

Then

$$\int_A (X_\tau - X_\sigma) = \int_\Omega (X_{\tau'} - X_{\sigma'}) \text{ and } \sigma' \in T,\ \tau' \in T(\sigma')$$

and also $\sigma' \geq \sigma$. Hence

$$\sup_{A\in \mathcal{F}_\sigma} \left\| \int_A (X_\tau - X_\sigma) \right\|$$

$$\leq \sup_{\substack{\tau' \geq \sigma' \\ \tau', \sigma' \in T(\sigma)}} \left\| \int_\Omega (X_{\tau'} - X_{\sigma'}) \right\|$$

Therefore

$$\limsup_{\sigma\in T} \sup_{\tau\in T(\sigma)} \left\| E^{\mathcal{F}_\sigma} X_\tau - X_\sigma \right\|_{Pe}$$

$$\leq \limsup_{\sigma\in T} \sup_{\substack{\tau' \geq \sigma' \\ \tau', \sigma' \in T(\sigma)}} \left\| \int_\Omega X_{\tau'} - \int_\Omega X_{\sigma'} \right\| = 0 \qquad \square$$

Examples V.2.3

1. Every uniform amart is an amart and hence, by V.1.2, every quasi-martingale is an amart.

2. Let E be a Banach lattice which is an AL-space and $(X_n, F_n)_{n \in \mathbb{N}}$ a positive submartingale such that

$$\sup_{n \in \mathbb{N}} \left\| \int_\Omega X_n \right\| < \infty \qquad (1)$$

Then $(X_n, F_n)_{n \in \mathbb{N}}$ is an amart.

Proof : Indeed, if $\sigma \in T$ and $\tau \in T(\sigma)$ then $0 \leqslant \int_\Omega X_\sigma \leqslant \int X_\tau$. So

$$\left\| \int_\Omega X_\sigma \right\| \leqslant \left\| \int_\Omega X_\tau \right\|$$

and this net $(\int_\Omega X_\sigma)_{\sigma \in T}$ is bounded above by (1). Hence it converges; see theorem III.1.7. From theorem V.2.2 it follows that $(X_n, F_n)_{n \in \mathbb{N}}$ is an amart. □

Now follows the Riesz-decomposition theorem for amarts, proved in Edgar and Sucheston [1976d] .

Theorem V.2.4 (Edgar and Sucheston) : Let E have (RNP) and let $(X_n, F_n)_{n \in \mathbb{N}}$ be an amart such that

$$\liminf_{n \in \mathbb{N}} \int_\Omega \|X_n\| < \infty \quad .$$

Then for each $n \in \mathbb{N}$, X_n can be uniquely written as

$$X_n = Y_n + Z_n$$

where $(Y_n, F_n)_{n \in \mathbb{N}}$ is an L^1_E-bounded martingale and $(Z_n, F_n)_{n \in \mathbb{N}}$ satisfies $\lim_{\tau \in T} \|Z_\tau\|_{Pe} = 0$. Furthermore if $(X_n)_{n \in \mathbb{N}}$ is uniformly integrable then $(Y_n)_{n \in \mathbb{N}}$ is also.

An adapted sequence $(Z_n, F_n)_{n \in \mathbb{N}}$ with the above property is called a potential.

Proof : The decomposition follows easily from the definition of amart, theorem IV.1.5 and the construction of $(Y_n, F_n)_{n\in\mathbb{N}}$. The uniqueness is seen as follows : Suppose $X_n = Y'_n + Z'_n$ is a second decomposition with the properties of the theorem. Then

$$\lim_{n\to\infty} \int_A Z_n = \lim_{n\to\infty} \int_A Z'_n = 0$$

for every $A \in \bigcup_n F_n$. Hence, for $m \in \mathbb{N}$ fixed but arbitrary in $\mathbb{N}$ and for $A \in F_m$

$$\int_A (Y'_m - Y_m) = \lim_{n\to\infty} \int_A (Y'_n - Y_n)$$

$$= \lim_{n\to\infty} \int_A (Z_n - Z'_n) = 0$$

So $Y_m = Y'_m$, a.e. and $Z_m = Z'_m$, a.e. for each $m \in \mathbb{N}$. □

Corollary V.2.5 (Uhl [1977]) : Let E have (RNP). Then every uniformly integrable amart $(X_n, F_n)_{n\in\mathbb{N}}$ converges in the Pettis norm to a function in L^1_E and conversely, E has (RNP) if already every uniformly bounded martingale is Pettis convergent to a function in L^1_E.

Proof : If E has (RNP) the convergence follows readily from the Riesz-decomposition theorem and theorem II.2.2.1. Conversely, suppose only that every uniformly bounded martingale $(X_n, F_n)_{n\in\mathbb{N}}$ is Pettis convergent in L^1_E say to X_∞. Then, as is readily seen, $E^{F_n} X = X_n$ for each $n \in \mathbb{N}$. But then, by theorem II.1.3, $(X_n)_{n\in\mathbb{N}}$ converges to X_∞ in the mean norm. Theorem II.2.2.1, (iii) ⇒ (i) now shows that E has (RNP). □

For a Pettis convergence result in Banach spaces which do not necessarily have (RNP), very analogous to theorem III.5.1 we refer to Słaby [1982], theorem 3.3 and corollaries. In these results, read "Bochner integrable" instead of "Pettis integrable" if you are not familiar with this last notion - you still reach the heart of the matter!

In the Riesz-decomposition theorem for uniform amarts (theorem V.1.4) one did not suppose E to have (RNP) nor $(X_n)_{n\in\mathbb{N}}$ to satisfy

$$\liminf_{n\in\mathbb{N}} \int_\Omega \|X_n\| < \infty \quad .$$

Both assumptions are however necessary in the amart RDT (see Edgar and Sucheston [1976d], p.89-91). If we restrict ourselves to amarts consisting of stepfunctions however, we do not need the two assumptions above as can be trivially seen from the proofs.

Also in theorem V.1.4, the conclusion is that $\lim_{n\to\infty} Z_n = 0$, strongly a.e. and $\lim_{n\to\infty} \|Z_n\|_1 = 0$. This is not possible in the amart case. As a matter of fact it will follow from chapter VII that this is false in any infinite dimensional Banach space. What about weak convergence a.e. then? We have the following result.

<u>Theorem V.2.6 (Chacon and Sucheston [1975])</u> : Let E have (RNP) and let E' be separable. Then every amart of class (B) is weakly convergent a.e.. We prefer to postpone the proof to section V.3. There the same result can be proved for class (B) weak sequential amarts which are more general as amarts. This generalization is due to Brunel and Sucheston [1976a] and will be proved in section V.3 using the Bellow-Egghe inequalities. The reader can study this proof now, if necessary.

From theorem V.2.6 it now follows that in theorem V.2.4, weak-$\lim_{n\to\infty} Z_n = 0$, a.e. if, in addition to the hypothesis of theorem V.2.4 we suppose E' to be separable and $(X_n, F_n)_{n\in\mathbb{N}}$ to be of class (B). We now see that if dim E $< \infty$, then $\lim_{n\to\infty} Z_n = 0$, under the hypothesis of theorem V.2.4. This can also be seen by the fact that if dim E $< \infty$, then every amart is a uniform amart, using theorem I.1.2.8 and then using theorem V.1.3 or V.1.4. In fact this is only true if dim E $< \infty$: in any infinitely dimensional Banach space E every $\|\cdot\|_{Pe}$-convergent sequence in L^1_E which is not $\|\cdot\|_1$-convergent - see theorem I.1.2.8 - is an example of an amart (w.r.t. constant σ-algebra) which is not a uniform amart. Thus we have

<u>Theorem V.2.7</u> : If dim E $< \infty$, then every L^1_E-bounded amart converges a.e..

We remark that from theorem V.2.7 the real positive submartingale convergence theorem immediately follows, since these are amarts, as follows from example V.2.3.(2). We also have, from theorem V.1.6

Theorem V.2.8 : If dim $E < \infty$, then for every L^1_E-bounded amart $(X_n, F_n)_{n \in \mathbb{N}}$

(i) $(\|X_n\|, F_n)_{n \in \mathbb{N}}$ is also an amart.

(ii) $(X_n, F_n)_{n \in \mathbb{N}}$ is of class (B).

Theorems V.2.7 and V.2.8 are extended in Fréchet spaces in Egghe [1980b] and Egghe [1982a]. Then the results are true if the Fréchet space is nuclear. For definition and properties of nuclear spaces, see Pietsch [1976]. These spaces can be infinite dimensional and have some importance in probability theory. When the space is a Banach space, nuclearity is equivalent with finite dimensionality. Nevertheless, the above mentioned results from Egghe [1980b] and Egghe [1982a] are more positive than theorems V.2.7 and V.2.8.

We also remark that the converse of theorems V.2.7 and V.2.8 (being in fact bad news for amarts!) is also true and will be proved in chapter VII. See example V.2.10 below for a first taste!

Corollary V.2.9 : If $E = \mathbb{R}$, then for all L^1-bounded amarts $(X_n, F_n)_{n \in \mathbb{N}}$ and $(Y_n, F_n)_{n \in \mathbb{N}}$, the adapted sequences $(X_n \vee Y_n, F_n)_{n \in \mathbb{N}}$ and $(X_n \wedge Y_n, F_n)_{n \in \mathbb{N}}$ are amarts. Hence so are $(X_n^+, F_n)_{n \in \mathbb{N}}$ and $(X_n^-, F_n)_{n \in \mathbb{N}}$.

Proof : This follows from theorem V.2.8 and the identities

$$X_n \vee Y_n = \frac{1}{2} (X_n + Y_n + |X_n - Y_n|)$$

$$X_n \wedge Y_n = \frac{1}{2} (X_n + Y_n - |X_n - Y_n|)$$

for each $n \in \mathbb{N}$. For the second assertion, just take $Y_n = 0$ for each $n \in \mathbb{N}$. □

Corollary V.2.9 was proved in Edgar and Sucheston [1976a] in a direct, but elaborate way (*).

(*) However, in their proof, to obtain (1.4') (p.198 in Edgar and Sucheston [1976a]), one has to define a second stopping time σ_2; σ_1 does not work in this case.

Thus the supremum of a finite number of real amarts is an amart. This is not so for a countable number of amarts.

Example V.2.10 (Chacon and Sucheston [1975]) : Let $E = \ell^2$ and give indices to the canonical basis of ℓ^2, thus : e_n^i with $n \in \mathbb{N}$ and $i \in \{1,\ldots,2^n\}$. Put

$$X_n = \sum_{i=1}^{2^n} e_n^i \chi_{A_n^i}$$

where $A_n^i \cap A_n^j = \phi$ $(i \neq j)$ and $P(A_n^i) = 2^{-n}$ and put $F_n = \sigma(X_1,\ldots,X_n)$ for every $n \in \mathbb{N}$. Hence

$$F_n = \sigma(\{A_k^i \| k = 1,\ldots,n;\ i = 1,\ldots,2^n\}),$$

for every $n \in \mathbb{N}$.

Put $Y_{2n} = X_n$ and $Y_{2n-1} = 0$ for each $n \in \mathbb{N}$. Now $(\int_\Omega X_\tau)_{\tau \in T} \to 0$. Indeed : there exists $A \subset \{1,\ldots,2^n\}$ such that

$$X_\tau = \sum_{n=\min \tau}^{\max \tau} \sum_{i \in A} e_n^i \chi_{A_n^i \cap \{\tau = n\}}$$

Hence if $\tau \in T(n_o)$, where $n_o \in \mathbb{N}$:

$$\|\int_\Omega X_\tau\|^2 = \Sigma\, P(A_n^i \cap \{\tau = n\})^2 \leqslant 2^{-n_o} \Sigma\, P(A_n^i \cap \{\tau = n\}) = 2^{-n_o}$$

Incidentally $(X_n, F_n)_{n \in \mathbb{N}}$ is an example of a uniformly bounded (by 1) amart in ℓ^2 such that $(X_n)_{n \in \mathbb{N}}$ does not converge. In fact $(X_n)_{n \in \mathbb{N}}$ diverges everywhere since for all $\omega \in \Omega$ and $n \in \mathbb{N}$.

$$\|X_{n+1}(\omega) - X_n(\omega)\| = \sqrt{2}$$

(cf. theorem V.2.7). Since $\lim_{\tau\ T} \int_\Omega X_\tau = 0$, we have that $(Y_n, F_n)_{n \in \mathbb{N}}$ is also an amart. However $(\|Y_n\|, F_n)_{n \in \mathbb{N}}$ is not since it is alternately 1 and 0; cf. theorem V.2.8. But for each $n \in \mathbb{N}$

$$\|Y_n\| = \sup_{x' \in D} |x'(Y_n)|$$

where D is a countable subset of B_{ℓ^2}, the closed unit ball of ℓ^2.
Since $(Y_n, F_n)_{n\in\mathbb{N}}$ is an amart, every $(x'(Y_n), F_n)_{n\in\mathbb{N}}$ is an amart.
So the countable supremum of amarts is not necessarily an amart.

The optional sampling theorem V.1.8 is also true for amarts.

Theorem V.2.11 (Edgar and Sucheston [1976a]) : Let E be any Banach space and $(X_n, F_n)_{n\in\mathbb{N}}$ be an amart. Let $(\tau_k)_{k\in\mathbb{N}}$ be an increasing sequence in T. Define $Y_k = X_{\tau_k}$ and $G_k = F_{\tau_k}$ for each $k \in \mathbb{N}$. Then $(Y_k, G_k)_{k\in\mathbb{N}}$ is an amart.

Proof : Let $\varepsilon > 0$. Choose $n_o \in \mathbb{N}$ such that for all $\tau, \tau' \in T(n_o)$

$$\|\int_\Omega X_\tau - \int_\Omega X_{\tau'}\| < \varepsilon$$

Put $\tau_\infty = \lim_{k\to\infty} \tau_k$. Since for any cofinal sequence $(\sigma_j)_{j\in\mathbb{N}}$ in T', the set of all bounded stopping times w.r.t. $(G_n)_{n\in\mathbb{N}}$, $\lim_{j\to\infty} X_{\tau_{\sigma_j} \wedge n_o} = X_{\tau_\infty \wedge n_o}$ and since

$$\int_\Omega \sup_{j\in\mathbb{N}} \|X_{\tau_{\sigma_j} \wedge n_o}\| \leq \int_\Omega \|X_1\| \vee \ldots \vee \|X_{n_o}\|$$

the dominated convergence theorem assures that

$$\lim_{j\to\infty} \int_\Omega X_{\tau_{\sigma_j} \wedge n_o} = \int_\Omega X_{\tau_\infty \wedge n_o}$$

So

$$\lim_{\sigma\in T'} \int_\Omega X_{\tau_\sigma \wedge n_o} = \int_\Omega X_{\tau_\infty \wedge n_o}$$

and so $(X_{\tau_k \wedge n_o}, G_k)_{k\in\mathbb{N}}$ is an amart. Hence, let $n_1 \in \mathbb{N}$ be such that $\sigma, \sigma' \in T'(n_1)$ implies

$$\|\int_\Omega X_{\tau_\sigma \wedge n_o} - \int_\Omega X_{\tau_{\sigma'} \wedge n_o}\| < \varepsilon$$

It is easily seen that if $\sigma \in T'$ then $\tau_\sigma \in T$. So, if $\sigma, \sigma' \in T'(n_1)$

$$\left\|\int_\Omega Y_\sigma - \int_\Omega Y_{\sigma'}\right\|$$

$$= \left\|\int_\Omega X_{\tau_\sigma} - \int_\Omega X_{\tau_{\sigma'}}\right\|$$

$$= \left\|\int_\Omega X_{\tau_\sigma \wedge n_o} + \int_\Omega X_{\tau_\sigma \vee n_o} - \int_\Omega X_{n_o}\right.$$

$$\left. - \int_\Omega X_{\tau_{\sigma'} \wedge n_o} - \int_\Omega X_{\tau_{\sigma'} \vee n_o} + \int_\Omega X_{n_o}\right\|$$

$$\leq \left\|\int_\Omega X_{\tau_\sigma \vee n_o} - \int_\Omega X_{\tau_{\sigma'} \vee n_o}\right\| + \left\|\int_\Omega X_{\tau_\sigma \wedge n_o} - \int_\Omega X_{\tau_{\sigma'} \wedge n_o}\right\|$$

$$\leq 2\varepsilon . \qquad \square$$

Remarks V.2.12 :

1. The same remark as in V.1.9 yields the optional stopping theorem for amarts.
2. An optional sampling theorem where finite but possibly unbounded stopping times appear is proved in Krengel and Sucheston [1978], theorem 1.12, p.212-214.

The next result is a useful boundedness property for amarts.

Theorem V.2.13 : Let $(X_n, F_n)_{n\in \mathbb{N}}$ be an amart in L_E^1 such that $\sup_{n\in\mathbb{N}} \int_\Omega \|X_n\| < \infty$. Then

$$\sup_{\tau\in T} \sup_{A\in F_\tau} \left\|\int_A X_\tau\right\| < \infty \quad .$$

If we do not have that $(X_n)_{n\in\mathbb{N}}$ is L_E^1-bounded, then we have

$$\sup_{\tau\in T} \left\|\int_\Omega X_\tau\right\| < \infty$$

Note that the first assertion extends (ii) in theorem V.2.8.

Proof : Choose $n_o \in \mathbb{N}$ such that if $\tau \in T(n_o)$

$$\left\| \int_\Omega X_\tau - \int_\Omega X_{n_0} \right\| \leq 1$$

applying theorem V.2.2. For $\tau \in T(n_0)$ and $A \in F_\tau$, define, as in the proof of theorem V.2.2 :

$$\tau' \begin{cases} = \tau & \text{on } A \\ = \max \tau & \text{on } \Omega \setminus A \end{cases}$$

Then $\tau' \in T(n_0)$ and $\int_A X_\tau = \int_\Omega X_{\tau'} - \int_{\Omega \setminus A} X_{\max \tau}$.
So

$$\sup_{\tau \in T(n_0)} \sup_{A \in F_\tau} \left\| \int_A X_\tau \right\| \leq 1 + 2 \sup_{n \in \mathbb{N}} \left\| \int_\Omega X_n \right\|$$

Let now $\tau \in T$ and $A \in F_\tau$ be arbitrary. Then

$$\int_A X_\tau = \int_{A \cap \{\tau \leq n_0\}} X_\tau + \int_{A \cap \{\tau > n_0\}} X_\tau$$

$$= \int_{A \cap \{\tau \leq n_0\}} X_{\tau \wedge n_0} + \int_{A \cap \{\tau > n_0\}} X_{\tau \vee n_0}$$

So

$$\left\| \int_A X_\tau \right\| \leq \int \sup_{1 \leq j \leq n_0} \|X_j\| + \sup_{\tau \in N(n_0)} \sup_{A \in F_\tau} \left\| \int_A X_\tau \right\|$$

since $A \cap \{\tau > n_0\} \in F_\tau$. Since this upper bound is independent of $A \in F_\tau$ or $\tau \in T$ we have

$$\sup_{\tau \in T} \sup_{A \in F_\tau} \left\| \int_A X_\tau \right\| < \infty \quad .$$

An inspection of the above proof shows that we have in any case, even without L^1_E-boundedness

$$\sup_{\tau \in T} \left\| \int_\Omega X_\tau \right\| < \infty \quad . \qquad \square$$

For real adapted sequences the following theorem gives another characterization of amarts.

Theorem V.2.14 (Edgar and Sucheston [1976d]) : Let $E = \mathbb{R}$, and $(X_n, F_n)_{n\in\mathbb{N}}$ be an adapted sequence. Then $(X_n, F_n)_{n\in\mathbb{N}}$ is an amart if and only if for each increasing sequence $(\tau_n)_{n\in\mathbb{N}}$ with $\tau_n \in T(n)$ for each $n \in \mathbb{N}$ one has

$$E^{F_n} X_{\tau_n} - X_n \to 0 \ , \qquad (1)$$

a.e. and in the L^1-sense, for $n \to \infty$.

Proof : If

Since

$$\lim_{n\to\infty} \| E^{F_n} X_{\tau_n} - X_n \|_1 = 0$$

it follows that $\lim_{n\to\infty} \int_\Omega (X_{\tau_n} - X_n) = 0$. So $(\int_\Omega X_\tau)_{\tau\in T}$ is Cauchy and hence convergent.

Only if

Let for each $n \in \mathbb{N}$, $X_n = Y_n + Z_n$ be the Riesz-decomposition of the amart $(X_n, F_n)_{n\in\mathbb{N}}$. We only have to prove (1) for $(Z_n, F_n)_{n\in\mathbb{N}}$. The optional sampling theorem V.2.11 shows that $(Z_{\tau_n}, F_{\tau_n})_{n\in\mathbb{N}}$ is an amart. Since $F_n \subset F_{\tau_n}$ for each $n \in \mathbb{N}$ we also have that $(E^{F_n} Z_{\tau_n}, F_n)_{n\in\mathbb{N}}$ is an amart. Since E^{F_n} is an L^1-contraction

$$\lim_{n\to\infty} \int_\Omega |E^{F_n} Z_{\tau_n}| \leq \lim_{n\to\infty} \int_\Omega |Z_{\tau_n}| = 0 \ .$$

Hence by theorem V.2.7, $\lim_{n\to\infty} E^{F_n} Z_{\tau_n} = 0$, a.e.. We conclude that $E^{F_n} Z_{\tau_n} - Z_n \to 0$, a.e. and in L^1. □

We define now the "Doob-decomposition" of a real amart and give an application of this notion and of theorem V.2.14 (see also lemma III.3.2).

<u>Definition V.2.15</u> : Let $E = \mathbb{R}$ and $(X_n, F_n)_{n \in \mathbb{N}}$ be an adapted sequence. The <u>Doob-decomposition</u> of $(X_n)_{n \in \mathbb{N}}$ is

$$X_n = M_n + A_n$$

where

$$M_1 = X_1$$

$$A_1 = 0$$

$$M_n - M_{n-1} = X_n - E^{F_{n-1}} X_n$$

$$A_n - A_{n-1} = E^{F_{n-1}} X_n - X_{n-1}$$

for each $n \in \mathbb{N}(2)$.

Note that $(M_n, F_n)_{n \in \mathbb{N}}$ is a martingale and that each A_n is F_{n-1} - measurable. Hence $(A_n, F_n)_{n \in \mathbb{N}}$ is predictable in the sense of definition III.3.1.

Using this formalism we can prove a convergence result for real amarts, namely Cesàro a.e. convergence (or the law of large numbers for amart difference sequences), extending the same well-known result for martingales, due to Chow [1963].

<u>Theorem V.2.16 (Krengel and Sucheston [1978])</u> : Suppose that $(X_n = \sum_{i=1}^{n} Y_i, F_n)_{n \in \mathbb{N}}$ is a real amart such that for some $\alpha \geq 1$

$$\sum_{i=1}^{\infty} \frac{\int_\Omega (Y_i^{2\alpha})}{i^{1+\alpha}} < \infty$$

Then $$\lim_{n \to \infty} \frac{X_n}{n} = 0, \text{ a.e.}.$$

<u>Proof</u> : Write the Doob-decomposition of $(X_n, F_n)_{n \in \mathbb{N}}$: $X_n = M_n + A_n$ as in definition V.2.15 above. We have

$$\int_\Omega |M_n - M_{n-1}|^{2\alpha}$$

$$\leq (\int_\Omega |Y_n|^{2\alpha} + \int_\Omega |A_n - A_{n-1}|^{2\alpha})^{2\alpha}$$

$$\leq 2^{2\alpha} \int_\Omega |Y_n|^{2\alpha}$$

So :

$$\sum_{i=1}^{\infty} \frac{\int_\Omega |M_i - M_{i+1}|^{2\alpha}}{i^{1+\alpha}} < \infty$$

and $(M_n, F_n)_{n \in \mathbb{N}}$ is also a martingale.
Hence, see Chow [1963] ,

$$\lim_{n\to\infty} \frac{M_n}{n} = 0 \ , \quad \text{a.e.} \qquad (1)$$

Furthermore, since $(X_n, F_n)_{n \in \mathbb{N}}$ is an amart, it follows from theorem V.2.14 that $(A_n - A_{n-1})_{n \in \mathbb{N}(2)}$ converges to zero a.e.. Hence also, defining $A_o = A_1 = 0$:

$$\frac{1}{n} A_n = \frac{1}{n} \sum_{i=0}^{n-1} (A_{i+1} - A_i)$$

Therefore

$$\lim_{n\to\infty} \frac{A_n}{n} = 0 \ , \quad \text{a.e.} \qquad (2)$$

From (1) and (2) we now have

$$\lim_{n\to\infty} \frac{X_n}{n} = 0 \ , \quad \text{a.e..} \qquad \square$$

For a special case of this theorem, see Edgar and Sucheston [1976a] .

So far only convergence theorems have been investigated. However, we have encountered in this section a norm which is incomplete, namely the Pettis norm $\|\cdot\|_{Pe}$ (see also theorem I.1.2.8). So one can expect that spaces in which every, say uniformly bounded, amart is Cauchy for the Pettis norm might differ from (RNP) spaces, because (RNP) spaces are exactly the spaces in which every uniformly bounded amart is convergent in L^1_E for the Pettis norm, by corollary V.2.5. This will be investigated now and we shall indeed prove that the class of Banach spaces in which every uniformly bounded amart is Cauchy for the Pettis norm is strictly larger than the class of (RNP) spaces. In fact an operator characterization can be given.

To work more easily we shall study occasionally amarts and martingales of the form $(X_i, F_i)_{i \in I}$ where I is an index set, $X_i \in L^1_E$ for each $i \in I$ and if $i \leq j$ then $F_i \subset F_j$. The definition of martingale has already been given (see (iii) $\Rightarrow$ (i) in theorem II.2.2.1). We call $(X_i, F_i)_{i \in I}$ an <u>amart</u> if the net

$$(\int_\Omega X_\tau)_{\tau \in T}$$

converges where T is the set of all finite stopping times, i.e. : functions $\tau : \Omega \to I$ such that $\{\tau = i\} \in F_i$ for every $i \in I$ and that take only finitely many values in I. We do not study index sets other than $\mathbb{N}$ except in those cases where the presentation can be given more elegantly by using these. This was already the case in theorem II.2.2.1. This is also the case here.

The reader interested in a more detailed treatment of adapted nets is referred to chapter VI, but at this stage we do not need more about adapted nets than the few notions we have introduced here.

We start with an individual characterization of amarts which are Cauchy for $\|\cdot\|_{Pe}$ due to J.J. Uhl Jr. [1977], but first we state (only part of) a result from Hoffmann-Jørgensen which we need in Uhl's proof. For details, see Hoffmann and Jørgensen [1971], theorem 9 and Dunford and Schwartz [1957], IV.13.19 :

<u>Lemma V.2.17 (Hoffmann-Jørgensen)</u> : Let $F_o \subset F$ be an algebra, and $F : F_o \to E$ be an additive set function where E is a Banach space. The following assertions are equivalent :

(i) $F(\mathcal{F}_o)$ is relatively norm compact

(ii) There exists a positive, finitely additive measure ν on $\mathcal{F}_o$ and a sequence $(f_n)_{n \in \mathbb{N}}$ of E-valued $\mathcal{F}_o$-stepfunctions such that

$$\lim_{n\to\infty} \sup_{A \in \mathcal{F}_o} \left\| \int_A f_n \, d\nu - F(A) \right\| = 0 .$$

Furthermore ν is absolutely continuous w.r.t. every element in

$$\{x'(F) \| x' \in E' \text{ and } \|x'\| \leq 1\} .$$

<u>Theorem V.2.18 (J.J. Uhl Jr.)</u> : Let $(X_i, \mathcal{F}_i)_{i \in I}$ be an amart. Then $(X_i)_{i \in I}$ is Pettis-Cauchy if and only if its limit measure F is P-continuous and has relatively norm compact range.

<u>Proof</u> : Suppose $(X_i)_{i \in I}$ is Pettis-Cauchy. Then by theorem I.1.2.7, the nets $(\int_A X_i)_{i \in I}$ are Cauchy, uniformly in $A \in \sigma(\bigcup_i \mathcal{F}_i)$. This uniformity together with the relative compactness and P-continuity of the range of a Bochner integral finishes the "only if" part of the proof.
Conversely, suppose first that I = $\mathbb{N}$. We apply lemma V.2.17 with $\mathcal{F}_o = \bigcup_n \mathcal{F}_n$. So there is a finitely additive measure ν on $\bigcup_n \mathcal{F}_n$ and a sequence of $\bigcup_n \mathcal{F}_n$ - stepfunctions such that

$$\lim_{n\to\infty} \sup_{A \in \bigcup_m \mathcal{F}_m} \left\| \int_A f_n \, d\nu - F(A) \right\| = 0$$

and such that ν is absolutely continuous w.r.t. every element in $\{x'(F) \| x' \in E' \text{ and } \|x'\| \leq 1\}$. Since $F \ll P$ we thus have $\nu \ll P$. By the classical theorem of Radon-Nikodym, also called Lebesgue-Nikodym, see e.g. Dinculeanu [1967], theorem 5, p.182, there exists $f \in L^1(\Omega, \mathcal{F}, P)$, such that

$$\nu(A) = \int_A f \, dP$$

for every $A \in \bigcup_m \mathcal{F}_m$. So

$$\lim_{n\to\infty} \sup_{A\in\cup_m F_m} \|\int_A f_n.f\,dP - F(A)\| = 0 ,$$

but since $\cup_m F_m$ - stepfunctions are dense in $L^1(\Omega,F,P)$, there exists a sequence $(\varphi_n)_{n\in\mathbb{N}}$ of $\cup_m F_m$ - stepfunctions, such that

$$\lim_{n\to\infty} \sup_{A\in\cup_m F_m} \|\int_A \varphi_n\,dP - F(A)\| = 0 .$$

So, for every $\varepsilon > 0$, there is a function which we now denote φ_ε of the form $\varphi_\varepsilon = \sum_{i=1}^{p} x_i \chi_{A_i}$, where $x_i \in E$ and $A_i \in \cup_m F_m$ for every $i = 1,\ldots,p$ such that

$$\|F(A) - \int_A \varphi_\varepsilon\| < \varepsilon \tag{1}$$

for all $A \in \cup_n F_n$. Now

$$\limsup_{n,m\to\infty} \|X_n - X_m\|_{Pe}$$

$$\leq 2 \limsup_{n,m\to\infty} \sup_{A\in\cup_n F_n} \|\int_A X_n - \int_A X_m\|$$

$$\leq 4 \limsup_{n\to\infty} \sup_{A\in\cup_n F_n} \|\int_A X_n - F(A)\| \tag{2}$$

If we knew that in the above sup we could restrict ourselves to one of the F_n, then theorem V.2.2 would finish the proof. That is where (1) comes in! Indeed,

$$(2) \leq 4 \limsup_{n\to\infty} \sup_{A\in\cup_n F_n} \|\int_A (X_n - \varphi_{\frac{\varepsilon}{16}})\| + \frac{\varepsilon}{4}$$

due to (1). There is an $n_o \in \mathbb{N}$ such that $\varphi_{\frac{\varepsilon}{16}}$ is F_{n_o}-measurable. Then

$$(2) = 4 \limsup_{n\to\infty} \sup_{A\in F_n} \|\int_A X_n - \int \varphi_{\frac{\varepsilon}{16}}\| + \frac{\varepsilon}{4}$$

since in the above lim sup we can suppose $n \geq n_o$.
Hence

$$(2) \leq 4 \lim_{n\to\infty} \sup_{A\in F_n} \|\int_A X_n - F(A)\| + 4 \sup_{A\in F_n} \|F(A) - \int \varphi_{\frac{\varepsilon}{16}}\| + \frac{\varepsilon}{4}$$

$$\leq 4 \lim_{n\to\infty} \sup_{A\in F_n} \|\int_A X_n - F(A)\| + \frac{\varepsilon}{2}$$

$$= \frac{\varepsilon}{2},$$

applying theorem V.2.2. Since this is true for every ε.

$$\lim_{n,m\to\infty} \|X_n - X_m\|_{Pe} = 0 \quad .$$

In the general case $(X_i, F_i)_{i\in I}$, we have by the above countable case that any countable subsequence $(X_{i_n})_{n\in \mathbb{N}}$ is $\|\cdot\|_{Pe}$-Cauchy, since the restriction of F, the limitmeasure of $(X_i)_{i\in I}$, to $\sigma(\bigcup_{n\in\mathbb{N}} F_{i_n})$ is certainly still P-continuous and has relatively norm compact range. This now implies that $(X_i)_{i\in I}$ itself is $\|\cdot\|_{Pe}$-Cauchy. This is seen by the following lemma : □

<u>Lemma V.2.19</u> : A net $(X_i)_{i\in I}$ in L^1_E is $\|\cdot\|_{Pe}$-Cauchy if and only if every countable subnet $(X_{i_n})_{n\in\mathbb{N}}$ is $\|\cdot\|_{Pe}$-Cauchy.

<u>Proof</u> : This is easy and in fact the proof is contained in the proof of (iii) ⇒ (i) in theorem II.2.2.1; just change $\|\cdot\|_1$ into $\|\cdot\|_{Pe}$. □

We could also have proved the converse part of theorem V.2.18 directly for nets, remarking that theorem V.2.2 remains valid for amarts indexed by directed sets, and by making obvious modifications to the proof of the case I = $\mathbb{N}$.

<u>Definition V.2.20 (Musiał, Egghe)</u> : We say that a Banach space has "<u>the compact range property</u>" (<u>CRP</u>) if every uniformly integrable amart is Cauchy for the Pettis topology (for every complete probability space (Ω, F, P)).

The reason for the choice of the term as well as the link with Musiał [1980], theorem 3 is seen in the next two results V.2.21 and V.2.23.

Lemma V.2.21 (Egghe [1980c]) : Every uniformly integrable amart is Cauchy for $\|\cdot\|_{Pe}$ (i.e. E has (CRP)) if and only if every uniformly integrable martingale consisting of stepfunctions is Cauchy for $\|\cdot\|_{Pe}$.

Proof : Let $(X_n, F_n)_{n\in\mathbb{N}}$ be a uniformly integrable amart. Write

$$X_n = \|\cdot\|_1 - \lim_{m\to\infty} X_{n,m}$$

where

$$X_{n,m} = \sum_{A\in\pi_{n,m}} \frac{\int_A X_n}{P(A)} \chi_A$$

and where $(\pi_{n,m})_{m\in\mathbb{N}}$ is an increasing sequence of finite partitions of Ω into elements of F_n. Then we have a table

$$(X_{n,m}; \sigma(\pi_{n,m}))_{n,m\in\mathbb{N}}$$

with $\pi_{n,m+1} \geqslant \pi_{n,m}$ for every $m,n \in \mathbb{N}$. We can of course also suppose that $\pi_{n+1,m} \geqslant \pi_{n,m}$ for every $n,m \in \mathbb{N}$.
The rows are trivially seen to be martingales. Take now an arbitrary column $(X_{n,m})_{n\in\mathbb{N}}$. Denote by T_m the set of all bounded stopping times w.r.t. the sequence $(\sigma(\pi_{n,m}))_{n\in\mathbb{N}}$. Let $\tau \in T_m$. Then

$$\int_\Omega X_{\tau,m} = \sum_{k=\min\tau}^{\max\tau} \int_{\{\tau=k\}} X_{k,m}$$

Now $\{\tau=k\}\in\sigma(\pi_{k,m}) \subset F_k$, $\int_A X_{k,m+1} = \int_A X_{k,m}$ for all $A \in \sigma(\pi_{k,m})$ and $\|\cdot\|_1 - \lim_{m\to\infty} X_{k,m} = X_k$. Hence, for all $A \in \sigma(\pi_{k,m})$, $\int_A X_k = \int_A X_{k,m}$, Consequently

$$\int_\Omega X_{\tau,m} = \int_\Omega X_\tau$$

Since $(X_n,F_n)_{n\in\mathbb{N}}$ is an amart and since T_m is cofinal in T (since $\mathbb{N} \subset T_m$) we see now that every column is an amart. Suppose now that $(X_n)_{n\in\mathbb{N}}$ is not $\|\cdot\|_{Pe}$-Cauchy, then we can construct a subsequence $(X_{n_j,m_j})_{j\in\mathbb{N}}$ of the table, which is not $\|\cdot\|_{Pe}$-Cauchy. Now $(X_{n_j,m_j};\sigma(\pi_{n_j,m_j}))_{j\in\mathbb{N}}$ is a uniformly integrable amart consisting of stepfunctions.

To simplify the notation call this $(X_j',G_j)_{j\in\mathbb{N}}$. From the remark after V.2.5, write the Riesz-decomposition of $(X_j',G_j)_{j\in\mathbb{N}}$

$$X_j' = Y_j + Z_j$$

where $\lim_{\tau\in T'} \|Z_\tau\|_{Pe} = 0$ and where T' is the set of bounded stopping times w.r.t. $(G_j)_{j\in\mathbb{N}}$. Then the uniformly integrable martingale $(Y_j,G_j)_{j\in\mathbb{N}}$ which consists of stepfunctions is not $\|\cdot\|_{Pe}$-Cauchy, contradicting the assumptions. □

Remark V.2.22 : In the definition of (CRP) or in any equivalent formulation using the $\|\cdot\|_{Pe}$-Cauchy property we may use nets or sequences of amarts or of martingales. This follows from lemma V.2.19.

Theorem V.2.23 (Musiał) : The following assertions are equivalent :

(i) E has (CRP).

(ii) For every complete probability space (Ω,F,P) and every P-continuous vectormeasure $F : F \to E$ of bounded variation we have that $F(F)$ is relatively norm compact.

Proof : (i) ⇒ (ii)

Let F be a P-continuous vectormeasure of bounded variation. For $\pi \in \Pi$ = {all finite partitions of Ω into elements of F}, put

$$X_\pi = \sum_{A\in\pi} \frac{F(A)}{P(A)} \chi_A .$$

Then, using (CRP) and remark V.2.22, the martingale $(X_\pi,\sigma(\pi))_{\pi\in\Pi}$ is Cauchy for $\|\cdot\|_{Pe}$. Its limit measure is of course F. So by theorem V.2.18, F has relatively norm compact range.

(ii) ⇒ (i)

Let $(X_n, \mathcal{F}_n)_{n \in \mathbb{N}}$ be a uniformly integrable amart. Then the limit measure F exists trivially on $\bigcup_n \mathcal{F}_n$, and due to the uniform integrability, F exists on $\sigma(\bigcup_n \mathcal{F}_n)$. Obviously $F \ll P$ and $|F|(\Omega) < \infty$. So $F(\mathcal{F})$ is relatively norm compact. By theorem V.2.18, $(X_n)_{n \in \mathbb{N}}$ is $\|\cdot\|_{Pe}$-Cauchy. Hence E has (CRP). □

The following result is handy in checking (CRP).

Theorem V.2.24 : In the definition of (CRP) as well as in lemma V.2.21 we may change "uniformly integrable" into "uniformly bounded".

Proof : For the definition this is seen from theorem V.2.23 and the fact that in (ii) of this theorem we may restrict ourselves to vector measures F with bounded average range (see Phillips [1943] , lemma 5.4). For lemma V.2.21, this is proved with the modified (equivalent) definition in exactly the same way. □

Applying lemma V.2.21 we get the following result (proved differently in Musiał [1980]).

Theorem V.2.25 : If E has the property that w.r.t. one fixed complete non atomic probability space $(\Omega, \mathcal{F}, P)$, every uniformly bounded martingale consisting of stepfunctions is $\|\cdot\|_{Pe}$-Cauchy, then E has (CRP).

An operator characterization of (CRP) is now given.
In this theorem we use the following result of Diestel :

Theorem V.2.26 (Diestel) : Let $T \in \mathcal{L}(L^\infty, E)$, where $L^\infty = L^\infty(\Omega, \mathcal{F}, P)$. Define

$$F : \mathcal{F} \to E$$

$$A \to F(A) =: T(\chi_A)$$

Then $F(\mathcal{F})$ is relatively norm compact iff T is a compact operator. See Diestel [1975a] , p.214 for the proof.

Theorem V.2.27 (Egghe [1980c]) : Let E be a Banach space and $(\Omega, \mathcal{F}, P)$ a fixed complete non-atomic probability space. Then the following assertions are equivalent :

(i) E has (CRP).

(ii) For every operator $T \in \mathcal{L}(L^1(\Omega,\mathcal{F},P),E)$, the restriction $T|_{L^\infty}$ is a compact operator.

(iii) Every operator $T \in \mathcal{L}(L^1(\Omega,\mathcal{F},P),E)$ is Dunford-Pettis i.e. : maps weakly convergent sequences into norm convergent sequences.

Proof : (i) ⇒ (ii)
Suppose $T \in \mathcal{L}(L^1,E)$ and put

$$F : \mathcal{F} \to E$$
$$A \to F(A) = T(\chi_A) .$$

Then $\|F(A)\| \leq \|T\| P(A)$, hence the average range of F is bounded. Put

$$X_\pi = \sum_{A \in \pi} \frac{F(A)}{P(A)} \chi_A$$

where $\pi \in \Pi$, the set of all finite partitions of Ω into elements of $\mathcal{F}$. Since $(X_\pi, \sigma(\pi))_{\pi \in \Pi}$ is a uniformly bounded martingale, it is $\|\cdot\|_{Pe}$-Cauchy, due to remark V.2.22. Hence, by theorem V.2.18, $F(\mathcal{F})$ is relatively norm compact. Hence, using theorem V.2.26, $T|_{L^\infty}$ is a compact operator.

(ii) ⇒ (iii)
See Musiał [1980] .

(ii) ⇒ (i)
Using theorem V.2.24 and lemma V.2.21 it is enough to suppose that $(X_n,\mathcal{F}_n)_{n\in\mathbb{N}}$ is a uniformly bounded martingale consisting of step-functions. Then we can write

$$X_n = \sum_{A \in \pi_n} \frac{F(A)}{P(A)} \chi_A$$

where $\mathcal{F}_n = \sigma(\pi_n)$ and π_n is a finite partition of Ω into elements of $\mathcal{F}_n$, for each $n \in \mathbb{N}$, and where F, as limitmeasure of $(X_n)_{n\in\mathbb{N}}$ is uniquely determined on $\mathcal{F} = \sigma(\bigcup_n \pi_n)$. Define

$$T : L^1 \to E$$

$$X \to \int_\Omega X \, dF = T(X)$$

where the integral is taken in the Dinculeanu sense (see Dinculeanu [1967]). T is linear and furthermore $\|T(X)\| \leq \int_\Omega |X| \, d|F|$ for every $X \in L^1$. From the scalar Radon-Nikodym-theorem, $|F| = \varphi.P$ where $\varphi \in L^\infty$. Hence

$$\|T(X)\| \leq \int_\Omega |X| \, d(\varphi.P)$$

$$\leq \|\varphi\| \, \|X\|_1$$

Hence $T \in \mathcal{L}(L^1,E)$. So $T|_{L^\infty}$ is a compact operator and so $F(F)$ is relatively norm compact by theorem V.2.26. From theorem V.2.18 we have now that $(X_n,F_n)_{n\in\mathbb{N}}$ is $\|\cdot\|_{Pe}$-Cauchy. □

Using some Banach space techniques beyond the scope of this book one can prove directly or using theorem V.2.27 :

Theorem V.2.28 (Bourgain) : Let E be a Banach space. Then E has no subspace isomorphic to ℓ^1 if and only if E' has (CRP).

Using theorem V.2.27 and V.2.28 we are in a position to give examples of spaces with or without (CRP).

V.2.29 Examples of spaces without (CRP)

(a) c_0 - see the remarks after II.2.2.1.

(b) ℓ^∞. This is so since $c_0 \subsetneq \ell^\infty$ and since (CRP) is a property hereditary for closed subspaces, as is easily seen.

(c) L^∞, $C(\Omega)'$ where $C(\Omega)$ is the space of continuous functions on a compact Hausdorff space. This follows from theorem V.2.28.

(d) $C(\Omega)$ since $c_0 \subsetneq C(\Omega)$.

(e) $L^1(\Omega,F,P)$ if (Ω,F,P) is atomless. Indeed, the identity operator I : $L^1(P) \to L^1(P)$ has no compact restriction to $L^\infty(P)$. For if it had, then $\{\chi_A \| A\in F\}$ would be relatively compact in $L^1(P)$, contradicting

Hoffmann and Jørgensen [1971], p.29. So theorem V.2.27 gives the result. For another proof, see also the remarks after II.2.2.1.

V.2.30 Examples of spaces with (CRP)

(a) All (RNP) spaces.
This follows from corollary V.2.5.

(b) All Schur spaces i.e. : spaces in which every weakly convergent sequence is strongly convergent.
Indeed, every operator $T \in \mathcal{L}(L^1, E)$ is also continuous w.r.t. the weak topologies on L^1 and on E. So it transforms weakly convergent sequences into weakly, hence strongly convergent sequences. Hence T is Dunford-Pettis. So theorem V.2.27 finishes the proof. □

(c) JT', JF' (the spaces from Lindenstrauss and Stegall [1975]).
This follows from theorem V.2.28. These spaces do not have (RNP). Hence we see here that (CRP) is a strictly weaker property than (RNP). In Egghe [1980c] some results previously proved under the assumption (RNP) are extended assuming only (CRP). These are beyond the scope of this work.

When working in Banach lattices, one may also define an amart-type adapted sequence by using order convergence. Indeed one can define, as Ghoussoub did, an orderamart as follows.

Definition V.2.31 (Ghoussoub [1979a]) : An adapted sequence $(X_n, F_n)_{n \in \mathbb{N}}$ with values in a Banach lattice E is called an orderamart if the net

$$(\int_\Omega X_\tau)_{\tau \in T}$$

is orderconvergent in E. Analogously an orderpotential is an adapted sequence $(Z_n, F_n)_{n \in \mathbb{N}}$ such that the net

$$(\int_\Omega |Z_\tau|)_{\tau \in T}$$

is orderconvergent to zero.

We shall present the theory of orderamarts but not the proofs since they are much in line with the amart proofs. For the detailed proofs we refer the reader to Ghoussoub [1979a], p.167-172; see also Ghoussoub [1982].

Theorem V.2.32 : Let E be a Banach lattice with (RNP) and let $(X_n, F_n)_{n\in \mathbb{N}}$ be an orderamart such that

$$\liminf_{n\in\mathbb{N}} \int_\Omega \|X_n\| < \infty$$

Then $(X_n, F_n)_{n\in\mathbb{N}}$ can be written uniquely as

$$X_n = Y_n + Z_n$$

where $(Y_n, F_n)_{n\in\mathbb{N}}$ is an L^1_E-bounded martingale and $(Z_n, F_n)_{n\in\mathbb{N}}$ is an orderpotential.

Theorem V.2.33 : Let E be a Banach lattice with (RNP) such that E' has a countable t.o.s. (see Schaefer [1974] or lemma V.3.19.2 where the concept is effectively used and where the proof is the same as the one needed here to prove this theorem). Then each E-valued orderamart of class (B) converges weakly a.e.. For a more detailed result, see theorem V.3.20.

Theorem V.2.34 : For a Banach lattice E the following assertions are equivalent :

(i) E is lattice isomorphic to $\ell^1(\Gamma)$, for a certain Γ.

(ii) Each orderamart $(X_n, F_n)_{n\in\mathbb{N}}$ for which $\sup_{n\in\mathbb{N}} \int_\Omega |X_n|$ exists is convergent weakly a.e..

(iii) Each L^1_E-bounded orderamart is convergent weakly a.e..

(iv) Each orderamart of class (B) converges strongly a.e..

Theorem V.2.35 : For a Banach lattice E, the following assertions are equivalent :

(i) E is lattice isomorphic to an AL-space.

(ii) Every orderamart is a uniform amart.

(iii) Every orderpotential converges to zero strongly a.e..

(iv) Every orderpotential of class (B) converges to zero strongly a.e..

For non-trivial applications of amarts we refer the reader to chapter VI, which is completely devoted to it.

V.3. Weak sequential amarts

In this section, one of the main objectives is to prove the theorem of Brunel and Sucheston [1976a], which is theorem V.2.6 but extended to a more general class than amarts, namely weak sequential amarts (see definition further on). We give the Bellow-Egghe proof of this result, using the inequalities proved in section IV.1, and yielding also a slight generalization of the Brunel-Sucheston theorem.

As one can see in this theorem, rather heavy assumptions on the Banach space and on the adapted sequence are made in order to obtain weak convergence a.e.. Examples show that these assumptions are necessary.

The Brunel-Sucheston characterization of separability (Brunel and Sucheston [1977]) of the dual E' of a Banach space in terms of weak sequential potentials is also included here, but with Edgar's proof (Edgar [1980]). It extends the Brunel-Sucheston result in an operator fashion and it adds also some further characterizations, f.i. in terms of strong potentials. They reveal the most important applications of this type of adapted sequences, and are the very reason to study them.

Also in this section the Ghoussoub-Talagrand theorem (Ghoussoub and Talagrand [1979a]) on the characterization of weak convergence a.e. of positive weak sequential potentials is proved in detail.

Definition V.3.1 : Let $(X_n,F_n)_{n\in\mathbb{N}}$ be an arbitrary adapted sequence in L^1_E. We say that $(X_n,F_n)_{n\in\mathbb{N}}$ is a "weak sequential amart", abbreviated WS amart, if for every increasing sequence $(\tau_n)_{n\in\mathbb{N}}$ in T (not necessarily cofinal), the sequence

$$(\int_\Omega X_{\tau_n})_{n \in \mathbb{N}}$$

converges weakly.

Theorem V.3.2 : Every amart is a WS amart.

Proof : If $(X_n, F_n)_{n \in \mathbb{N}}$ is an amart and $(\tau_n)_{n \in \mathbb{N}}$ is an increasing sequence in T, then

$$(\int_\Omega X_{\tau_n})_{n \in \mathbb{N}}$$

converges strongly. Indeed, using the optional sampling theorem for amarts (theorem V.2.11) we see that $(X_{\tau_n}, F_{\tau_n})_{n \in \mathbb{N}}$ is an amart. So $(X_n, F_n)_{n \in \mathbb{N}}$ is a WS amart. □

We are now going to prove theorem V.2.6 but for WS amarts. In view of theorem V.3.2 this is a more general result. But first we prove even a more general result :

Theorem V.3.3 (Bellow-Egghe) : Let E have (RNP). Suppose that $(X_n, F_n)_{n \in \mathbb{N}}$ is a WS amart such that there is a sequence $(\tau_n)_{n \in \mathbb{N}}$ in T such that $n \leq \tau_n \leq \tau_{n+1}$ for each $n \in \mathbb{N}$ and such that

$$\sup_{n \in \mathbb{N}} \int_\Omega \|X_{\tau_n}\| < \infty$$

(f.i. L^1_E-boundedness suffices!). Then $(X_n)_{n \in \mathbb{N}}$ converges scalarly a.e. to a Bochner integrable function.

Proof : We apply theorem IV.1.4, case II, for $T = \sigma(E,E')$. Condition (B_T) there is valid. Indeed : for each $m \in \mathbb{N}$ and $A \in F_m$, define :

$$\sigma_n \begin{cases} = \tau_n & \text{on } A \\ = m & \text{on } \Omega \setminus A \end{cases}$$

for $n \in \mathbb{N}(m)$. Then $(\sigma_n)_{n \in \mathbb{N}}$ is an increasing sequence in T. Hence, since $(X_n, F_n)_{n \in \mathbb{N}}$ is a WS amart, $(\int_\Omega X_{\sigma_n})_{n \in \mathbb{N}}$ converges weakly. But $\int_\Omega X_{\sigma_n} = \int_A X_{\tau_n} + \int_{\Omega \setminus A} X_m$ for each $n \in \mathbb{N}$. Hence

$$(\int_A X_{\tau_n})_{n\in\mathbb{N}}$$

converges weakly. Since $(X_{\tau_n})_{n\in\mathbb{N}}$ is also L^1_E-bounded, condition (B_T) is verified. From theorem IV.1.4, case II, there is an L^1_E-bounded martingale $(Y_n,F_n)_{n\in\mathbb{N}}$ such that for each $x'\in E'$, there are sequences $(\sigma_n^{x'})_{n\in\mathbb{N}}$ and $(\tau_n^{x'})_{n\in\mathbb{N}}$ in T such that $n\leqslant\sigma_n^{x'}\leqslant\tau_n^{x'}$ for each $n\in\mathbb{N}$ and such that a.e.

$$\limsup_{n\in\mathbb{N}} |x'(Y_n(\omega) - X_n(\omega))|$$

$$\leqslant \liminf_{n\in\mathbb{N}} |x'(E^{F_{\sigma_n^{x'}}} X_{\tau_n^{x'}}(\omega) - X_{\sigma_n^{x'}}(\omega))|$$

Since E has (RNP), $(Y_n,F_n)_{n\in\mathbb{N}}$ converges a.e. to a Bochner integrable function Y. So certainly a.e.

$$\limsup_{n\in\mathbb{N}} |x'(Y(\omega) - X_n(\omega))|$$

$$\leqslant \liminf_{n\in\mathbb{N}} |x'(E^{F_{\sigma_n^{x'}}} X_{\tau_n^{x'}}(\omega) - X_{\sigma_n^{x'}}(\omega)|$$

Integrating and using Fatou's lemma yields :

$$\int_\Omega \limsup_{n\in\mathbb{N}} |x'(Y(\omega) - X_n(\omega))|$$

$$\leqslant \liminf_{n\in\mathbb{N}} \int_\Omega |x'(E^{F_{\sigma_n^{x'}}} X_{\tau_n^{x'}}(\omega) - X_{\sigma_n^{x'}}(\omega))|$$

Since, for each $x'\in E'$, $(x'(X_n),F_n)_{n\in\mathbb{N}}$ is a uniform amart and since the sequences $(\sigma_n^{x'})_{n\in\mathbb{N}}$ and $(\tau_n^{x'})_{n\in\mathbb{N}}$ are cofinal in T, the right hand side of the last inequality is zero, which finishes the proof. □

Corollary V.3.4 (Brunel and Sucheston) : Let E be a Banach space with (RNP) and such that E' is separable. Let $(X_n, F_n)_{n \in \mathbb{N}}$ be a WS amart of class (B). Then $(X_n)_{n \in \mathbb{N}}$ converges weakly a.e. to a Bochner integrable function.

Proof : Let $Y \in L^1_E$ be the function such that

$$\lim_{n \to \infty} x'(X_n(\omega)) = x'(Y(\omega))$$

a.e., for each $x' \in E'$, using the above theorem. Now applying lemma II.1.5(i) we have for every $\lambda > 0$:

$$P(\sup_{n \in \mathbb{N}} \|X_n\| > \lambda) \leq \frac{1}{\lambda} \sup_{\tau \in T} \int \|X_\tau\|$$

Hence $\sup_{n \in \mathbb{N}} \|X_n(\omega)\| < \infty$, a.e.. Due to the separability of E' and the boundedness a.e. of $\bigcup_{n \in \mathbb{N}} X_n(\Omega)$, the scalar convergence a.e. is in fact a weak convergence a.e.. □

If E' is separable then E' has (RNP). This is well known - see f.i. Diestel and Uhl [1977]. In fact it suffices to require "E' (RNP)" in corollary V.3.4 above :

Corollary V.3.5 : Corollary V.3.4 holds if "E' separable" is replaced by "E' (RNP)".

Proof : Let E_1 be a separable subspace of E such that

$$X_n(\omega) \in E_1$$

for every $n \in \mathbb{N}$, a.e.. Now apply the result of Stegall (see also Diestel and Uhl [1977]) stating that E' has (RNP) if and only if for every separable Banach subspace E_1 of E, E_1' is separable. Then we can apply the previous corollary in E_1. Hence there is an integrable function Y such that a.e

$$\sigma(E_1, E_1') - \lim_{n \to \infty} X_n(\omega) = Y(\omega)$$

Hence also a.e.

$$\sigma(E,E') - \lim_{n\to\infty} X_n(\omega) = Y(\omega) ,$$

using the well-known fact that the weak topology on E, restricted to E_1 is the weak topology on E_1. □

If one looks at the proof of theorem V.3.3 one sees that the results V.3.3, V.3.4 and V.3.5 remain trivially valid for the following type of adapted sequence, called "uniform weak amart", introduced by Schmidt [1981b] : an adapted sequence $(X_n,\mathcal{F}_n)_{n\in\mathbb{N}}$ is called a <u>uniform weak amart</u> if the net $(\int_\Omega X_\tau)_{\tau\in T}$ weakly converges and if for every $A \in \bigcup_n \mathcal{F}_n$, the sequence $(\int_A X_n)_{n\in\mathbb{N}}$ weakly converges. That every WS amart is a uniform weak amart is easy to see and in fact is contained in the proofs of theorem V.3.3 and theorem V.4.2 (trivial to check at this point!).

We now show that the above results are "sharp" in the sense that the minimal conditions (E has only (RNP), $(X_n)_{n\in\mathbb{N}}$ is only L^1_E-bounded) do not suffice and that strong convergence a.e. is not obtained.

<u>Examples V.3.6</u>

1. Strong convergence need not hold (Chacon and Sucheston). For this, take example V.2.10. There $(X_n,\mathcal{F}_n)_{n\in\mathbb{N}}$ is a uniformly bounded amart in ℓ^2 which diverges everywhere strongly!

2. L^1_E-boundedness is not enough (Chacon and Sucheston).
Let E be ℓ^2 again. Also let e^i_n and A^i_n be as in example V.2.10.
For $k = 1,\dots,2^n$, define

$$Y^k_n = \sum_{i=1}^{2^n} \alpha_i\, e^i_n\, \chi_{A^i_n}$$

where $\alpha_i = 1$ if $i \neq k$ and $\alpha_k = n$. Define $m = 2^{n-1} + k - 1$ and $X_m = Y^k_n$ for $k = 1,\dots,2^{n-1}$. Now

$$\int_\Omega \|Y^k_n\| = \sum_{i=1}^{2^n} \alpha_i\, P(A^i_n) = 1 - 2^{-n} + n2^{-n} < 2 .$$

Hence

$$\sup_{m \in \mathbb{N}} \int_\Omega \|X_m\| \leqslant 2 < \infty .$$

Let $\tau \in T$. Then X_τ is of the form

$$X_\tau = \sum_{i,n} \beta_n^i e_n^i \chi_{B_n^i} \tag{1}$$

where the B_n^i are disjoint, $B_n^i \subset A_n^i$ and β_n^i has the values 1 or n. Fix $N \in \mathbb{N}$. If $\tau \in T(2^N)$, then, since $\tau(\omega)$ is an $m = 2^{n-1} + k-1$ ($k \in \{1,\dots,2^{n-1}\}$), we see from $2^{n-1} + k - 1 \geqslant 2^N + 1$ that $2^{n-1} \geqslant 2^N + 2 - k \geqslant 2^N + 2 - 2^{n-1}$. Hence $2^n \geqslant 2^N + 2$. So $n > N$. From this

$$\left\|\int_\Omega X_\tau\right\|^2 \leqslant \sum_{\substack{n \in \mathbb{N}(N) \\ i}} n^2 P(B_n^i)^2 \leqslant \sup_{n \in \mathbb{N}(N)} (n^2 2^{-n}) \sum_{i,n} P(B_n^i)$$

$$\leqslant \sup_{n \in \mathbb{N}(N)} (n^2 2^{-n}) \to 0$$

if $N \to \infty$. So $(X_n, F_n)_{n \in \mathbb{N}}$ is an amart where $F_n = \sigma(X_1,\dots,X_n)$ for each $n \in \mathbb{N}$. But $\sup_{n \in \mathbb{N}} \|X_n(\omega)\| = \infty$ everywhere. Therefore $(X_n(\omega))_{n \in \mathbb{N}}$ diverges weakly everywhere.

3. The condition that E has (RNP) cannot be omitted, even when E' is separable (Edgar and Sucheston).
Indeed, take $E = c_o$. We have remarked already that c_o lacks (RNP) although its dual ℓ^1 is separable. Order the canonical basis of c_o as $\{e_n^i \| n \in \mathbb{N},\ 1 \leqslant i \leqslant 2^n\}$ and take A_n^i as in the above example. Define

$$Y_j = \sum_{i=1}^{2^j} e_j^i \chi_{A_j^i}$$

$$X_n = \sum_{j=1}^{n} Y_j$$

If $\tau \in T$, then $X_\tau = \sum_{i,n} e_n^i \chi_{B_n^i}$ where $B_n^i = A_n^i \cap \{\tau \geqslant n\}$.
Hence $\|X_\tau\| = 1$ everywhere. Also if we fix $N \in \mathbb{N}$, and $\tau \in T(N)$, then

$$\int_\Omega X_\tau - \int_\Omega X_N = \sum_{i,n} P(B_n^i)\, e_n^i$$

where now $n \in \mathbb{N}(N)$. Then

$$\|\int_\Omega X_\tau - \int_\Omega X_N\| = \max_{n,i} P(B_n^i) \leq 2^{-N}$$

So, $(X_n, F_n)_{n \in \mathbb{N}}$ is an amart. Now $(X_n)_{n \in \mathbb{N}}$ obviously converges to $\sum_{n=1}^{\infty} Y_n$ in the $\sigma(\ell^\infty, \ell^1)$-topology and $\sum_{n=1}^{\infty} Y_n(\omega) \notin c_o$ for any ω. Hence $(X_n(\omega))_{n \in \mathbb{N}}$ diverges weakly everywhere.

4. The assumption that E' has (RNP) cannot be omitted, even when E has (RNP) (Brunel and Sucheston).
Indeed, take $E = \ell^1$ which has (RNP), being a separable dual. Then $E' = \ell^\infty$ which lacks (RNP). Let $(Y_n)_{n \in \mathbb{N}}$ be a sequence of independent real integrable functions such that $Y_n(\Omega) = \{-1,+1\}$ with $P(Y_n = 1) = P(Y_n = -1) = \frac{1}{2}$. Define

$$X_n = \frac{1}{n} \sum_{i=1}^{n} Y_i \, e_i$$

where $(e_n)_{n \in \mathbb{N}}$ is the canonical basis of ℓ^1. Now, for every $\omega \in \Omega$, there is $x' \in E' = \ell^\infty$ such that $\lim_{n\to\infty} x'(X_n(\omega)) = 1$. Indeed : choose x' to be a sequence whose terms are ± 1 according to the values of Y_i. But by the strong law of large numbers, $\lim_{n\to\infty} x'(X_n(\omega)) = 0$, a.e., for each $x' \in E'$. So trivially $\lim_{\tau \in T} x'(X_\tau(\omega)) = 0$ a.e., for each $x' \in E'$. By uniform boundedness, $\lim_{\tau \in T} \int_\Omega x'(X_\tau) = 0$. So for every $x' \in E'$, $(x'(X_n), F_n)_{n \in \mathbb{N}}$ is an amart. Using theorem V.2.11 we see that for every increasing sequence $(\tau_n)_{n \in \mathbb{N}}$ in T, $\lim_{n\to\infty} \int_\Omega x'(X_{\tau_n}) = 0$. Hence for every increasing sequence $(\tau_n)_{n \in \mathbb{N}}$ in T,

$$\left(\int_\Omega X_{\tau_n}\right)_{n \in \mathbb{N}}$$

is weakly Cauchy. Now $E = \ell^1$ is a Schur space, hence the sequence $(\int_\Omega X_{\tau_n})_{n \in \mathbb{N}}$ strongly converges. Therefore $(X_n, F_n)_{n \in \mathbb{N}}$ is a uniformly bounded amart in ℓ^1, diverging weakly everywhere.

This finishes the theorem of Brunel and Sucheston and its extensions.

Another result of Brunel and Sucheston [1977] describes in a probabilistic way the separability of the dual E' of a separable Banach space E. Indeed in this result, the separability of E' is characterized in terms of weak convergence a.e. of WS potentials (see definition further on). However a considerable extension of this theorem is given by Edgar [1980] where he characterises Asplund operators in terms of weak convergence a.e. of weak or strong amart potentials. It is this elegant proof we shall present. Let us first fix some terminology.

Definition V.3.7 (Brunel and Sucheston) : An adapted sequence $(X_n, F_n)_{n \in \mathbb{N}}$ is called a weak sequential potential, abbreviated WS potential, if $(X_n, F_n)_{n \in \mathbb{N}}$ is a WS amart satisfying

$$\text{weak}-\lim_{n \to \infty} \int_A X_n = 0$$

for every $A \in \bigcup_{n \in \mathbb{N}} F_n$.

The theory of Asplund spaces and Asplund operators is beyond the scope of the present work. Hence we indicate only the necessary definitions and theorems to understand the main result. The presentation of the Brunel-Sucheston characterization of the separability of the dual in the framework of Asplund theory is indeed very elegant and yields also considerable extensions of this result.

Definition V.3.9 : Let E and F be Banach spaces and $T \in \mathcal{L}(E,F)$, i.e. T is a continuous linear operator from E into F. We say that T is a Radon-Nikodym-operator if for any probability space (Ω, F, P) and any vector measure $F : F \to E$ with bounded average range, the measure $T \circ m$: $F \to F$ has a Bochner integrable Radon-Nikodym derivative. Here we denote of course

$$(T \circ m)(A) = T(m(A))$$

for every $A \in F$.

Definition V.3.10 (Edgar) : Let E and F be Banach spaces and $T \in \mathcal{L}(E,F)$. We say that T is an Asplund operator if for every continuous and convex function $\varphi : F \to \mathbb{R}$, $\varphi \circ T$ is Fréchet differentiable on a dense subset of E (see Diestel [1975b]).

The definitions V.3.9 and V.3.10 have found their origin in the fact that if E = F and $T = ID_E$, the identity operator in E, T is a Radon-Nikodym operator (Asplund operator) if and only if E has (RNP) (is an Asplund space). For (RNP), this is clear; for Asplund space, see also Namioka and Phelps [1975].

We conclude this sequence of definitions with the notions of Haar function and Haar operator.

Definition V.3.11 : Let Δ be the Cantor set and μ be Haar measure on Δ. Let $\{\Delta_{ni} \| n \in \mathbb{N},\ 1 \leq i \leq 2^n\}$ be clopen (= closed and open) disjoint intervals in Δ such that $\mu(\Delta_{ni}) = 2^{-n}$, $\Delta_{n+1,2i-1} \cup \Delta_{n+1,2i} = \Delta_{ni}$ for each i,n. The Haar functions $h_{ni} : \Delta \to \mathbb{R}$ are defined as :

$$h_{ni} = \chi_{\Delta_{n+1,2i-1}} - \chi_{\Delta_{n+1,2i}}$$

Definition V.3.12 : Let $\{e_{n_i} \| n \in \mathbb{N},\ 1 \leq i \leq 2^n\}$ be an enumeration of the canonical ℓ^1-basis. We define the Haar operator $H : \ell^1 \to L^\infty(\Delta,\mu)$ as

$$H(e_{n_i}) = h_{n_i}$$

for every i,n.

We have the following result of Stegall [1972].

Theorem V.3.13 (Stegall) : Let E and F be Banach spaces and $T \in \mathcal{L}(E,F)$. The following assertions are equivalent :

(i) T', the adjoint of T, is a Radon-Nikodym operator.

(ii) T is an Asplund operator.

(iii) T is not a factor of the Haar operator H, i.e. the following scheme cannot be true.

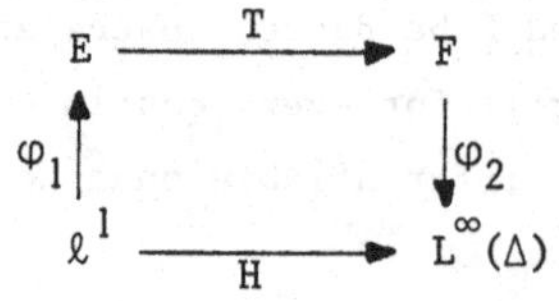

(So $H \neq \varphi_2 \circ T \circ \varphi_1$ where $\varphi_1 \in \mathcal{L}(\ell^1, E)$ and $\varphi_2 \in \mathcal{L}(F, L^\infty(\Delta))$).

(iv) T factors through a space G, where G has the property that

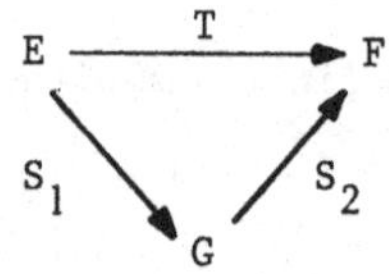

every separable subspace of G has a separable dual.

The proof of this theorem is not the subject of this book and hence is omitted.

We now come to Edgar's theorem (Edgar [1980]).

Theorem V.3.14 (Edgar) : Let E and F be Banach spaces and $T \in \mathcal{L}(E,F)$. The following assertions are equivalent :

(i) T is an Asplund operator.

(ii) Let $(\Omega, \mathcal{F}, P)$ be any probability space. Let $X_n \in L^1_E(\Omega, \mathcal{F}, P)$ for every $n \in \mathbb{N}$. Suppose that $\sup_{n \in \mathbb{N}} \|X_n(\omega)\| < \infty$, a.e. and that $(X_n)_{n \in \mathbb{N}}$ converges scalarly a.e. to zero. Then $(TX_n)_{n \in \mathbb{N}}$ converges weakly a.e. to zero in F.

(iii) Let $(X_n, \mathcal{F}_n)_{n \in \mathbb{N}}$ be a WS potential of class (B) in E. Then $(TX_n)_{n \in \mathbb{N}}$ converges weakly a.e. to zero in F.

(iv) Let $(X_n, \mathcal{F}_n)_{n \in \mathbb{N}}$ be a $\|\cdot\|_\infty$-bounded potential in E. Then $(TX_n)_{n \in \mathbb{N}}$ converges weakly a.e. to zero in F.

Proof : (i) ⇒ (ii)
Use theorem V.3.13 to obtain a space G such that $T = S_2 \circ S_1$

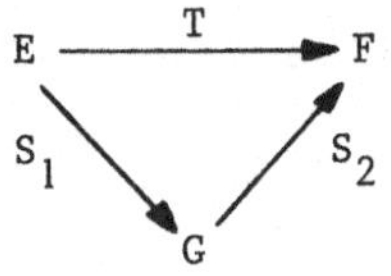

where $S_1 \in \mathcal{L}(E,G)$, $S_2 \in \mathcal{L}(G,F)$ and such that every separable subspace of G has a separable dual. Call G_1 the separable subspace of G such that $S_1X_n(\omega) \in G_1$ a.e. for all $n \in \mathbb{N}$. Since G_1' is separable and $\bigcup_{n\in\mathbb{N}} S_1Y_n(\Omega)$ is a.e. bounded in G_1 it follows that $(S_1X_n)_{n\in\mathbb{N}}$ converges weakly a.e. to zero. Hence, since S_2 is also in $\mathcal{L}((G,\sigma(G,G')),(F,\sigma(F,F')))$ and since the restriction of $\sigma(G,G')$ to G_1 is $\sigma(G_1,G_1')$ we see now that $(TX_n)_{n\in\mathbb{N}}$ converges weakly a.e..

<u>(ii) ⇒ (iii)</u>

Certainly $(x'(X_n),F_n)_{n\in\mathbb{N}}$ is a potential in $\mathbb{R}$, so $\lim_{n\to\infty} x'(X_n) = 0$ for every $x' \in E'$. Hence $(X_n)_{n\in\mathbb{N}}$ converges scalarly to zero. Using lemma II.1.5 we also have $\sup_{n\in\mathbb{N}} \|X_n\| < \infty$, a.e.. So from (ii), $(TX_n)_{n\in\mathbb{N}}$ converges weakly a.e..

<u>(iii) ⇒ (iv)</u>

Indeed, a potential is a WS potential, see theorem V.3.2.
The main part of the proof now follows.

<u>(iv) ⇒ (i)</u>

Suppose that T is not an Asplund operator. Then by theorem V.3.13 T is a factor of the Haar operator H.

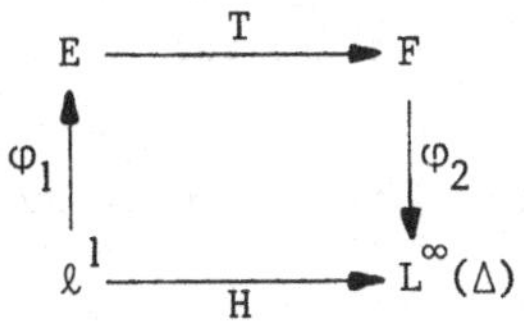

If we can construct a $\|\cdot\|_\infty$-bounded potential $(X_n,F_n)_{n\in\mathbb{N}_\infty}$ in ℓ^1 such that $(HX_n)_{n\in\mathbb{N}}$ does not converge to zero weakly a.e. in L^∞ then $(\varphi_1(X_n),F_n)_{n\in\mathbb{N}}$ will be a $\|\cdot\|_\infty$-bounded potential in E such that $(T\varphi_1X_n)_{n\in\mathbb{N}}$ does not converge to zero weakly a.e. in F. This would contradict (iii). So let us define $(X_n,F_n)_{n\in\mathbb{N}}$. Choose $\Omega = \Delta$, $P = \mu$, and define $X_n : \Delta \to \ell^1$ as

$$X_n(\omega) = \frac{1}{n} \sum_{m=1}^{n} \sum_{i=1}^{2^m} h_{mi}(\omega)\, e_{mi}$$

Then we have that $(X_n, \sigma(X_1,\dots,X_n))_{n\in\mathbb{N}}$ satisfies all the required properties

(1) $(X_n)_{n\in\mathbb{N}}$ is $\|\cdot\|_\infty$-bounded : for every $\omega \in \Omega$,

$$\|X_n(\omega)\|_{\ell^1} = \frac{1}{n} \sum_{m=1}^{n} \sum_{i=1}^{2^m} |h_{mi}(\omega)| = \sum_{m=1}^{n} \frac{1}{n} = 1 .$$

(2) $(HX_n)_{n\in\mathbb{N}}$ does not converge weakly a.e. to zero :

$$(*) \quad ((HX_n)(\omega))(\omega) = \frac{1}{n} \sum_{m=1}^{n} \sum_{i=1}^{2^m} h_{mi}(\omega)\, h_{mi}(\omega)$$

$$= \frac{1}{n} \sum_{m=1}^{n} 1 = 1$$

Now every $(HX_n)(\omega)$ is a continuous function on Δ. Call the space of continuous functions on Δ, $C(\Delta)$.
So $\bigcup_{n\in\mathbb{N}} (HX_n)(\Omega) \subset C(\Delta)$. Now (*) means that $(HX_n)(\omega)$ is not $\sigma(C(\Delta),C(\Delta)')$-convergent, for every $\omega \in \Omega(=\Delta)$ (for the Dirac-measure $\delta_\omega \in C(\Delta)'$ we have indeed $\delta_\omega[(HX_n)(\omega)] = ((HX_n)(\omega))(\omega))$. Again using that $\sigma(L^\infty(\Delta),L^{\infty\prime}(\Delta))$ restricted to $C(\Delta)$ is $\sigma(C(\Delta),C(\Delta)')$ we have shown that $(HX_n)_{n\in\mathbb{N}}$ does not converge weakly a.e. to zero.

$(X_n,\mathcal{F}_n)_{n\in\mathbb{N}}$ is a potential. Indeed we shall show that

$$\lim_{\tau\in T} \|X_\tau\|_{Pe} = 0 \tag{3}$$

Here of course, $\|X_\tau\|_{Pe} = \sup_{\substack{\|x'\|\leq 1 \\ x'\in \ell^{1\prime} = \ell^\infty}} \int |x'(X_\tau)|$.

Take $x' \in \ell^{1'}$ with $\|x'\| \leq 1$ and put $\alpha_{ni} = x'(e_{ni})$ for every i,n. Hence $|\alpha_{ni}| \leq 1$, for every i,n. Let $p \in \mathbb{N}$, $p > 1$. The classical Khintchine inequality, see e.g. Lindenstrauss and Tzafriri [1977], theorem 2.b.3, p.66, implies now

$$\left\| \frac{1}{n} \sum_{m=1}^{n} \sum_{i=1}^{2^m} \alpha_{mi} h_{mi} \right\|_p^p \leq \frac{C_p}{n^{\frac{p}{2}}}$$

where C_p is a constant only depending on p. So, for every $\lambda > 0$

$$P\left(\left|\frac{1}{n} \sum_{m=1}^{n} \sum_{i=1}^{2^m} \alpha_{mi} h_{mi}\right| > \lambda\right) \leq \frac{C_p}{n^{\frac{p}{2}} \lambda^p}$$

Now let $\tau \in T(N)$, where $N \in \mathbb{N}$. Then :

$$\int_\Omega |x'(X_\tau)|$$

$$= \sum_{n=N}^{\infty} \int_{\{\tau = n\}} \left|\frac{1}{n} \sum_{m=1}^{n} \sum_{i=1}^{2^m} \alpha_{mi} h_{mi}\right|$$

$$(**) \qquad = \sum_{n=N}^{\infty} \int_0^{\infty} P\left(\left|\frac{1}{n} \sum_{m=1}^{n} \sum_{i=1}^{2^m} \alpha_{mi} h_{mi}\right| \chi_{\{\tau=n\}} > \lambda\right) d\lambda$$

using Fubini. Let $\varepsilon_n = n^{-\alpha}$ where $\alpha > 0$ will be specified later. Then

$$\int_\Omega |x'(X_\tau)| = \sum_{n=N}^{\infty} \left(\int_0^{\varepsilon_n} + \int_{\varepsilon_n}^{\infty}\right)$$

where the same integrand as in (**) is used

$$\leq \sum_{n=N}^{\infty} \varepsilon_n P(\{\tau = n\}) + \sum_{n=N}^{\infty} \frac{C_p}{n^{\frac{p}{2}}} \int_{\varepsilon_n}^{\infty} \frac{d\lambda}{\lambda^p}$$

$$= \sum_{n=N}^{\infty} n^{-\alpha} P(\{\tau = n\}) + \sum_{n=N}^{\infty} \frac{C_p}{n^{\frac{p}{2}} \varepsilon_n^{p-1} (p-1)}$$

$$\leq \sum_{n=N}^{\infty} N^{-\alpha} P(\{\tau = n\}) + \frac{C_p}{p-1} \sum_{n=N}^{\infty} \frac{1}{n^{\frac{p}{2}-\alpha(p-1)}}$$

$$= N^{-\alpha} + \frac{C_p}{p-1} \sum_{n=N}^{\infty} \frac{1}{n^{\frac{p}{2}-\alpha(p-1)}}$$

Since this is independent of x', we have

$$(***) \qquad \sup_{\substack{x' \in \ell^{1'} \\ \|x'\| \leq 1}} \int_\Omega |x'(X_\tau)| \leq N^{-\alpha} + \frac{C_p}{p-1} \sum_{n=N}^{\infty} \frac{1}{n^{\frac{p}{2}-\alpha(p-1)}}$$

If we choose $\alpha > 0$ and p so that

$$\frac{p}{2} - \alpha(p-1) > 1$$

then (***)converges to zero for N going to ∞. Hence

$$\lim_{\tau \in T} \|X_\tau\|_{Pe} = 0$$

which implies that $(X_n, F_n)_{n \in \mathbb{N}}$ is a potential. □

Corollary V.3.15 (Brunel and Sucheston) : Let E be a separable Banach space. Then E' is separable if and only if every $\|\cdot\|_\infty$-bounded WS potential converges weakly a.e..

Proof : If we take E = F in theorem V.3.14 and T = Id_E, the identity operator on E, we have that Id_E is an Asplund operator, i.e. E is an Asplund space, if and only if every $\|\cdot\|_\infty$-bounded WS potential converges weakly a.e.. Now E is an Asplund space if and only if E' has (RNP) (see theorem V.3.13). Finally, for E separable, E' has (RNP) if and only if E' is separable, as is well-known (cf. Diestel and Uhl [1977]). □

Edgar's theorem V.3.14 and the method developed in the proof of theorem V.3.3 now yield another extension of theorem V.3.4.

Corollary V.3.16 (Edgar) : Let E and F be Banach spaces and $T \in \mathcal{L}(E,F)$. The following assertions are equivalent :

(i) T and T' are Radon-Nikodym-operators.

(ii) For every WS amart $(X_n, F_n)_{n\in\mathbb{N}}$ of class (B) in E, the sequence $(TX_n)_{n\in\mathbb{N}}$ converges weakly a.e..

(iii) For every $\|\cdot\|_\infty$-bounded amart $(X_n, F_n)_{n\in\mathbb{N}}$ in E the sequence $(TX_n)_{n\in\mathbb{N}}$ converges weakly a.e..

Proof : (i) ⇒ (ii)

Following the proof and notations of theorem V.3.3 we have a martingale $(Y_n, F_n)_{n\in\mathbb{N}}$ such that for every $x' \in E'$

$$\int_\Omega \limsup_{n\in\mathbb{N}} |x'(Y_n(\omega) - X_n(\omega))|$$

$$\leq \liminf_{n\in\mathbb{N}} \int_\Omega |x'(E^{F_{\sigma_n^{x'}}} X_{\tau_n^{x'}}(\omega) - X_{\sigma_n^{x'}}(\omega))|$$

a.e.. Hence

$$X_n = Y_n + Z_n$$

where $(Y_n, F_n)_{n\in\mathbb{N}}$ is an L^1_E-bounded martingale and $(Z_n, F_n)_{n\in\mathbb{N}}$ is a WS amart of class (B). Indeed, $(X_n, F_n)_{n\in\mathbb{N}}$ as well as $(Y_n, F_n)_{n\in\mathbb{N}}$ are of class (B). Also $(Z_n)_{n\in\mathbb{N}}$ converges scalarly to zero. From the maximal inequality II.1.5, $\sup_{n\in\mathbb{N}} \|Z_n(\omega)\| < \infty$, a.e.. From theorem V.3.14 (i) ⇔ (ii) and theorem V.3.13 (i) ⇔ (ii) we now have that $(TZ_n)_{n\in\mathbb{N}}$ weakly converges a.e.. Now, $(TY_n, F_n)_{n\in\mathbb{N}}$ is an L^1_F-bounded martingale having a limitmeasure with a Radon-Nikodym density, since T is a Radon-Nikodym operator. So the Chatterji-proof of theorem II.2.4.3 shows that $(TY_n)_{n\in\mathbb{N}}$ converges strongly a.e.. Hence $(TX_n)_{n\in\mathbb{N}}$ converges weakly a.e..

(ii) ⇒ (iii)

Is obvious.

(iii) ⇒ (i)

Certainly T' is a Radon-Nikodym operator due to theorem V.3.14 (iv) ⇔ (i) and theorem V.3.13 (i) ⇔ (ii). Since (iii) now implies that for every $\|\cdot\|_\infty$-bounded martingale $(X_n, F_n)_{n\in\mathbb{N}}$, the sequence $(TX_n)_{n\in\mathbb{N}}$ converges weakly a.e., it is easy to see that T must be a Radon-Nikodym operator (cf. the proof of theorem II.2.2.1 (iii) ⇒ (i)).

Remark V.3.17 : In the above proof the following result was remarked : Let E and F be Banach spaces and $T \in \mathcal{L}(E,F)$. Then T is a Radon-Nikodym operator if and only if for every $\|\cdot\|_\infty$- or L^1_E-bounded martingale $(X_n, F_n)_{n\in\mathbb{N}}$ in E, the sequence $(TX_n)_{n\in\mathbb{N}}$ converges weakly (or strongly) a.e..

We end this section on WS amarts with a characterization of weak convergence a.e. analogous as that of corollary V.3.15, but now for positive WS potentials in a Banach lattice. This result now can be used to prove the earlier mentioned result in remark III.4.2 as we shall see. Let E be a Banach lattice. Let $A \subset E$. Denote by H_A the closed ideal generated by A. If $u \in E^+$ is such that $H_{\{u\}} = E$ then we say that u is a quasi-interior point. The cardinal number dens (E) is the smallest cardinality of a dense subset of E. Let $(X_n, F_n)_{n\in\mathbb{N}}$ be a positive adapted sequence with values in E.

Definition V.3.18 : $(X_n, F_n)_{n\in\mathbb{N}}$ is called a Doob potential (resp. strong Doob potential) if $(\int_\Omega X_\tau)_{\tau\in T}$ decreases to zero (resp. and converges in norm to zero).

We remark that if $E = \mathbb{R}$, then every Doob potential is a strong Doob potential. Indeed if $\lim_{\tau\in T} \int_\Omega X_\tau = 0$ and $X_n \geq 0$ for each $n \in \mathbb{N}$, then it follows from corollary I.3.5.2 that

$$\int_\Omega \limsup_{n\in\mathbb{N}} X_n = \int_\Omega \lim_{n\to\infty} X_{\tau_n}$$

where $(\tau_n)_{n\in\mathbb{N}}$ is an increasing cofinal sequence in T. So

$$\int_\Omega \limsup_{n\in\mathbb{N}} X_n \leq \liminf_{n\in\mathbb{N}} \int_\Omega X_{\tau_n} \qquad \text{(Fatou)}$$

$$\leq \lim_{\tau\in T} \int_\Omega X_\tau = 0$$

Since $X_n \geq 0$ for each $n \in \mathbb{N}$, $\lim_{n\to\infty} |X_n| = 0$, a.e..

For a set A, we denote by $\#A$ its cardinality. The analogous result to corollary V.3.15 is as follows :

Theorem V.3.19 (Ghoussoub and Talagrand [1979a]) : For a Banach lattice, the following assertions are equivalent :

(i) Every positive WS potential of class (B) converges weakly a.e..

(ii) Every uniformly bounded strong Doob potential converges weakly a.e..

(iii) For every sublattice F of E, there is a subset A of F' such that $\# A \leqslant \text{dens } F$ for which $H_A = F'$.

(iv) Every separable sublattice F of E has a quasi-interior point in its dual.

Proof : (i) ⇒ (ii)

Is done in the same way as the proof of theorem V.3.2.

(ii) ⇒ (iii)

This is the most difficult part of the proof. We can put E = F. We first need a lemma.

Lemma V.3.19.1 : Suppose (iii) is not valid. Let $W = \cup\{V \| V$ is a w^*-open subset of $K = \{x \in E'^+ \| \|x\| \leqslant 1\}$ for which there exists $A \subset E'$, $\# A \leqslant \text{dens } E$ such that $V \subset H_A + \frac{1}{2} K\}$. Then $L = K \setminus W$ is non-empty and for every w^*-open non-empty set U of L and every $v \in E'$ we have that $V \setminus (H_{\{v\}} + \frac{1}{2} K) \neq \phi$.

Proof : $W \subset H_A + \frac{1}{2} K$ for a set A for which $\# A \leqslant \text{dens }(E)$. This follows from the fact that K (and hence W) has a basis (in the w^*-topology) of cardinality dens (E) and from $\bigcup_{i \in I} H_{A_i} \subset H_{\bigcup_{i \in I_1} A_i}$ for any family $(A_i)_{i \in I}$ of sets. Suppose now that $L = \phi$. Then $K \subset H_A + \frac{1}{2} K$. Hence $K \subset H_A + \frac{1}{2}(H_A + \frac{1}{2} K) = H_A + \frac{1}{4} K$. Hence, by induction, $K \subset H_A + \frac{1}{2^n} K$ for every $n \in \mathbb{N}$ and so $K \subset H_A$. Consequently $E' = H_A$ which is impossible since we supposed that (iii) is not valid. So $L \neq \phi$. If V is a w^*-open subset of K such that $L \cap V \subset H_{\{v\}} + \frac{1}{2} K$ for a certain $v \in E'$, then

$$V \subset (H_{\{v\}} + \frac{1}{2} K) \cup W$$

$$\subset (H_{\{v\}} + \frac{1}{2} K) \cup (H_A + \frac{1}{2} K)$$

$$\subset H_{\{v\} \cup A} + \frac{1}{2} K \quad .$$

So $V \cap L = \phi$. This finishes the proof of the lemma. We now continue the proof of (ii) ⇒ (iii), supposing that (iii) is not valid.
Put $\Omega = [0,1]$, λ the Lebesguemeasure on Ω. We shall construct an adapted sequence $(X_n, F_n)_{n\in\mathbb{N}}$ on Ω with values in E^+ and also, for every $t \in \Omega$, closed sets $V_n(t)$ in L such that

(a) $\quad s_n = \inf_{B\in F_n} \lambda(B) > 0$

(b) $\quad \| E^{F_n} X_{n+1}(t) \| \leq \frac{s_n}{2^n}$

(c) $\quad \| X_n(t) \| \leq 1$

(d) $\quad \overset{\circ}{V}_n(t) \neq \phi$ and $V_{n+1}(t) \subset V_n(t)$

(e) $\quad$ for every $h \in V_n(t)$, $h\ (V_n(t)) \geq \frac{1}{3}$

(f) $\quad V_n(t)$ is constant on every atom of F_n.

Suppose we have constructed already $V_1(t),\ldots,V_n(t)$, $X_1,\ldots,X_n$ and $F_1,\ldots,F_n$. The inductive step is as follows (this can also be used to define $V_1(t)$, X_1 and F_1) : Let Z be an atom of F_n and denote by V the common value of $V_n(t)$ for $t \in Z$. Let $B = \{x\in E^+ \| \|x\| \leq 1\}$. Put

$$D = \{x\in B \| \exists h \in \overset{\circ}{V} \text{ with } h(x) \geq \frac{1}{2}\}$$

and $C = \overline{\text{con}}\ D$.
Then $d(0,C) = 0$ ($d(0,C)$ denotes the distance between 0 and C). For suppose, to obtain a contradiction, that $\beta = d(0,C) > 0$. Now $d(0,C) = d(0,\overline{C+E^+})$ since $\geq$ is trivial and since $\leq$ follows from the fact

that $\|\cdot\|$ is increasing on E^+. Using Hahn-Banach we get $f \in E'$, $\|f\| \leq 1$ such that $f \geq \beta$ on $C + E^+$. So for every $x \in C$ and $y \in E^+$, $f(x+y) \geq \beta$. Hence $f(y) \geq \beta - \inf_{x\in C} f(x)$ for every $y \in E^+$. Now $\beta \geq \inf_{x\in C} f(x)$ since $d(0,C) = \beta$. Hence $f \in E'^+$ and $f \geq \beta$ on D. From lemma V.3.19.1 we have an $h \in \overset{\circ}{V}$ such that $h \notin H_{\{f\}} + \frac{1}{2} K$.
Choose $n \geq \frac{2}{\beta}$. Certainly $h \notin n[0,f] + \frac{1}{2} K$ ($[0,f]$ denotes the order interval from 0 to f). So, since $h \geq 0$, $h \notin n[0,f] + \frac{1}{2} K - E'^+$. This last set is w*-closed. Using Hahn-Banach again, there is $x \in E$, $\|x\| = 1$ such that for every $k \in K$, $g \in [0,nf]$ and $\ell \in E'^+$:

$$g(x) + \frac{k}{2}(x) - \ell(x) \leq h(x) \qquad (1)$$

Suppose there exists $\ell \in E'^+$ such that $\ell(x) < 0$. Then $\lim_{p\to\infty} p.\ell(x) = -\infty$ and $p.\ell \in E'^+$. This contradicts (1). So $\ell(x) \geq 0$ and hence $x \in E^+$. Since $h(x) \geq \frac{k}{2}(x)$ for every $k \in K$ and since there is a sequence k_n in K such that $\lim_{n\to\infty} k_n(x) = \|x\| = 1$, it follows that $h(x) \geq \frac{1}{2}$, so $x \in D$. Also $nf(x) \leq h(x) \leq 1$. So $f(x) \leq \frac{1}{n} \leq \frac{\beta}{2}$. This is not possible for $x \in D$, a contradiction. Therefore $d(0,C) = 0$.
Consequently we can choose $\{x_1,\ldots,x_k\}$ in D and $\alpha_1,\ldots,\alpha_k > 0$ such that $\sum_{i=1}^{k} \alpha_i = 1$, such that $\left\| \sum_{i=1}^{k} \alpha_i x_i \right\| \leq \frac{s_n}{2^n}$. Divide Z into k measurable sets Z_i of measure $\alpha_i\, \lambda(Z)$. Define X_{n+1} on Z by $X_{n+1}(t) = x_i$ for $t \in Z_i$ and

$$V_{n+1}(t) = \{h \in V_n(t) \| h(X_{n+1}(t)) \geq \tfrac{1}{3}\}$$

Then $\overset{\circ}{V}_{n+1}(t) \neq 0$ since $\overset{\circ}{V}_n(t) \neq 0$ (easy w*-argument). This must of course be done on every atom Z and then the results glued together to obtain X_{n+1}, F_{n+1} and $V_{n+1}(t)$ for $t \in \Omega$. Certainly (a) - (f) are trivially verified. Put $a_n = \int_\Omega X_n$ and $b_n = \sum_{p=n}^{\infty} \frac{a_p}{s_{p-1}}$. Hence

$$\|a_n\| = \int_\Omega E^{F_{n-1}} X_n \leq \frac{s_{n-1}}{2^{n-1}} \quad \text{and} \quad \|b_n\| \leq \sum_{p=n}^{\infty} \frac{\|a_p\|}{s_{p-1}} \leq 2^{-n+2} .$$

The final adapted sequence which does the job is $(Y_n, F_n)_{n\in\mathbb{N}}$ where $Y_n = X_n + b_n$. Certainly $\|Y_n(t)\| \leq \frac{1}{3}$ for every $t \in \Omega$. So $(Y_n)_{n\in\mathbb{N}}$ is uniformly bounded. Furthermore

$$E^{\mathcal{F}_n} Y_{n+1} = b_{n+1} + E^{\mathcal{F}_n} X_{n+1}$$

On Z, an atom of $\mathcal{F}_n$ we have

$$s_n(E^{\mathcal{F}_n} X_{n+1})(t) \leq \lambda(Z)(E^{\mathcal{F}_n} X_{n+1})(t)$$

$$= \int_Z X_{n+1} \leq \int_\Omega X_{n+1} = a_{n+1} .$$

Then

$$E^{\mathcal{F}_n} Y_{n+1} \leq b_{n+1} + \frac{a_{n+1}}{s_n} = b_n \leq Y_n .$$

So $(Y_n, \mathcal{F}_n)_{n \in \mathbb{N}}$ is a supermartingale and also $\|\int_\Omega Y_n\| = \|a_n + b_n\| \leq 2^{-n+3}$. Hence $(Y_n, \mathcal{F}_n)_{n \in \mathbb{N}}$ is a strong Doob potential. But $(Y_n)_{n \in \mathbb{N}}$ does not converge weakly a.e.. To see this, take $h \in V_n(t)$. Then $h(Y_n(t)) \geq h(X_n(t)) \geq \frac{1}{3}$ for every $n \in \mathbb{N}$. Hence for $h \in \bigcap_{n \in \mathbb{N}} V_n(t) \neq \phi$ (due to (d)) we have $h(Y_n(t)) \geq \frac{1}{3}$ for every $n \in \mathbb{N}$. So $w - \lim_{n \to \infty} Y_n$ - if this exists - is non zero. Denote $F = \overline{\text{span}} \{Y_n(\Omega) \| n \in \mathbb{N}\}$, F being separable, and let D be a countable norming subset in the unit ball of F'. For each $f \in D$, $(f(Y_n))_{n \in \mathbb{N}}$ converges a.e. to zero since $(f^+(Y_n), \mathcal{F}_n)_{n \in \mathbb{N}}$ and $(f^-(Y_n), \mathcal{F}_n)_{n \in \mathbb{N}}$ are positive supermartingales and since $\lim_{n \to \infty} \int_\Omega Y_n = 0$. If $(Y_n)_{n \in \mathbb{N}}$ converges weakly a.e., say to Y, then of course $f(Y) = 0$, a.e. for every $f \in D$, so $\|Y\| = 0$, a.e., a contradiction.

<u>(iii) ⇒ (iv)</u>

Let A be a subset of F' such that $\# A \leq \text{dens}(F)$ and such that $H_A = F'$. Since F is separable, A is countable : $A = \{a_n \| n \in \mathbb{N}\}$. Put

$$u = \sum_{n=1}^{\infty} \frac{1}{2^n} \frac{a_n}{\|a_n\|}$$

Then $H_{\{u\}} = H_A = F'$. Hence u is a quasi-interior point in F'.

(iv) ⇒ (i)

We can suppose E to be separable. Hence E' has a quasi-interior point. For later use, we call the next argument a lemma (it is more general than we need right now) :

Lemma V.3.19.2 : Suppose that there is a countable orthogonal set $A \subset E'$ such that $H_A = E'$ (such a set A is called a countable topological orthogonal system (t.o.s.)). Then every positive E-valued WS potential of class (B) is weakly convergent a.e..

Proof of the lemma : Denote $A = \{a_i \| i \in \mathbb{N}\}$. Suppose that $(Z_n, F_n)_{n \in \mathbb{N}}$ is a positive WS potential of class (B). Then, for every $x' \in E'$, $(x'(Z_n), F_n)_{n \in \mathbb{N}}$ is a scalar potential of class (B), hence converges to zero a.e. : see f.i. theorem V.3.3 applied scalarly - the limit-function being zero (this follows from the proof of theorem V.3.3). Also

$$\sup_{n \in \mathbb{N}} \|Z_n\| < \infty \text{ , a.e.}$$

as follows from lemma II.1.5. For each $a_i \in A$, call Ω_n the null set in Ω on which $(a_i(Z_n))_{n \in \mathbb{N}}$ does not converge to zero. We recall the well known result (Schaefer [1974]) that A is a t.o.s. if and only if for each $x' \in E'^+$,

$$x' = \lim_{m,H} \sum_{i \in H} x' \wedge ma_i$$

where the limit runs through $m \in \mathbb{N}$ and H in a finite subset of $\mathbb{N}$ (ordered in the natural way). Call $\Omega' = \bigcup_n \Omega_n$ and Ω'' the null set on which $\sup_{n \in \mathbb{N}} \|Z_n\| = \infty$. For each $x' \in E'^+$, $m,n \in \mathbb{N}$ and H a finite subset of $\mathbb{N}$ we have, since $Z_n \geq 0$ for every $n \in \mathbb{N}$,

$$0 \leq \sum_{i \in H} (x' \wedge ma_i)(Z_n) \leq \sum_{i \in H} ma_i(Z_n) \qquad (1)$$

Since for every $\omega \in \Omega \setminus (\Omega' \cup \Omega'')$, $\lim_{m,H} (x' - \sum_{i \in H} x' \wedge ma_i)(Z_n(\omega)) = 0$, uniformly in $n \in \mathbb{N}$, and since $\sum_{i \in H} ma_i(Z_n(\omega))$ converges to zero on this set, it follows from (1) that

$$x'(Z_n(\omega)) = (x' - \sum_{i\in H} (x' \wedge ma_i))(Z_n(\omega)) + \sum_{i\in H} (x' \wedge ma_i)(Z_n(\omega))$$

converges to zero for $n \to \infty$. This finishes the proof of the lemma. This also finishes the proof of (iv) ⇒ (i) since we may take $A = \{u\}$, u being the quasi-interior point. This in turn finishes the whole proof. □

From theorem V.3.19 we now derive :

Theorem V.3.20 (Ghoussoub and Talagrand [1979a]) : Let E be a Banach lattice. The following assertions are equivalent :

(i) E has (RNP) and every separable sublattice F of E has a quasi-interior point in F'.

(ii) Every orderamart of class (B) in E converges weakly a.e..

(iii) Every supermartingale of class (B) in E converges weakly a.e..

(iv) Every positive supermartingale of class (B) in E converges weakly a.e..

Proof : (i) ⇒ (ii)

Since E has (RNP), it follows from theorem V.2.32 that, if $(X_n, \mathcal{F}_n)_{n\in\mathbb{N}}$ denotes the orderamart,

$$X_n = Y_n + Z_n$$

where $(Y_n, \mathcal{F}_n)_{n\in\mathbb{N}}$ is an L^1_E-bounded martingale and $(Z_n, \mathcal{F}_n)_{n\in\mathbb{N}}$ is an orderpotential. Hence theorem V.3.19 implies that $(Z_n)_{n\in\mathbb{N}}$ converges weakly a.e. to zero, since $(|Z_n|, \mathcal{F}_n)_{n\in\mathbb{N}}$ is a WS potential. So, theorem II.2.4.3 finishes the proof.

(ii) ⇒ (iii)

Indeed, every supermartingale of class (B) is an orderamart : First remark that (ii) implies that E does not contain c_o as an isomorphic subspace, by the remarks after II.2.2.1 or by the proof of III.2.5. Hence, by III.1.3, E is weakly sequentially complete. Thus every decreasing norm bounded sequence is norm convergent. Since a supermartingale $(X_n, \mathcal{F}_n)_{n\in\mathbb{N}}$ of class (B) is norm bounded due to lemma II.1.5, and since $(\int_\Omega X_\sigma)_{\sigma\in T}$ decreases, it so follows that $(\int_\Omega X_\sigma)_{\sigma\in T}$ converges in norm. Since E^+ is closed,

$$\|\cdot\|-\lim_{\tau\in T} \int_\Omega X_\tau \leq \int_\Omega X_\sigma$$

for every $\sigma \in T$. So $(\int_\Omega X_\sigma)_{\sigma\in T}$ decreases and has an order lower bound. Hence it must be orderconvergent, since E is Dedekind σ-complete, and therefore $(X_n, F_n)_{n\in\mathbb{N}}$ is an orderamart.

(iii) ⇒ (iv)

Is trivial.

(iv) ⇒ (i)

(iv) implies that every strong Doob potential of class (B) converges weakly a.e.. So from theorem V.3.19 it already follows that every separable sublattice F of E has a quasi-interior point in F'. To prove (RNP), remark that (iv) implies the weak a.e. convergence of positive uniformly bounded martingales. But this implies that they are of the form

$$X_n = E^{F_n} X_\infty \ ,$$

for every $n \in \mathbb{N}$ where $X_\infty \in L^1_E$, as is easily seen. Hence, by theorem II.1.6, $(X_n)_{n\in\mathbb{N}}$ converges strongly a.e.. Now (RNP) follows from theorem III.2.5. □

To relate property (i) in theorem V.3.20 to certain other properties we mention :

Proposition V.3.21 : Consider the following properties of a separable Banach lattice E

(a) E is reflexive.

(b) E has (RNP) and E' has a quasi-interior point.

(c) E has (RNP) and E' has a countable t.o.s. (see lemma V.3.19.2).

(d) E is ordercontinuous and E' has a countable t.o.s..

Then (a) ⇒ (b) ⇔ (c) ⇒ (d).

Proof : (a) ⇒ (b)

This follows from Schaefer [1974] (theorem II.6.6), its corollary and from theorem III.1.2 here.

(b) ⇒ (c)

Is trivial.

(c) ⇒ (b)

This follows from theorem V.3.19 (i) ⇔ (iii) ⇔ (iv), lemma V.3.19.2 and the separability of E.

(c) ⇒ (d)

Since E has (RNP), $c_o \not\hookrightarrow E$. Now apply theorem III.1.2. □

Proposition V.3.21 has been stated only for separable Banach lattices since this condition can obviously be assumed when studying adapted sequences of strongly measurable functions, as we do in this book.

Thus we have completed the important convergence result, which was announced in remark III.4.2. We finish by remarking that the convergence properties described in theorem V.3.20 cannot be improved.

Examples V.3.22 (Benyamini)

1. In theorem V.3.20, the class (B) assumption cannot be weakened : choose $\Omega_i = [0,1]$, λ_i the Lebesguemeasure and F_i the Borelsets on Ω_i. Put

$$\Omega = \prod_{i \in \mathbb{N}} \Omega_i , \qquad F = \prod_{i \in \mathbb{N}} F_i , \qquad P = \prod_{i \in \mathbb{N}} \lambda_i .$$

An element of Ω will be denoted by $\omega = (\omega_n)_{n \in \mathbb{N}}$. Take $E = \ell^2$. Fix $n \in \mathbb{N}$ and $k \in \{1,\dots,2^n\}$. Define X_m by indicating every coordinate in ℓ^2 : for the $(2^n+k)^{th}$ coordinate

$$(X_m(\omega))_{2^n+k} = \begin{cases} \dfrac{1}{2^n} & \text{if } m < n \\ m & \text{if } m = n \text{ and } \dfrac{k-1}{m2^m} \leqslant \omega_m \leqslant \dfrac{k}{m2^m} \\ 0 & \text{otherwise} \end{cases}$$

$X_m(\omega)$ is completely defined in this way for every $\omega \in \Omega$. Now we see that :

a) $(X_n)_{n\in \mathbb{N}}$ is $L^1_{\ell^2}$-bounded :

$$\|X_m(\omega)\| = \begin{cases} [m^2 + \sum_{n>m} 2^n (\frac{1}{2^n})^2]^{1/2} \leqslant m+1 & \text{if} \quad \omega_m \leqslant \frac{1}{m} \\ [\sum_{n>m} 2^n (\frac{1}{2^n})^2]^{1/2} \leqslant 1 & \text{if} \quad \omega_m > \frac{1}{m} \end{cases}$$

This is easily seen from the definition of X_m. So

$$\int_\Omega \|X_m\| = \int_{\{\omega\|\omega_m \leqslant \frac{1}{m}\}} \|X_m\| + \int_{\{\omega\|\omega_m > \frac{1}{m}\}} \|X_m\|$$

$$\leqslant \frac{m+1}{m} + 1 \leqslant 3$$

b) $(X_n, F_n)_{n\in \mathbb{N}}$ is a positive supermartingale where $F_n = \sigma(X_1,\ldots,X_n)$ for every $n \in \mathbb{N}$. Indeed, the X_n's are obviously independent and $X_m \geqslant \int_\Omega X_{m+1}$ for every $m \in \mathbb{N}$ since for every $n \in \mathbb{N}$ and $k \in \{1,\ldots,2^n\}$

$$(\int_\Omega X_{m+1})_{2^n+k} = \begin{cases} \frac{1}{2^n} & \text{if} \quad m+1 < n \\ \frac{1}{2^{n+1}} & \text{if} \quad m+1 = n \\ 0 & \text{if} \quad m+1 > n \end{cases}$$

$$\leqslant (X_m)_{2^n+k}$$

c) $(X_n)_{n\in \mathbb{N}}$ does not converge weakly a.e. : Indeed we shall show that $\{\|X_m(\omega)\| \| m \in \mathbb{N}\}$ is not bounded for almost every $\omega \in \Omega$. Since $P(\{\omega\| 0 \leqslant \omega_m \leqslant \frac{1}{m}\}) = \frac{1}{m}$ it follows that $\sum_{m=1}^{\infty} P(\{\omega\| 0 \leqslant \omega_m \leqslant \frac{1}{m}\}) = \infty$.
Using the Borel-Cantelli result, almost all $\omega \in \Omega$ are in infinitely many sets $\{\omega\| 0 \leqslant \omega_m \leqslant \frac{1}{m}\}$. But on these sets $\|X_m(\omega)\| = m$.

2. In theorem V.3.20, strong convergence a.e. cannot be obtained. We use the same construction as in example 1; only the values of X_m (still in ℓ^2) change :

$$(X_m(\omega))_{2^n+k} = \begin{cases} \frac{1}{2^n} & \text{if} \quad m < n \\ 1 & \text{if} \quad m = n \quad \text{and} \quad \frac{k-1}{2^m} \leq \omega_m \leq \frac{k}{2^m} \\ 0 & \text{otherwise} \end{cases}$$

As in example 1 we also see that $(X_n, F_n)_{n \in \mathbb{N}}$ is a positive supermartingale. Now $(X_n)_{n \in \mathbb{N}}$ is uniformly bounded, hence of class (B). Hence $(X_n)_{n \in \mathbb{N}}$ is weakly converging (to zero) a.e., since we can apply theorem V.3.20 (using proposition V.3.21). Obviously $\|X_n(\omega)\|_2 \geq 1$ for all $\omega \in \Omega$. So $(X_n)_{n \in \mathbb{N}}$ cannot converge strongly a.e..

The reader interested in $\|\cdot\|_{Pe}$-Cauchy and convergence properties of WS amarts is referred to the next section since all $\|\cdot\|_{Pe}$-results for WS amarts are also valid for W amarts which are studied in the next section.

V.4. Weak amarts

As the numbering of the sections in this chapter increases, the type of adapted sequence studied becomes weaker. This section is devoted to the notion of "weak amart" with values in a Banach space.

Definition V.4.1 : An adapted sequence $(X_n, F_n)_{n \in \mathbb{N}}$ is called a weak amart (abbreviated W amart) if the net

$$\left(\int_\Omega X_\tau\right)_{\tau \in T}$$

converges weakly.

The weaker the notion, the heavier the hypotheses must be in order to get convergence. These hypotheses concern Banach space properties or boundedness properties on the adapted sequence. In this section we show that in reflexive Banach spaces, W amarts of class (B) converge weakly a.e. and that this result is the best possible in the sense that neither class (B), nor reflexivity can be removed from the hypothesis. Relations with amarts and WS amarts are given. Also Pettis convergence of W amarts and WS amarts is described.

Obviously every amart is a W amart. We can say more :

Theorem V.4.2 : Every WS amart is a W amart.

Proof : For every $x' \in E'$, $(x'(X_n), F_n)_{n \in \mathbb{N}}$ is obviously an amart, so

$$(\int_\Omega X_\tau)_{\tau \in T}$$

is weakly Cauchy. Since

$$(\int_\Omega X_n)_{n \in \mathbb{N}}$$

weakly converges it now follows that $(X_n, F_n)_{n \in \mathbb{N}}$ is a W amart. □

The converse is not true in general :

Theorem V.4.3 (Egghe [1982a]) : The following assertions are equivalent :

(i) E is weakly sequentially complete.

(ii) Every W amart is a WS amart.

(iii) For every uniformly bounded W amart $(X_n, F_n)_{n \in \mathbb{N}}$, and every $A \in \bigcup_n F_n$ the sequence

$$(\int_A X_n)_{n \in \mathbb{N}}$$

converges weakly.

Proof : (i) ⇒ (ii)

For every $x' \in E'$, $(x'(X_n), F_n)$ is an amart. Hence (theorem V.3.2), for every increasing sequence $(\tau_n)_{n \in \mathbb{N}}$ in T, the sequence

$$(\int_\Omega x'(X_{\tau_n}))_{n \in \mathbb{N}}$$

converges. So $(\int_\Omega X_{\tau_n})_{n \in \mathbb{N}}$ is weakly Cauchy, hence weakly convergent.

<u>(ii) ⇒ (iii)</u>

Let $n_o \in \mathbb{N}$ be such that $A \in F_{n_o}$. Define for $n \in \mathbb{N}(n_o)$

$$\tau_n = \begin{cases} n & \text{on } A \\ n_o & \text{on } \Omega \setminus A \end{cases}$$

Then $\tau_n \in T$ for every $n \in \mathbb{N}(n_o)$ and $(\tau_n)_{n\in\mathbb{N}(n_o)}$ increases. Since $(X_n, F_n)_{n\in\mathbb{N}(n_o)}$ is a WS amart, $(\int_\Omega X_{\tau_n})_{n\in\mathbb{N}(n_o)}$ converges weakly. But

$$\int_\Omega X_{\tau_n} = \int_A X_{\tau_n} + \int_{\Omega\setminus A} X_{n_o}$$

for every $n \in \mathbb{N}(n_o)$. Hence $(\int_A X_n)_{n\in\mathbb{N}}$ converges weakly.

<u>(iii) ⇒ (i)</u>

Suppose $(x_n)_{n\in\mathbb{N}}$ is a weak Cauchy sequence which is not weakly convergent. Choose $X : \Omega \to \{-1,+1\}$, for (Ω, F, P) an arbitrary probability space, such that $P(\{X=1\}) = P(\{X=-1\}) = \frac{1}{2}$. Put

$$X_n = X.x_n$$

for each $n \in \mathbb{N}$. For each $n \in \mathbb{N}$, put $F_n = \sigma(\pi_n)$ where $\pi_n = \{\{X=-1\},\{X=1\}\}$. Now $(X_n, F_n)_{n\in\mathbb{N}}$ is a W amart, but for $A = \{X=1\}$, we have $X_n\big|_A = x_n \chi_A$. So $\int_A X_n = \frac{x_n}{2}$, which is not weakly convergent. □

<u>Corollary V.4.4 (Brunel and Sucheston [1976b])</u> : Suppose the Banach space E does not contain a subspace isomorphic to ℓ^1. Then E is reflexive if and only if every W amart is a WS amart.

<u>Proof</u> : This follows from theorem V.4.3 and the celebrated theorem of Rosenthal [1974] on the characterization of Banach spaces not containing ℓ^1, saying that a bounded sequence has a subsequence equivalent with the ℓ^1-basis or has a weak Cauchy-subsequence. □

The main result on weak convergence a.e. of W amarts will now be proved. It is a result of Brunel and Sucheston [1976b], and shows the importance of the W amart notion.

Theorem V.4.5 (Brunel and Sucheston) : A Banach space is reflexive if and only if every class (B) W amart converges weakly a.e..

Proof : a) Suppose that E is reflexive. Then the W amart $(X_n, F_n)_{n \in \mathbb{N}}$ is a WS amart by corollary V.4.4. Hence, using corollary V.3.4, $(X_n)_{n \in \mathbb{N}}$ converges weakly a.e..

b) Suppose that E is not reflexive. Hence let $(x_n)_{n \in \mathbb{N}}$ be a sequence in E such that $\|x_n\| = 1$ for every $n \in \mathbb{N}$ such that $(x_n)_{n \in \mathbb{N}}$ does not have a weakly convergent subsequence. We apply Rosenthal's theorem (Rosenthal [1974]) on $(x_n)_{n \in \mathbb{N}}$ which says that there is a subsequence $(y_n)_{n \in \mathbb{N}}$ of $(x_n)_{n \in \mathbb{N}}$ such that (1) or (2) below is satisfied :

(1) $(y_n)_{n \in \mathbb{N}}$ is weakly Cauchy

(2) $(y_n)_{n \in \mathbb{N}}$ is equivalent with the canonical basis of ℓ^1.

In case (1), define $X_n = Xy_n$ where, for an arbitrary probability space (Ω, F, P), $X : \Omega \to \{-1,+1\}$ is a function such that $P(\{X=1\}) = P(\{X=-1\}) = \frac{1}{2}$ (cf. the proof of theorem V.4.3 : the argument originates from Brunel and Sucheston). For every $\omega \in \Omega$ we have $X_n(\omega) = y_n$ for every $n \in \mathbb{N}$ or $X_n(\omega) = -y_n$ for every $n \in \mathbb{N}$. Hence $(X_n)_{n \in \mathbb{N}}$ does not converge weakly a.e.. However, $(X_n, F_n)_{n \in \mathbb{N}}$ is a W amart, where $F_n = \sigma(X_1, \ldots, X_n)$, for every $n \in \mathbb{N}$. Indeed, let $\tau \in T$ and $x' \in E'$. We have :

$$\begin{aligned}
|x'(\int_\Omega X_\tau)| &= |x'(\sum_{k=\min \tau}^{\max \tau} \int_{\{\tau = k\}} X_k)| \\
&= |\sum_{k=\min \tau}^{\max \tau} x'(y_k)(P(\{\tau=k\} \cap \{X=1\}) - P(\{\tau=k\} \cap \{X=-1\}))| \\
&\leq 2 \sum_{k=\min \tau}^{\max \tau} |x'(y_k)| \; P(\{\tau=k\}) \\
&\leq 2 \max_{k \geq \min \tau} |x'(y_k)| \xrightarrow{\min \tau \to \infty} 0
\end{aligned}$$

Hence $\text{w-lim}_{\tau \in T} \int_\Omega X_\tau = 0$.

For case (2), we may as well suppose that $(y_n)_{n\in\mathbb{N}}$ is the canonical basis of ℓ^1. The example V.3.6.4 yields the contradiction in this case. □

Part (a) of the proof of theorem V.4.5 is also a corollary of a general convergence property of adapted sequences :

Theorem V.4.6 (Brunel and Sucheston [1976b]) : Let $(X_n,F_n)_{n\in\mathbb{N}}$ be an adapted sequence such that for $\omega \in \Omega$ a.e. the sequence $(X_n(\omega))_{n\in\mathbb{N}}$ is weakly relatively compact. Suppose also that for every $x' \in E'$

$$\lim_{n\to\infty} x'(X_n(\omega)) \tag{1}$$

exists and is finite a.e.. Then $(X_n)_{n\in\mathbb{N}}$ converges weakly a.e..

Proof : Let Ω_o be a measurable set in Ω such that $P(\Omega_o) = 1$, and such that for every $\omega \in \Omega_o$, $(X_n(\omega))_{n\in\mathbb{N}}$ is weakly relatively compact. Let $x' \in E'$ and let $\Omega_o^{x'}$ be a measurable set in Ω such that $P(\Omega_o^{x'}) = 1$ and such that $\lim_{n\to\infty} x'(X_n(\omega))$ exists and is finite for every $\omega \in \Omega_o^{x'}$. By the theorem of Eberlein (see e.g. Schaefer [1971]), for every $\omega \in \Omega_o$, there is a sequence $(k_n)_{n\in\mathbb{N}}$ in $\mathbb{N}$ such that

$$\text{w-}\lim_{n\to\infty} X_{k_n}(\omega) =: X_\infty(\omega) \tag{2}$$

exists. Then for every $x' \in E'$ and every $\omega \in \Omega_o \cap \Omega_o^{x'}$

$$\lim_{n\to\infty} x'(X_n(\omega)) = x'(X_\infty(\omega))$$

Hence X_∞ is scalarly measurable. Due to (2), $X_\infty(\Omega)$ is weakly relatively compact.

Now invoke the classical theorem of Phillips to ensure the existence of a strongly measurable function X'_∞ such that for every $x' \in E'$

$$x'(X'_\infty) = x'(X_\infty)$$

a.e.. Thus, in our theorem, we may as well suppose that X_∞ itself is strongly measurable. It is already clear that X_∞ is a weak accumulation point of $(X_n)_{n\in\mathbb{N}}$. If we can show it is the only one, we are done. In order to obtain this, we proceed as follows. Since $\sup_{n\in\mathbb{N}} \|X_n(\omega)\|$ is a.e. finite, for every $\varepsilon > 0$, there is a constant $M \in \mathbb{R}^+$ such that

$$\sup_{n\in\mathbb{N}} \|X_n(\omega)\| \leq M$$

for all $\omega \in \Omega_1$, where Ω_1 is a measurable part of Ω_0 such that $P(\Omega_1) > 1-\varepsilon$. So we can and do assume that $(X_n)_{n\in\mathbb{N}}$ is uniformly bounded. If a function Y is another weak accumulation point of $(X_n)_{n\in\mathbb{N}}$ then, for every $x' \in E'$ and every $A \in F$

$$\lim_{n\to\infty} \int_A x'(X_n) = \int_A x'(X_\infty) = \int_A x'(Y)$$

due to the Lebesgue dominated convergence theorem. Hence, since by (1) Y is necessarily scalarly measurable we have that

$$x'(X_\infty) = x'(Y)$$

a.e.. So, if Y is a strongly measurable weak limit of $(X_n)_{n\in\mathbb{N}}$, $Y = X_\infty$, a.e. necessarily, since two strongly measurable functions, which are scalarly equal a.e., are equal a.e., as is well known (see e.g. Diestel and Uhl [1977], corollary 7, p.48). □

Remark V.4.7 : Theorem V.4.6, together with lemma II.1.5 reproves (a) in theorem V.4.5 without using corollary V.3.4 on WS amarts. Indeed, in reflexive Banach spaces, bounded sets are weakly relatively compact and if $(X_n)_{n\in\mathbb{N}}$ is a W amart, then $(x'(X_n), F_n)_{n\in\mathbb{N}}$ is an amart for every $x' \in E'$. Another application of theorem V.4.6 is given in section VII.2 below (theorem VII.2.28) which is in fact a generalization of (a) in theorem V.4.5.

We close this section by studying $\|\cdot\|_{Pe}$-Cauchy or -convergence properties of W amarts. They are proved in Egghe [1980c]. First a characterization of Schur spaces : i.e. spaces in which weak sequential convergence implies strong sequential convergence.

Theorem V.4.8 (Egghe) : The following assertions are equivalent :

(i) X is a Schur space.

(ii) Every W amart is an amart.

(iii) Every WS amart is an amart.

Proof : (i) ⇒ (ii)

Let (X_n, F_n) be the W amart. Suppose that $(\int_\Omega X_\tau)_{\tau \in T}$ does not converge strongly. Then let $(\tau_n)_{n \in \mathbb{N}}$ be a cofinal sequence in T such that $(\int_\Omega X_{\tau_n})_{n \in \mathbb{N}}$ does not converge strongly. By (i), $(\int_\Omega X_{\tau_n})_{n \in \mathbb{N}}$ is not weakly Cauchy and hence, so is $(\int_\Omega X_\tau)_{\tau \in T}$, which is a contradiction.

(ii) ⇒ (iii)

This follows from theorem V.4.2.

(iii) ⇒ (i)

If E is not a Schur space, then there is a sequence $(x_n)_{n \in \mathbb{N}}$ in E which is weakly but not strongly convergent. Put trivially :

$$X_n = x_n \chi_\Omega$$

$F_n = \{\Omega\}$ for (Ω, F, P) any probability space. Then $(X_n, F_n)_{n \in \mathbb{N}}$ is readily seen to be a WS amart which is not an amart. □

This result has two corollaries, based on $\|\cdot\|_{Pe}$-properties of amarts :

Corollary V.4.9 (Egghe) : The following assertions are equivalent :

(i) E is a Schur space.

(ii) Every uniformly integrable W amart is Cauchy for $\|\cdot\|_{Pe}$.

(iii) Every uniformly bounded WS amart is Cauchy for $\|\cdot\|_{Pe}$.

Proof : (i) ⇒ (ii)

This follows from theorem V.4.8 and the fact that a Schur space has (CRP) by V.2.30(b).

(ii) ⇒ (iii)

This follows from theorem V.4.2.

(iii) ⇒ (i)

If E is not a Schur space, then the same example as in the proof (iii) ⇒ (i) of theorem V.4.8 does the job. □

Corollary V.4.10 (Egghe) : The following assertions are equivalent :

(i) E is a Schur space and has (RNP).

(ii) Every uniformly integrable W amart is $\|\cdot\|_{Pe}$-convergent.

(iii) Every uniformly bounded WS amart is $\|\cdot\|_{Pe}$-convergent.

Proof : (i) ⇒ (ii)

This follows from corollary V.4.9 and corollary V.2.5.

(ii) ⇒ (iii)

This follows from theorem V.4.2.

(iii) ⇒ (i)

This also follows from corollaries V.4.9 and V.2.5. □

V.5. Semiamarts

So far, in this chapter, we have studied adapted sequences $(X_n, F_n)_{n\in\mathbb{N}}$ such that

$$(E^{F_\sigma} X_\tau - X_\sigma)_{\substack{\tau\in T(\sigma)\\ \sigma\in T}}$$

goes to zero in a certain way, thus extending the martingale property.

Of a different nature is the extension where

$$(E^{F_\sigma} X_\tau - X_\sigma)_{\substack{\tau\in T(\sigma)\\ \sigma\in T}}$$

is bounded in a certain way. We cannot expect the convergence of such a kind of adapted sequence but nevertheless it is an interesting type to investigate since boundedness is often valid without having convergence to zero. Let us now state more precisely the (uniform) semiamart notion.

Definition V.5.1 : Let E be a Banach space and $(X_n,\mathcal{F}_n)_{n\in\mathbb{N}}$ an adapted sequence. We say that $(X_n,\mathcal{F}_n)_{n\in\mathbb{N}}$ is a semiamart if

$$\sup_{\sigma\in T}\ \sup_{\tau\in T(\sigma)} \| E^{\mathcal{F}_\sigma} X_\tau - X_\sigma \|_{Pe} < \infty$$

We say that $(X_n,\mathcal{F}_n)_{n\in\mathbb{N}}$ is a uniform semiamart if

$$\sup_{\sigma\in T}\ \sup_{\tau\in T(\sigma)} \| E^{\mathcal{F}_\sigma} X_\tau - X_\sigma \|_1 < \infty$$

(cf. the analogy with the definitions of amart and uniform amart).
As for amarts we have

Theorem V.5.2 : For an adapted sequence $(X_n,\mathcal{F}_n)_{n\in\mathbb{N}}$ the following assertions are equivalent :

(i) $(X_n,\mathcal{F}_n)_{n\in\mathbb{N}}$ is a semiamart

(ii) For any fixed $\sigma \in T$, the net

$$\left(\int_A X_\tau\right)_{\substack{\tau\in T\\ A\in\mathcal{F}_\sigma}}$$

is bounded in E.

(iii) The net $(\int_\Omega X_\tau)_{\tau\in T}$ is bounded in E.

Proof : (i) ⇒ (ii)

$$\left\|\int_A X_\tau\right\| = \left\|\int_A E^{\mathcal{F}_\sigma} X_\tau\right\|$$

$$\leq \left\|\int_A (E^{\mathcal{F}_\sigma} X_\tau - X_\sigma)\right\| + \left\|\int_A X_\sigma\right\|$$

$$\leq \|E^{F_\sigma} X_\tau - X_\sigma\|_{Pe} + \|X_\sigma\|_{Pe} < \infty$$

(ii) ⇒ (iii)

Is trivial.

(iii) ⇒ (i)

This is carried out in exactly the same way as the proof of (iii) ⇒ (i) in theorem V.2.2. □

Of course, every uniform semiamart is a semiamart and the converse is true in finite dimensional Banach spaces, due to theorem I.1.2.8.

An interesting result on real semiamarts is the Riesz-decomposition theorem of Krengel and Sucheston [1978]. Before giving the result and its proof we need some permanence properties of uniform semiamarts in general Banach spaces.

Theorem V.5.3 : Let E be any Banach space. An L^1_E-bounded adapted sequence is a uniform semiamart if and only if it is of class (B).

Proof : (a) Let $(X_n, F_n)_{n \in \mathbb{N}}$ be an L^1_E-bounded uniform semiamart. Let $\sigma \in T$ and choose $n \in \mathbb{N}(\sigma)$. Then

$$\int_\Omega \|X_\sigma\|$$

$$\leq \int_\Omega \|E^{F_\sigma} X_n - X_\sigma\| + \int_\Omega \|E^{F_\sigma} X_n\|$$

$$\leq \int_\Omega \|E^{F_\sigma} X_n - X_\sigma\| + \int_\Omega \|X_n\|$$

Hence

$$\sup_{\sigma \in T} \int_\Omega \|X_\sigma\| \leq \sup_{\sigma \in T} \sup_{n \in \mathbb{N}(\sigma)} \int_\Omega \|E^{F_\sigma} X_n - X_\sigma\| + \sup_{n \in \mathbb{N}} \int_\Omega \|X_n\| < \infty$$

(b) If $(X_n, F_n)_{n \in \mathbb{N}}$ is of class (B) then obviously

$$\sup_{\sigma \in T} \sup_{\tau \in T(\sigma)} \int_\Omega \| E^{F_\sigma} X_\tau - X_\sigma \| \leq 2 \sup_{\sigma \in T} \int_\Omega \| X_\sigma \| < \infty \qquad \square$$

Corollary V.5.4 : An L^1_E-bounded adapted sequence $(X_n, F_n)_{n \in \mathbb{N}}$ is a uniform semiamart if and only if $(\|X_n\|, F_n)_{n \in \mathbb{N}}$ is a (uniform) semiamart.

Corollary V.5.5 : Let $(X_n, F_n)_{n \in \mathbb{N}}$ and $(Y_n, F_n)_{n \in \mathbb{N}}$ be real L^1-bounded semiamarts. Then $(X_n \vee Y_n, F_n)_{n \in \mathbb{N}}$ and $(X_n \wedge Y_n, F_n)_{n \in \mathbb{N}}$ are L^1-bounded semiamarts.

Proof : This follows from corollary V.5.4 and the identities

$$X_n \vee Y_n = \frac{1}{2} (X_n + Y_n + |X_n - Y_n|)$$

$$X_n \wedge Y_n = \frac{1}{2} (X_n + Y_n - |X_n - Y_n|)$$

for each $n \in \mathbb{N}$. $\square$

Theorem V.5.6 : Let $(X_n, F_n)_{n \in \mathbb{N}}$ be a real semiamart where $F_n = F_m$ for every $n, m \in \mathbb{N}$. Then $\sup_{n \in \mathbb{N}} |X_n| \in L^1$.

Proof : Suppose first that $\int_\Omega \sup_{n \in \mathbb{N}} X_n = \infty$. Then for each $m \in \mathbb{N}$, choose $N(m) \in \mathbb{N}$ such that

$$\int_\Omega \max_{n \leq N(m)} X_n > m$$

For each $\omega \in \Omega$, let $\tau_m(\omega)$ be the first natural number k for which $X_k(\omega) = \max_{n \leq N(m)} X_n(\omega)$. Then $\int_\Omega X_{\tau_m} > m$. Hence $(X_n, F_n)_{n \in \mathbb{N}}$ is not a semiamart, a contradiction. So $\int_\Omega \sup_{n \in \mathbb{N}} X_n < \infty$. Analogously $\int_\Omega \inf_{n \in \mathbb{N}} X_n > -\infty$. Hence $\int_\Omega \sup_{n \in \mathbb{N}} |X_n| < \infty$. $\square$

The classical notion of "Banach limit" is needed now. This is natural since we are dealing with semiamarts : since we have no indication of the existence of certain limits we must deal with generalized limits. However the notion "Banach limit" is not often used, so we shall repeat the definition and main theorem. For further information, see Sucheston [1967] .

Definition V.5.7 : Let B be the space of all bounded real sequences $(x_n)_{n\in\mathbb{N}}$. A linear functional L on B is called a Banach limit if

(i) $L((x_n)_{n\in\mathbb{N}}) \geqslant 0$ if $x_n \geqslant 0$ for every $n \in \mathbb{N}$

(ii) $L((x_{n+1})_{n\in\mathbb{N}}) = L((x_n)_{n\in\mathbb{N}})$

(iii) $L(\overline{1}) = 1$, where $\overline{1} = (1,1,1,1,\ldots)$.

Theorem V.5.8 (Sucheston) :

(i) Banach limits exist.

(ii) The maximal value of Banach limits on a bounded sequence $(x_n)_{n\in\mathbb{N}}$ is

$$L^*((x_n)_{n\in\mathbb{N}}) = \lim_{n\to\infty} (\sup_{j\in\mathbb{N}} \frac{1}{n} \sum_{i=0}^{n-1} x_{i+j})$$

(iii) The minimal value of Banach limits on a bounded sequence $(x_n)_{n\in\mathbb{N}}$ is

$$L_*((x_n)_{n\in\mathbb{N}}) = \lim_{n\to\infty} (\inf_{j\in\mathbb{N}} \frac{1}{n} \sum_{i=0}^{n-1} x_{i+j})$$

(iv) A necessary and sufficient condition in order that all Banach limits on a bounded sequence $(x_n)_{n\in\mathbb{N}}$ are equal (with value denoted by s) is that

$$\lim_{n\to\infty} \frac{1}{n} \sum_{i=0}^{n} x_{i+j} = s$$

uniformly in j. So, on a convergent sequence, the Banach limit is the limit.

For the proof, see Sucheston [1967] , p.309-311.

We can now state and prove the Riesz-decomposition theorem for semiamarts of Krengel and Sucheston [1978] :

Theorem V.5.9 (Krengel and Sucheston) : Every real semiamart $(X_n, F_n)_{n\in\mathbb{N}}$ can be written as

$$X_n = Y_n + Z_n$$

where $(Y_n, F_n)_{n\in\mathbb{N}}$ is a martingale and $(Z_n, F_n)_{n\in\mathbb{N}}$ is what is called a <u>semipotential</u>, i.e. : an L^1-bounded semiamart such that for each $A \in \bigcup_n F_n$:

$$\liminf_{n\in\mathbb{N}} \frac{1}{n} \sum_{i=1}^{n} \int_A Z_i \leq 0 \leq \limsup_{n\in\mathbb{N}} \frac{1}{n} \sum_{i=1}^{n} \int_A Z_i \qquad (1)$$

One has, for each $m \in \mathbb{N}$ and $A \in \bigcup_n F_n$:

$$\liminf_{n\in\mathbb{N}} \frac{1}{n} \sum_{i=1}^{n} \int_A X_i \leq \int_A Y_m \leq \limsup_{n\in\mathbb{N}} \frac{1}{n} \sum_{i=1}^{n} \int_A X_i \qquad (2)$$

<u>Proof</u> : For each $m \in \mathbb{N}$, $(E^{F_m} X_n, F_m)_{n\in\mathbb{N}(m)}$ is a semiamart since there are fewer stopping times w.r.t. the constant sequence $F_m, F_m, F_m, \ldots$ than w.r.t. $(F_n)_{n\in\mathbb{N}}$ and by theorem V.5.2. Hence (theorem V.5.6),

$$\sup_{n\in\mathbb{N}(m)} |E^{F_m} X_n| \in L^1 \qquad (3)$$

Let L be an arbitrary Banach limit and for $A \in F_m$ put :

$$\mu_n(A) = \int_A X_n = \int_A E^{F_m} X_n$$

$$\mu(A) = L\left(\frac{1}{n} \sum_{i=1}^{n} \mu_i(A)\right)$$

(denote from now on $L(x_n)$ instead of $L((x_n)_{n\in\mathbb{N}})$).
Due to (3), for each $\varepsilon > 0$, there is a $\delta > 0$ such that $P(A) < \delta$ implies $\sup_{n\in\mathbb{N}} |\mu_n(A)| < \varepsilon$. Since L is linear, μ is certainly finitely additive.
But also

$$|\mu(A)| \leq L\left(\frac{1}{n} \sum_{i=1}^{n} |\mu_i(A)|\right)$$

$$\leq \sup_{n\in\mathbb{N}} |\mu_n(A)|$$

using (i) and (iii) in V.5.7. Hence μ is a measure such that $\mu \ll P$. Denote $Y_m = \frac{d\mu}{dP}$, the Radon-Nikodym-derivative of μ (on F_m) w.r.t. P (or more exactly w.r.t. $P|F_m$). Put $Z_m = X_m - Y_m$ for each $m \in \mathbb{N}$. Obviously $(Y_n, F_n)_{n\in\mathbb{N}}$ is a martingale. From theorem V.5.8 it follows that any Banach limit on a bounded sequence lies between the lim inf and the lim sup of this sequence. So (2) follows. Hence also (1) since, in replacing X_i by $Y_i + Z_i$ in (2) we see that $\liminf_{n\in\mathbb{N}} \frac{1}{n} \sum_{i=1}^{n} \int_A Y_i = \int_A Y_m$, the lim inf being a limit.

$(Z_n)_{n\in\mathbb{N}}$ is L^1-bounded : Let $(\varepsilon_n)_{n\in\mathbb{N}}$ be a sequence of numbers decreasing to zero. Choose, for each $m \in \mathbb{N}$, $N(m) \in \mathbb{N}(m)$ such that

$$\int_{\Omega\setminus\{Z_m<0\}} Z_{N(m)} < \varepsilon_m$$

This is possible, due to (1). Define

$$\tau_m \begin{cases} = m & \text{on } \{Z_m < 0\} \\ = N(m) & \text{on } \Omega\setminus\{Z_m < 0\} \end{cases}$$

So

$$\int_\Omega Z_{\tau_m} - \int_{\{Z_m<0\}} Z_m = \int_{\Omega\setminus\{Z_m<0\}} Z_{N(m)} < \varepsilon_m$$

$(\int_\Omega Z_\tau)_{\tau\in T}$ is bounded since $(Z_n, F_n)_{n\in\mathbb{N}}$ is a semiamart and by theorem V.5.2. It follows that

$$-\int_{\{Z_m<0\}} Z_m = \int_\Omega Z_m^-$$

is bounded above when m varies in $\mathbb{N}$. The same argument for $(\int_\Omega Z_m^+)_{m\in\mathbb{N}}$. So $\sup_{m\in\mathbb{N}} \int_\Omega |Z_m| < \infty$. □

As one can see from the above proof using Banach limits, the decomposition given in theorem V.5.9 is not unique. It is unique in case $(X_n, F_n)_{n\in\mathbb{N}}$ is an amart but then the theorem reduces to the known Riesz-decomposition theorem for amarts (theorem V.2.4).

As an application of theorem V.5.9 we can include the following criterion for L^1-boundedness of a semiamart.

<u>Theorem V.5.10 (Krengel and Sucheston)</u> : Let $(X_n, F_n)_{n\in\mathbb{N}}$ be a real semiamart such that

(1) For each $A \in \bigcup_n F_n$

$$\lim_{n\to\infty} \left(\frac{1}{n} \sum_{i=1}^{n} \int_A X_i - P(A) \frac{1}{n} \sum_{i=1}^{n} \int X_i\right) = 0$$

("mixing" notion in ergodic theory).

Then $(X_n)_{n\in\mathbb{N}}$ is L^1-bounded.

<u>Proof</u> : The proof is very simple : In this case one sees, due to (1) and the definition of Y_m in theorem V.5.9, that $(Y_m)_{n\in\mathbb{N}}$ is constant, Y_m being $L\left(\frac{1}{n} \sum_{i=1}^{n} \int_\Omega X_i\right)$, for each $m \in \mathbb{N}$. Hence since $(Z_n)_{n\in\mathbb{N}}$ is L^1-bounded, so is $(X_n)_{n\in\mathbb{N}}$. □

Condition (1) in the above theorem is for instance satisfied if

$$X_n = \frac{1}{c_n} \sum_{i=1}^{n} Y_i$$

where $(c_n)_{n\in\mathbb{N}}$ is a positive sequence increasing to ∞ and where $(Y_n)_{n\in\mathbb{N}}$ is independent.

We now come to another application of theorem V.5.9. It is the semiamart version of theorem V.2.16 on amarts :

<u>Theorem V.5.11 (Krengel and Sucheston)</u> : Let $(X_n = \sum_{i=1}^{n} Y_i, F_n)_{n\in\mathbb{N}}$ be a semiamart such that for some $\alpha \geq 1$

$$\sum_{i=1}^{\infty} \frac{\int_\Omega (Y_i^{2\alpha})}{i^{1+\alpha}} < \infty$$

Then

$$\sup_{n\in\mathbb{N}} \frac{|X_n|}{n} < \infty \quad , \text{ a.e..}$$

Proof : Write the Doob decomposition of $(X_n,F_n)_{n\in\mathbb{N}}$: $X_n = M_n + A_n$ (see definition V.2.15). As in the proof of theorem V.2.16 we have that

$$\lim_{n\to\infty} \frac{M_n}{n} = 0 \text{ , a.e..}$$

So we only have to show that

$$\sup_{n\in\mathbb{N}} \frac{|A_n|}{n} < \infty, \text{ a.e..}$$

To see this, remark that $A_n - A_{n-1} = E^{F_{n-1}} X_n - X_{n-1}$ for each $n \in \mathbb{N}(2)$ and that

$$(E^{F_{n-1}} X_n - X_{n-1}, F_{n-1})_{n\in\mathbb{N}(2)}$$

is a semiamart. This is due to theorem V.5.2 and the fact that every bounded stopping time w.r.t. $(G_n)_{n\in\mathbb{N}(2)}$ where $G_n = F_{n-1}$, is in T. To this semiamart apply theorem V.5.9 : here we have

$$\lim_{n\to\infty} \frac{1}{n} \sum_{i=1}^{n} \int_A (A_{i+1} - A_i) = 0 \tag{1}$$

for every $A \in F_m$ and $m \in \mathbb{N}$. Indeed

$$\{\int_A X_n \| n > m\}$$

is bounded by theorem V.5.2. From (2) in theorem V.5.9 and (1) above it follows that $Y_m = 0$ for every $m \in \mathbb{N}$. So $(E^{F_{n-1}} X_n - X_{n-1}, F_n)_{n\in\mathbb{N}}$ is a semipotential and hence an L^1-bounded semiamart.
From theorem V.5.3, for $E = \mathbb{R}$, we see that

$$\sup_{\tau\in T'} \int \|A_\tau - A_{\tau-1}\| < \infty \quad ,$$

where T' denotes the set of bounded stopping times w.r.t. $(G_n)_{n\in\mathbb{N}(2)}$ where $G_n = F_{n-1}$. Lemma II.1.5 implies that $\sup_{n\in\mathbb{N}(2)} |A_n - A_{n-1}| < \infty$, a.e. Hence $\sup_{n\in\mathbb{N}} \frac{|A_n|}{n} < \infty$, a.e.. □

In this work, as we mentioned in the introduction, we start from the knowledge of the real martingale and semimartingale theory. A characterization of amarts and semiamarts in terms of these notions is given by Ghoussoub and Sucheston [1978], using theorem V.5.9 above.

Theorem V.5.12 (Ghoussoub and Sucheston) :

(i) Let $(X_n,F_n)_{n\in\mathbb{N}}$ be a real adapted sequence. $(X_n,F_n)_{n\in\mathbb{N}}$ is a semiamart if and only if $(X_n)_{n\in\mathbb{N}}$ can be written as

$$X_n = Y_n + Z_n$$

where $(Y_n,F_n)_{n\in\mathbb{N}}$ is a martingale and $(Z_n)_{n\in\mathbb{N}}$ is dominated by a positive supermartingale.

(ii) Let $(X_n,F_n)_{n\in\mathbb{N}}$ be an E-valued adapted sequence. $(X_n,F_n)_{n\in\mathbb{N}}$ is a uniform amart if and only if $(X_n)_{n\in\mathbb{N}}$ can be written as

$$X_n = Y_n + Z_n$$

where $(Y_n,F_n)_{n\in\mathbb{N}}$ is a martingale and $(Z_n)_{n\in\mathbb{N}}$ is dominated in norm by a Doob potential.

Proof : (i) Due to theorem V.5.9 we have that

$$X_n = Y_n + Z_n$$

where $(Y_n,F_n)_{n\in\mathbb{N}}$ is a martingale and $(Z_n,F_n)_{n\in\mathbb{N}}$ is an L^1-bounded semiamart. Then $(\|Z_n\|,F_n)_{n\in\mathbb{N}}$ is a positive semiamart of class (B), by V.5.3 and V.5.4. Let $M = \sup_{\tau\in T} \int_\Omega |Z_\tau|$. Put

$$S_n = \sup_{\sigma\in T(n)} E^{F_n}|Z_\sigma|$$

S_n is measurable since every element in T has its values in $\mathbb{N}$; otherwise we need to write essential sup - see Neveu [1975] or chapter VI. Applying the same argument as in lemma I.3.5.4 we see that there is a sequence $(\sigma_k^{(n)})_{k\in\mathbb{N}}$ in T(n) such that

$$S_n = \sup_{k \in \mathbb{N}} E^{F_n} |Z_{\sigma_k^{(n)}}|$$

Hence a classical argument - in fact exactly the same argument as in the proof of lemma I.3.5.4 - shows the existence of a sequence $(\tau_k^{(n)})_{k \in \mathbb{N}}$ in $T(n)$ such that $(E^{F_n}|Z_{\tau_k^{(n)}}|)_{k \in \mathbb{N}}$ increases to S_n. From this we now see that $\sup_{n \in \mathbb{N}} \int_\Omega S_n \leqslant M$, hence $S_n \in L^1$ for each $n \in \mathbb{N}$! So

$$E^{F_{n-1}} S_n = \lim_{k \to \infty} E^{F_{n-1}} |Z_{\tau_k^{(n)}}|$$

$$\leqslant \sup_{\sigma \in T(n-1)} E^{F_{n-1}} |Z_\sigma| = S_{n-1}$$

(ii)

If $(X_n, F_n)_{n \in \mathbb{N}}$ is an E-valued uniform amart, then, due to theorem V.1.4, using also the notation of theorem V.1.4, we see that $(\|Z_n\|, F_n)_{n \in \mathbb{N}}$ is a positive potential. Hence, for $(\|Z_n\|, F_n)_{n \in \mathbb{N}}$ we can define $(S_n)_{n \in \mathbb{N}}$ as above and we have already that $(\|Z_n\|)_{n \in \mathbb{N}}$ is dominated by $(S_n)_{n \in \mathbb{N}}$, a positive supermartingale. So, $\lim_{\tau \in T} \int_\Omega S_\tau$ exists since it decreases and is positive. Thus :

$$\lim_{\gamma \in T} \int_\Omega S_\gamma = \lim_{\gamma \in T} \int_\Omega \lim_{k \to \infty} E^{F_\gamma} \|Z_{\tau_k^{(\gamma)}}\|$$

where $\tau_k^{(\gamma)}$ is the stopping time $\tau_k^{(n)}$ on $\{\gamma = n\}$ for each $n \in \mathbb{N}$. So

$$\lim_{\gamma \in T} \int_\Omega S_\gamma = \lim_{\gamma \in T} \lim_{k \to \infty} \int_\Omega \|Z_{\tau_k^{(\gamma)}}\|$$

since $(E^{F_\gamma}\|Z_{\tau_k^{(\gamma)}}\|)_{k \in \mathbb{N}}$ increases to S_γ; indeed it does so on every set $\{\gamma = n\}$. For each $\gamma \in T$, choose $\tau_{k(\gamma)}^{(\gamma)} \in T(\gamma)$ such that

$$\left| \lim_{k \to \infty} \int_\Omega \|Z_{\tau_k^{(\gamma)}}\| - \int_\Omega \|Z_{\tau_{k(\gamma)}^{(\gamma)}}\| \right| < \frac{1}{\min \gamma} .$$

Then, if we put $\sigma_\gamma = \tau^{(\gamma)}_{k(\gamma)} \in T(\gamma)$

$$\lim_{\gamma\in T} \int_\Omega S_\gamma \leq \lim_{\gamma\in T} \int_\Omega \|Z_{\sigma_\gamma}\| + \lim_{\gamma\in T} \frac{1}{\min \gamma} = 0$$

since $(\|Z_n\|,F_n)_{n\in\mathbb{N}}$ is a potential. Hence from definition V.3.18. $(S_n,F_n)_{n\in\mathbb{N}}$ is a (strong) Doob potential (dominating $(\|Z_n\|)_{n\in\mathbb{N}}$. □

We now come to a general construction of semiamarts. Within this class we shall characterise amarts. From the construction we can also disprove some possible speculations of the reader - ideas that are very plausible, such as "Is every semiamart which converges a.e. and in the L^1-sense necessarily an amart?" and other things!

General Construction V.5.13 (Krengel and Sucheston [1978]) : Consider three sequences of numbers $(p_n)_{n\in\mathbb{N}}$ in $\mathbb{N}$, $(\alpha_n)_{n\in\mathbb{N}}$ in $]0,1]$ and $(\rho_n)_{n\in\mathbb{N}}$ in $]0,1]$. They will be specified according to the result we want to prove. Let $\Omega = [0,1]$, F = the Borel-σ-algebra on $[0,1]$, P = Lebesguemeasure on $[0,1]$. The construction is by induction. Divide $[0,\alpha_1]$ into p_1 disjoint intervals of equal length $\frac{\alpha_1}{p_1}$, say $A_1^1,\ldots,A_1^{p_1}$. Put $X_1 = 0$, $X_i = h_1 \chi_{A_1^{i-1}}$ $(i = 2,\ldots,p_1+1)$ where $h_1 = \frac{\rho_1}{\alpha_1}$.

Suppose we have constructed in the nth-step measurable functions $X_{g_n}, X_{g_n+1},\ldots,X_{g_{n+1}-1}$, where $g_1 = 1$, $g_{j+1} = g_j + 1 + \prod_{i=1}^{j} p_i$ $(j\in\mathbb{N})$, and where these functions take only the value 0 and $h_n = \prod_{i=1}^{n} \frac{\rho_i}{\alpha_i}$. Now comes step n+1 : Let $B_n^{i_1,\ldots,i_n}$ be a sub-interval of $A_n^{i_1,\ldots,i_n}$ of length

$$\alpha_{n+1} P(A_n^{i_1,\ldots,i_n}) = \frac{\prod_{i=1}^{n+1} \alpha_i}{\prod_{i=1}^{n} p_i} .$$

Divide each set $B_n^{i_1,\ldots,i_n}$ into p_{n+1} disjoint subintervals of equal length. For $i_{n+1} \in \{1,\ldots,p_{n+1}\}$, denote by $A_{n+1}^{i_1,\ldots,i_{n+1}}$ the i_{n+1}-st subinterval of $B_n^{i_1,\ldots,i_n}$. Put $X_{g_{n+1}} = 0$ and

Let $n \leq t \leq s$, $A = A_t^{i_1,\dots,i_t}$ and $\sigma \in T$ which stops on A during 1 period $\{g_s,\dots,g_{s+1} - 1\}$. Let $\sigma' \in T$ which stops on A during 1 period $\{g_t,\dots,g_{t+1} - 1\}$. Then

$$\int_A X_\sigma \leq \int_A X_{\sigma'}$$

So $\int_A X_\sigma \leq v_t\, q_t^{-1}$.

Assume that we have proved

$$\int_A X_\sigma \leq v_t\, q_t^{-1}$$

for all $t \in \mathbb{N}(n)$ and for $\sigma \in T$ which stop on A during at most r periods (the starting point g_s where $s \geq t$ being unimportant). Now let $\tau \in T(t)$ be such that τ stops on A during r + 1 periods, say $\{g_s,\dots,g_{s+r+1} - 1\}$. Fix $(i_1,\dots,i_t)$. With respect to this $(i_1,\dots,i_t)$ the sets $A_s^{i_1,\dots,i_s}$ can be subdivided into two groups : the first group contains those sets $A_s^{i_1,\dots,i_s}$ on which τ stops during the period $\{g_s,\dots,g_{s+1} - 1\}$ and the second group contains those sets $A_s^{i_1,\dots,i_s}$ on which τ stops later. If $A' = A_s^{i_1,\dots,i_s}$ belongs to group 1, then

$$\int_{A'} X_\tau \leq v_s\, q_s^{-1}$$

by the preceding argument (the case r = 1).

If $A'' = A_s^{i_1,\dots,i_s}$ belongs to group 2, then, by induction

$$\int_{A''} X_\tau \leq v_s\, q_s^{-1}$$

On the complement of the union of both groups $\Omega \setminus \bigcup_{i_1,\dots,i_s} A_s^{i_1,\dots,i_s}$, $X_\tau = 0$ by the definition of $(X_n)_{n \in \mathbb{N}}$ and the property of τ. In $A = A_t^{i_1,\dots,i_t}$ we are then dealing with $\prod_{i=t+1}^{s} p_i$ sets of type $A_s^{i_1,\dots,i_s}$, by construction. Hence

$$\int_A X_\tau \leq v_s\, q_s^{-1} \quad p_{t+1} \cdots p_s \leq v_t\, q_t^{-1}$$

Thus we have dealt with all $\tau \in T(t)$ and all $t \geq n$.
So $\int_A X_\tau \leq v_n\, q_n^{-1}$ for all $\tau \in T(n)$.
There are by construction precisely q_n sets of type $A = A_n^{i_1,\ldots,i_n}$. So

$$\sup_{\tau \in T(n)} \int_\Omega X_\tau \leq v_n$$

The conclusion is now : $\sup_{\tau \in T(n)} \int_\Omega X_\tau = v_n$ for each $n \in \mathbb{N}$. Therefore $(X_n, F_n)_{n \in \mathbb{N}}$ is a semiamart.

<u>Theorem V.5.13.1</u> : $(X_n, F_n)_{n \in \mathbb{N}}$ is an amart if and only if $\prod_{n=1}^{\infty} \rho_n = 0$.

<u>Proof</u> : If $(X_n, F_n)_{n \in \mathbb{N}}$ is an amart then $\lim_{\tau \in T} \int_\Omega X_\tau = 0$; indeed since $\int_\Omega X_{g_n} = 0$ for every $n \in \mathbb{N}$, this limit (which exists, since $(X_n, F_n)_{n \in \mathbb{N}}$ is an amart) has to be zero. This implies

$$\lim_{n\to\infty} v_n = \prod_{n=1}^{\infty} \rho_n = 0 .$$

Conversely, if $\prod_{n=1}^{\infty} \rho_n = 0$ then

$$\lim_{n\to\infty} \sup_{\tau \in T(n)} \int_\Omega X_\tau = 0 .$$

Hence $(X_n, F_n)_{n \in \mathbb{N}}$ is an amart, due to theorem V.2.2. □

<u>Theorem V.5.13.2</u> : $\lim_{n\to\infty} X_n = 0$ if and only if $\prod_{n=1}^{\infty} \alpha_n = 0$ or $\prod_{n=1}^{\infty} \rho_n = 0$.

<u>Proof</u> : We have that $P\big(\bigcup_{i_1,\ldots,i_n} A_n^{i_1,\ldots,i_n}\big) = \prod_{i=1}^{n} \alpha_i$.

Note that $\big(\bigcup_{i_1,\ldots,i_n} A_n^{i_1,\ldots,i_n}\big)_{n \in \mathbb{N}}$ decreases and that $X_t = 0$ on the complement of $\bigcup_{i_1,\ldots,i_n} A_n^{i_1,\ldots,i_n}$ for $t \geq g_n$. So $\prod_{n=1}^{\infty} \alpha_n = 0$ implies $\lim_{n\to\infty} X_n = 0$, a.e..

Conversely, assume $\prod_{n=1}^{\infty} \alpha_n > 0$. Then $\lim_{n\to\infty} h_n = 0$ if and only if $\prod_{n=1}^{\infty} \rho_n = 0$. Now $\lim_{n\to\infty} h_n \neq 0 \Leftrightarrow \lim_{n\to\infty} X_n \neq 0$ on $\bigcap_{n=1}^{\infty} [\bigcup_{i_1,\ldots,i_n} A_n^{i_1,\ldots,i_n}]$ which has positive measure precisely since $\prod_{n=1}^{\infty} \alpha_n > 0$. □

Example V.5.13.3 : There exists a semiamart which converges a.e. and in L^1, but which is not an amart.

Proof : Take $\rho_n = 1$, $\alpha_n = \frac{1}{2}$, $p_n = 2$ for all $n \in \mathbb{N}$. Then from the preceding results : $(X_n, \mathcal{F}_n)_{n\in\mathbb{N}}$ is a semiamart converging to zero and is not an amart. Furthermore for $k \in \{1,\ldots,g_{n+1} - 1 - g_n\}$

$$\int_\Omega |X_{g_n+k}| = \int_\Omega h_n \chi_{A_n^{i_1,\ldots,i_n}}$$

$$= \frac{\prod_{i=1}^{n} \rho_i \prod_{i=1}^{n} \alpha_i}{\prod_{i=1}^{n} \alpha_i \prod_{i=1}^{n} p_i}$$

$$= \frac{1}{2^n}$$

and

$$\int_\Omega |X_{g_n}| = 0$$

Hence $\lim_{n\to\infty} \int_\Omega |X_n| = 0$ also. □

We remark that from the above example it follows that the Riesz-decomposition theorem for amarts does not characterise amarts!

And another negative result is to follow.

Example V.5.13.4 : There exists an L^1-bounded positive semiamart $(X_n,F_n)_{n\in\mathbb{N}}$ such that in the Riesz-decomposition theorem V.5.9, the martingale part $Y_n = 0$ and such that $(Z_n)_{n\in\mathbb{N}} = (X_n)_{n\in\mathbb{N}}$ is not uniformly integrable.

Proof : Take $\alpha_n = \frac{1}{2}$, $p_n = \rho_n = 1$ for all $n \in \mathbb{N}$. Then as in V.5.13.3 we can see that $\sup_{n\in\mathbb{N}} \int_\Omega |X_n| \leq 1$. In fact in this case we have $X_k = 0$ for $k \neq 2^n$ for a certain $n \in \mathbb{N}$ and $X_k = 2^n \chi_{[0,\frac{1}{2^n}]}$ for $k = 2^n$ $(n \in \mathbb{N})$. Since it is now trivial that

$$\lim_{n\to\infty} \frac{1}{n} \sum_{i=1}^{n} \int_\Omega X_i = 0$$

we see that $Y_n = 0$ for every $n \in \mathbb{N}$ by the construction of $(Y_n,F_n)_{n\in\mathbb{N}}$ in V.5.9. Obviously $(X_n)_{n\in\mathbb{N}}$ is not uniformly integrable. □

V.6. Notes and remarks

V.6.1. Our definition of uniform amart is different from the way it was first defined in Bellow [1978a], p.179. If (Ω,F,P) is the probability space and G is a sub-algebra of F, denote by $\mathcal{E}_E(\Omega,G)$ the set of all G-stepfunctions with values in E, and by $\mathcal{E}^1_E(\Omega,G) = \{g\in\mathcal{E}_E(\Omega,G) \| \|g(\omega)\| \leq 1$, for every $\omega \in \Omega\}$.
In Bellow [1978a] the definition of uniform amart is the following :
An adapted sequence $(X_n,F_n)_{n\in\mathbb{N}}$ is called a uniform amart if :

(i) $\lim_{\sigma\in T} \int_\Omega g(X_\sigma) =: L(g)$ exists for each $g \in \mathcal{E}_{E'}(\Omega, \bigcup_{n\in\mathbb{N}} F_n)$

(ii) The above limit is uniform in the sense that

$$\lim_{\sigma\in T} \sup_{g\in\mathcal{E}^1_{E'}(\Omega,F_\sigma)} \left|\int_\Omega g(X_\sigma) - L(g)\right| = 0$$

Based on our definition we have that if $(X_n,F_n)_{n\in\mathbb{N}}$ is a uniform amart, then it is an amart and hence (i) is satisfied (see V.2.2). (ii) follows also since :

$$\lim_{\sigma\in T} \sup_{g\in \&^1_{E'}(\Omega,F_\sigma)} \sup_{\tau\in T(\sigma)} \left|\int_\Omega g(X_\sigma) - \int_\Omega g(X_\tau)\right|$$

$$= \lim_{\sigma\in T} \sup_{g\in \&^1_{E'}(\Omega,F_\sigma)} \sup_{\tau\in T(\sigma)} \left|\int_\Omega g(X_\sigma) - \int_\Omega g(E^{F_\sigma} X_\tau)\right|$$

$$\leqslant \lim_{\sigma\in T} \sup_{\tau\in T(\sigma)} \int_\Omega \| E^{F_\sigma} X_\tau - X_\sigma \| .$$

Conversely, suppose (i) and (ii) are valid. Then certainly

$$0 = \lim_{\sigma\in T} \sup_{g\in \&^1_{E'}(\Omega,F_\sigma)} \sup_{\tau\in T(\sigma)} \left|\int_\Omega g(X_\sigma) - \int_\Omega g(X_\tau)\right|$$

$$= \lim_{\sigma\in T} \sup_{\tau\in T(\sigma)} \sup_{g\in \&^1_{E'}(\Omega,F_\sigma)} \left|\int_\Omega g(E^{F_\sigma} X_\tau - X_\sigma)\right| \quad (1)$$

Now for every sub-σ-algebra G of F and every $X \in L^1_E(\Omega,G,P|G)$ we have

$$\sup_{g\in \&^1_{E'}(\Omega,G)} \left|\int_\Omega g(X)\right| = \int_\Omega \|X\| \quad (2)$$

<u>Proof</u> : We suffice to show this for stepfunctions

$$X = \sum_{i=1}^n x_i \chi_{A_i}$$

$(A_1,\ldots,A_n \in G)$. For this, choose, for every $\varepsilon > 0$ and $i \in \{1,\ldots,n\}$, elements $x_i' \in E'$ such that $\|x_i'\| = 1$ and such that

$$\|x_i\| - \varepsilon > x_i'(x_i)$$

Put

$$g = \sum_{i=1}^{n} x_i' \chi_{A_i}$$

Then $g \in \mathcal{E}^1_{E'}(\Omega, \mathcal{G})$ and

$$\int_\Omega g(X) = \int_\Omega \sum_{i=1}^{n} x_i'(x_i) \chi_{A_i}$$

$$\geq \int_\Omega \|X\| - \varepsilon. \qquad \square$$

Applying (2) with $\mathcal{G} = \mathcal{F}_\sigma$ to (1) we get

$$\lim_{\sigma \in T} \sup_{\tau \in T(\sigma)} \int_\Omega \|E^{\mathcal{F}_\sigma} X_\tau - X_\sigma\| = 0$$

Hence $(X_n, \mathcal{F}_n)_{n \in \mathbb{N}}$ is a uniform amart in our sense. The definition of uniform amart as given in this book is the one which is used most.

V.6.2. As noted in section V.2 the name "amart" comes from the term "asymptotic martingale". This name "amart" was proposed by Sucheston but for quite some time the two terms were used. The second one was especially used in the works of Bellow (see Bellow [1976] and [1977a]), until a certain time when the battle was won in favor of the name "amart"!

V.6.3. For an optional sampling theorem for real amarts using T_f = {all finite stopping times} we refer to Krengel and Sucheston [1978], p.212.

V.6.4. Also for real amarts, we mention an interesting theorem of A. Bellow [1977a], theorem 2, p.284-288 :

Theorem (A. Bellow) : For a function $G : \mathbb{R} \to \mathbb{R}$, the following assertions are equivalent :

(i) G satisfies (a) G is continuous

(b) $\lim_{t\to+\infty} \frac{G(t)}{t}$ and $\lim_{t\to-\infty} \frac{G(t)}{t}$

exist and are finite.

(ii) For every L^1-bounded amart $(X_n,\mathcal{F}_n)_{n\in\mathbb{N}}$, the adapted sequence $(G(X_n),\mathcal{F}_n)_{n\in\mathbb{N}}$ is an L^1-bounded amart.

For the proof, see Bellow [1977a].

V.6.5. As we mentioned earlier, the proof of corollary V.3.15 of Brunel and Sucheston is not the original one but is the proof of Edgar. The original proof uses some Ramsey theory (work of J. Stern [1978]), see Brunel and Sucheston [1977].

V.6.6. The case where the index set $\mathbb{N}$ in $(X_n,\mathcal{F}_n)_{n\in\mathbb{N}}$ is replaced by $-\mathbb{N}$ is not described in this book. We refer to Edgar and Sucheston [1976a] and Krengel and Sucheston [1978] for information on this topic (see also Gut [1982]). Generally speaking, adapted sequences $(X_n,\mathcal{F}_n)_{n\in-\mathbb{N}}$ are easier to handle and boundedness conditions (to get convergence) can be weakened or even deleted. The case where the index set is $\mathbb{R}$ is also not described (continuous parameter case) (see primarily Edgar and Sucheston [1976b]).

Also cartesian products $\mathbb{N}\times\mathbb{N}$ or $\mathbb{R}\times\mathbb{R}$ as index sets can be studied; see f.i. Walsh [1979], Cairoli [1970], Millet and Sucheston [1980d], Millet and Sucheston [1981b], Millet and Sucheston [1981c], Millet [1980], Millet [1982], Edgar [1982a], Fouque and Millet [1980], Fouque [1980a], Fouque [1980b] and others.

Although these notions are not studied in detail in this book, the next chapter will discuss some results in this area. The purpose is to exhibit applications of the amart theory. These applications were discovered rather recently. If the reader is not interested in these more general index sets for adapted processes, the next chapter may as well be skipped without losing continuity. In chapter VII we shall discuss the disadvantages of amarts in infinite dimensional Banach spaces and we give alternative extensions of the martingale concept which behave better (i.e. which are more suited to obtain strong convergence a.e.).

V.6.7. An abstract example of a class of adapted sequences in which amarts coincide with semiamarts is given as follows (Millet and Sucheston [1980a]) :

<u>Theorem V.6.7.1 (Millet and Sucheston)</u> : Let $(X_n)_{n\in\mathbb{N}}$ be a sequence in L^1 with disjoint supports $(A_n)_{n\in\mathbb{N}}$. Put

$$A = \bigcup_{n=1}^{\infty} A_n,$$

and put

$$F_n = \sigma(X_1,\ldots,X_n,A)$$

Then for $(X_n,F_n)_{n\in\mathbb{N}}$ the following assertions are equivalent :

(i) $(X_n,F_n)_{n\in\mathbb{N}}$ is an amart.

(ii) $(X_n,F_n)_{n\in\mathbb{N}}$ is a semiamart.

(iii) $\sum_{n=1}^{\infty} \int_\Omega |X_n| < \infty$.

<u>Proof</u> : <u>(i) ⇒ (ii)</u>

Is obvious.

<u>(ii) ⇒ (iii)</u>

For every $n \in \mathbb{N}$, define

$$\sigma_n \begin{cases} = \text{first } k\in\{1,\ldots,n\} \text{ such that } X_k > 0, \text{ if it exists} \\ = n+1 \text{ if such a k does not exist} \end{cases}$$

$$\tau_n \begin{cases} = \text{first } k\in\{1,\ldots,n\} \text{ such that } X_k < 0, \text{ if it exists} \\ = n+1 \text{ if such a k does not exist.} \end{cases}$$

Now :

$$\int_\Omega X_{\sigma_n} = \sum_{i=1}^{n} \int_{\{\sigma_n = k\}} X_k + \int_{\Omega \setminus \{\sigma_n \leq n\}} X_{n+1}$$

$$= \sum_{k=1}^{n} \int_{\{\sigma_n = k\}} X_k + \int_{\Omega \setminus \{\sigma_n \leq n\} \setminus \{\tau_n \leq n\}} X_{n+1}$$

since $X_{n+1}\Big|_{\{\tau_n \leq n\}} = 0$

Analogously for τ_n. So

$$\int_\Omega X_{\sigma_n} - \int_\Omega X_{\tau_n} = \sum_{k=1}^{n} \int_{\{\sigma_n = k\}} X_k - \sum_{k=1}^{n} \int_{\{\tau_n = k\}} X_k$$

$$= \sum_{k=1}^{n} \int_{A_k} |X_k| = \sum_{k=1}^{n} \int_\Omega |X_k|$$

Hence (ii) implies (iii).

(iii) ⇒ (i)

For each $\varepsilon > 0$, choose $M > 0$ such that $\sum_{n \in \mathbb{N}(M)} \int_\Omega |X_n| \leq \varepsilon$.
For $\tau \in T(M)$ we thus have $\int_\Omega |X_\tau| \leq \varepsilon$. So $(X_n, F_n)_{n \in \mathbb{N}}$ is an amart. □

V.6.8. In the previous remark, a class of adapted sequences in which amarts coincide with semiamarts was given. Of course example V.5.13.3 shows that this is not the case in general. A weaker but easier example, only showing the existence of an a.e. convergent semiamart that fails to be an amart, is found in Gut [1982]. Also in Gut [1982], there is a simple example of an L^1-convergent and a.e. convergent adapted sequence which is not a semiamart. We present it here. Let $p > \frac{1}{2}$. Define

$$Y_i^n(\omega) \begin{cases} = n^{1/p} & \text{if } \omega \in [\frac{i-1}{n^2}, \frac{i}{n^2}) \\ = 0 & \text{otherwise} \end{cases}$$

for $i \in \{1,\ldots,n\}$ and $n \in \mathbb{N}$. Order $(j,m) \leq (i,n)$ if $j < i$ or if $j = i$ and $m \leq n$. In this order, the sequence $(Y_i^n)_{\substack{i \in \{1,\ldots,n\} \\ n \in \mathbb{N}}}$

is denoted by $(X_n^{(p)})_{n\in\mathbb{N}}$. So $\lim_{n\to\infty} X_n = 0$, a.e. and in the L^1-sense. For each $n \in \mathbb{N}$, define τ_n as the index k of $X_k^{(p)}$ which corresponds to the first $Y_i^n \in \{Y_1^n,\ldots,Y_n^n\}$ which is nonzero, if it exists; otherwise, τ_n is the index k of $X_k^{(p)}$ corresponding to Y_n^n. Hence $\tau_n \in T$ for every $n \in \mathbb{N}$ and

$$\int_\Omega X_{\tau_n}^{(p)} = \int_\Omega \max_{i=1,\ldots,n} Y_i^n = n^{\frac{1}{p}-1}$$

Thus, for $p \in \left]\frac{1}{2}, 1\right[$,

$$\lim_{n\to\infty} \int_\Omega X_{\tau_n}^{(p)} = +\infty$$

Hence $(X_n, F_n)_{n\in\mathbb{N}}$ where, as usual, $F_n = \sigma(X_1,\ldots,X_n)$ for every $n \in \mathbb{N}$, is not a semiamart.

V.6.9. So far we have not considered scalar convergence a.e. of adapted sequences $(X_n, F_n)_{n\in\mathbb{N}}$ to an integrable function (cf. I.3.1).

For a scalar convergence result for W amarts, see corollary IX.1.2. As a matter of fact, the result is very easy to prove here; it is postponed until chapter IX where all scalar convergence results are collected together with one proof.

Chapter VI : GENERAL DIRECTED INDEX SETS AND APPLICATIONS OF AMART THEORY

This chapter is optional and hence can be skipped without losing continuity with the rest of the text. It deals with adapted nets : i.e. processes $(X_i,F_i)_{i\in I}$ where I is a directed index set filtering to the right, $X_i \in L_E^1$ is F_i-measurable, for every $i \in I$ and $(F_i)_{i\in I}$ is a stochastic basis w.r.t. I; this means that every F_i is a sub-σ-algebra of F and that $F_i \subset F_j$ whenever $i,j \in I$, $i \leq j$. With filtering to the right we mean that for every $i_1,i_2 \in I$, there exists an $i_3 \in I$ such that $i_3 \geq i_1$ and $i_3 \geq i_2$. For $i \in I$, denote

$$I(i) = \{j \in I \| j \geq i\}$$

A net $(A_i)_{i\in I}$ of sets is called adapted w.r.t. $(F_i)_{i\in I}$ if $(\chi_{A_i},F_i)_{i\in I}$ is an adapted net.

It is not our intention to develop the whole theory so far known concerning these adapted nets. We only intend to indicate the main trends in the theory and to give the reader a feeling of what properties can be expected and of what properties cannot. In this setting some applications of the amart theory are given showing the importance of it. In the next chapter however, shortcomings of the vector-valued amart theory are shown.

VI.1. Convergence of adapted nets

Let $(X_i,F_i)_{i\in I}$ be an adapted net. We say that $(X_i,F_i)_{i\in I}$ is a martingale if

$$E^{F_i} X_j = X_i$$

for every $i \in I$ and every $j \in I(i)$. Doob's fundamental theorem that states that real L^1-bounded martingales are convergent a.e., valid when the index set is $\mathbb{N}$, is no longer valid in this general case. For a counterexample, see Dieudonné [1950], see also Cairoli [1970] where we see that even L^1-bounded martingales indexed by $\mathbb{N} \times \mathbb{N}$ may diverge!

In Krickeberg [1956], Krickeberg has put a condition on $(\mathcal{F}_i)_{i \in I}$ in order to guarantee that Doob's theorem will be valid. It is generally called the Vitali condition V. Before introducing this condition we need to extend, when $E = \mathbb{R}$, our concept of "convergence a.e.". These extensions are classical but we include their definitions for the sake of completeness.

<u>Definition VI.1.1</u> : Let $(X_i)_{i \in I}$ be a net of real measurable functions. The <u>essential supremum of $(X_i)_{i \in I}$</u>, denoted by $\underset{i \in I}{\text{e sup}}\, X_i$ is the unique a.e. smallest measurable function such that for every $j \in I$, $\underset{i \in I}{\text{e sup}}\, X_i \geq X_j$, a.e.. The <u>essential infimum of $(X_i)_{i \in I}$</u>, denoted by $\text{e inf}\, X_i$, is defined as

$$\underset{i \in I}{\text{e inf}}\, X_i = -\underset{i \in I}{\text{e sup}}\, (-X_i)$$

For a net $(A_i)_{i \in I}$ of measurable sets we define the <u>essential supremum of $(A_i)_{i \in I}$</u>, denoted by $\text{e sup}\, A_i$, resp. the <u>essential infimum of $(A_i)_{i \in I}$</u>, denoted by $\underset{i \in I}{\text{e inf}}\, A_i$, by

$$\chi_{\underset{i \in I}{\text{e sup}}\, A_i} = \underset{i \in I}{\text{e sup}}\, \chi_{A_i}$$

resp.

$$\chi_{\underset{i \in I}{\text{e inf}}\, A_i} = \underset{i \in I}{\text{e inf}}\, \chi_{A_i}$$

which are defined a.e..

Let $(X_i)_{i \in I}$ be as above. The <u>stochastic upper limit of $(X_i)_{i \in I}$</u>, denoted by $\underset{i \in I}{\text{s lim sup}}\, X_i$, is defined as :

$$\operatorname{s\,lim\,sup}_{i \in I} X_i = \operatorname{e\,inf} \{Y \| \lim_{i \in I} P(Y < X_t) = 0\}$$

and correspondingly the _stochastic lower limit of $(X_i)_{i \in I}$_, denoted by $\operatorname{s\,lim\,inf}_{i \in I} X_i$, as

$$\operatorname{s\,lim\,inf}_{i \in I} X_i = -\operatorname{s\,lim\,sup}_{i \in I} (-X_i)$$

We say that _$(X_i)_{i \in I}$ converges stochastically to X_∞_ (or _converges in probability to X_∞_), a measurable function necessarily, if

$$\operatorname{s\,lim\,inf}_{i \in I} X_i = \operatorname{s\,lim\,sup}_{i \in I} X_i = X_\infty$$

X_∞ is called the _stochastic limit of $(X_i)_{i \in I}$_ and denoted by $X_\infty = \operatorname{s\,lim}_{i \in I} X_i$. The _essential upper limit of $(X_i)_{i \in I}$_, denoted by $\operatorname{e\,lim\,sup}_{i \in I} X_i$ is defined as

$$\operatorname{e\,lim\,sup}_{i \in I} X_i = \operatorname{e\,inf}_{i \in I} (\operatorname{e\,sup}_{j \in I(i)} X_j)$$

The essential lower limit of $(X_i)_{i \in I}$, denoted by $\operatorname{e\,lim\,inf}_{i \in I} X_i$ is defined as

$$\operatorname{e\,lim\,inf}_{i \in I} X_i = -\operatorname{e\,lim\,sup}_{i \in I} (-X_i)$$

We say that _$(X_i)_{i \in I}$ converges essentially to X_∞_, a measurable function necessarily, if

$$\operatorname{e\,lim\,sup}_{i \in I} X_i = \operatorname{e\,lim\,inf}_{i \in I} X_i = X_\infty$$

X_∞ is called the _essential limit of $(X_i)_{i \in I}$_ and denoted by $X_\infty = \operatorname{e\,lim}_{i \in I} X_i$. For a net $(A_i)_{i \in I}$ of sets we only need to define the _essential upper_ limit of $(A_i)_{i\in I}$, denoted by $e \limsup_{i\in I} A_i$ which is determined by

$$e \limsup_{i\in I} A_i = e \inf_{i\in I} (e \sup_{j\in I(i)} A_j)$$

or by

$$\chi_{e \limsup_{i\in I} A_i} = e \limsup_{i\in I} \chi_{A_i}$$

It is obvious that, in case $I = \mathbb{N}$, essential convergence is convergence a.e.. Indeed these definitions above have been given in this way so as to obtain a measurable function which resembles $\lim_{i\in I} X_i$, which might be non measurable.

In the case of Banach space valued functions $(X_i)_{i\in I}$, we say that $(X_i)_{i\in I}$ converges stochastically (resp. essentially) to a measurable function X_∞ if $s\text{-}\limsup_{i\in I} \|X_i - X_\infty\| = 0$ (resp. $e\text{-}\limsup_{i\in I} \|X_i - X_\infty\| = 0$).

Definition VI.1.2 : Let $(\mathcal{F}_i)_{i\in I}$ be a stochastic basis. We say that it satisfies the Vitali condition V if for every set A in $\mathcal{F}$, for every adapted net $(A_i)_{i\in I}$ such that $A \subset e \limsup_{i\in I} A_i$ and for every $\varepsilon > 0$, there exist finitely many indices $i_1 \dots i_n \in I$ and pairwise disjoint sets $B_j \in \mathcal{F}_{i_j}$, $B_j \subset A_{i_j}$ $(j = 1,\dots,n)$ such that

$$P(A \setminus \bigcup_{j=1}^{n} B_j) \leq \varepsilon \quad .$$

Now we have the following well-known theorem of Krickeberg [1956].

Theorem VI.1.3 (Krickeberg) : Let $(X_i,\mathcal{F}_i)_{i\in I}$ be a real L^1-bounded martingale such that $(\mathcal{F}_i)_{i\in I}$ satisfies condition V. Then $(X_i)_{i\in I}$ converges essentially.

This is an old theorem known since 1956. A new proof is contained in theorem VI.1.5.1 below. With L^1-boundedness we mean of course

$$\sup_{i \in I} \int_\Omega |X_i| < \infty .$$

<u>Definitions VI.1.4</u> : Let $(X_i, F_i)_{i \in I}$ be an adapted net with values in a Banach space E. A function $\tau : \Omega \to I$ is called a <u>stopping time</u> if $\{\tau = i\} \in F_i$, for every $i \in I$. Denote by T the set of all finitely valued stopping times. For $\tau \in T$, X_τ and F_τ are defined as in the case $I = \mathbb{N}$.

It is only a technical matter to see that the Vitali condition V is equivalent with the following properties (i) or (ii) (see Millet and Sucheston [1980a]) :

(i) For every adapted net $(A_i)_{i \in I}$ and for every $\varepsilon > 0$ there is a $\tau \in T$ such that

$$P\,(e \limsup_{i \in I} A_i \,\Delta\, A_\tau) \leq \varepsilon$$

(Here Δ denotes the symmetric difference and $A_\tau = \bigcup_{i \in \tau(\Omega)} [A_i \cap \{\tau = i\}]$) .

(ii) For every real adapted sequence $(X_i, F_i)_{i \in I}$ and every sequence $(i_n)_{n \in \mathbb{N}}$ in I, there exists an increasing sequence $(\sigma_n)_{n \in \mathbb{N}}$ in T with $\sigma_n \in T(i_n)$ for every $n \in \mathbb{N}$ and such that

$$e \limsup_{i \in I} X_i = \lim_{n \to \infty} X_{\sigma_n} , \text{ a.e.}$$

(cf.I.3.5.4.).

<u>VI.1.5</u> : With this knowledge we can see that the basic results I.3.5.5, IV.1.4, IV.1.9 and their corollaries are valid under the Vitali condition V, if we replace "$\limsup_{i \in I}$" by "$e \limsup_{i \in I}$". This is already trivial for I.3.5.5. For IV.1.4 and IV.1.9 this is easy too if we can make use of theorem II.2.4.3 adapted to our case. This is indeed the case, as is easy to prove :

<u>Theorem VI.1.5.1 (Millet and Sucheston [1980a])</u> : Let E have (RNP) and suppose that $(X_i, F_i)_{i \in I}$ is an L^1_E-bounded martingale such that $(F_i)_{i \in I}$ satisfies V. Then $(X_i)_{i \in I}$ converges essentially.

Proof : We only need to show the convergence in probability of $(X_\tau)_{\tau \in T}$ to a measurable function X_∞, due to I.3.5.5, trivially modified as indicated above, using the Vitali condition V. By I.3.4 we see that this type of convergence is in fact induced by a complete metric. Hence it suffices to prove that $(X_\tau)_{\tau \in T}$ is Cauchy in probability, but of course for this we may work with increasing sequences. If $(\tau_n)_{n \in \mathbb{N}}$ is an increasing sequence in T, then $(X_{\tau_n}, F_{\tau_n})_{n \in \mathbb{N}}$ is an L^1_E-bounded martingale. So the theorem of Ionescu-Tulcea applies.This finishes the proof too. □

As remarked, this proof yields also a new proof of Krickeberg's theorem VI.1.3 above.

So using VI.1.5.1 it is a relatively easy matter to reprove the results IV.1.4 and IV.1.9 in this setting, using the Vitali condition V. Of course, since we could apply these inequalities in chapter V in order to prove convergence results of certain adapted sequences $(X_n, F_n)_{n \in \mathbb{N}}$, we can do the same here for adapted nets $(X_i, F_i)_{i \in I}$ where $(F_i)_{i \in I}$ satisfies V. We do not intend to write down every result that can be obtained in this way since we do not need them later on and also since it is easy to obtain them. We only mention the trivial application on uniform amarts, hence on amarts in case $E = \mathbb{R}$.

Theorem VI.1.6 (Egghe [1980d]) : Let E have (RNP) and let $(X_i, F_i)_{i \in I}$ be an L^1_E-bounded uniform amart such that $(F_i)_{i \in I}$ satisfies V. Then $(X_i)_{i \in I}$ converges essentially to a Bochner integrable function.
Here uniform amarts are defined in exactly the same way as in section V.1.

Remark VI.1.7 : Most convergence results mentioned above are valid without V but with essential convergence replaced by stochastic convergence; see e.g. Millet and Sucheston [1980a] , Millet and Sucheston [1979a] , Egghe [1980d] .

VI.2. Applications of amart convergence results

We have now collected enough material to give some applications of real amart convergence theory. They are not included here for their own sake but to indicate the interest of amarts. But it must be said that the results are very important and interesting in themselves.

VI.2.1. First application

This application is concerned with adapted nets indexed by $I = Z \times Z$ (or $\mathbb{N} \times \mathbb{N}$). Order I in the usual way : $\overline{m} = (m_1,m_2) \leq \overline{n} = (n_1,n_2)$ for $n_1,n_2,m_1,m_2 \in Z$ if $m_1 \leq n_1$ and $m_2 \leq n_2$. For a stochastic basis $(F_{\overline{n}})_{\overline{n} \in Z \times Z}$ we put

$$F^1_{(n_1,n_2)} = \sigma(\bigcup_p F_{(n_1,p)})$$

$$F^2_{(n_1,n_2)} = \sigma(\bigcup_p F_{(p,n_2)})$$

and

$$F = \sigma(\bigcup_{\overline{n} \in Z \times Z} F_{\overline{n}}) \quad .$$

Definitions VI.2.1.1 : An adapted net $(X_{\overline{n}},F_{\overline{n}})_{\overline{n} \in Z \times Z}$ is called a 1-martingale if it is a martingale and if

$$E^{F^1_{\overline{m}}} X_{\overline{n}} = X_{(m_1,n_2)}$$

whenever $\overline{m} = (m_1,m_2) \leq \overline{n} = (n_1,n_2)$. Analogously, $(X_{\overline{n}},F_{\overline{n}})_{\overline{n} \in Z \times Z}$ is a 2-martingale if it is a martingale and if

$$E^{F^2_{\overline{m}}} X_{\overline{n}} = X_{(n_1,m_2)}$$

whenever $\overline{m} \leq \overline{n}$.

The set of all finite stopping times w.r.t. $(F^1_{\overline{n}})_{\overline{n} \in Z \times Z}$ is denoted by T^1. Its elements are called simple 1-stopping times. An adapted net $(X_{\overline{n}},F_{\overline{n}})_{\overline{n} \in Z \times Z}$ is called a 1-amart if the net

$$(\int_\Omega X_\tau)_{\tau \in T^1}$$

converges. Analogous definitions for simple 2-stopping times and 2-amarts.

We say that an adapted net $(X_{\overline{n}}, F_{\overline{n}})_{\overline{n} \in Z \times Z}$ is <u>L log L-bounded</u> if

$$\sup_{\overline{n} \in Z \times Z} \int_{\Omega} |X_{\overline{n}}| \log^{+} |X_{\overline{n}}| < \infty$$

This property implies uniform integrability, as is easily seen.

The first application goes as follows :

<u>Theorem VI.2.1.2 (Millet and Sucheston [1979a])</u> : Let $(X_{\overline{n}}, F_{\overline{n}})_{\overline{n} \in Z \times Z}$ be an L log L-bounded real 1-martingale. Then $(X_{\tau})_{\tau \in T^1}$ converges in L^1 and so, $(X_{\overline{n}}, F_{\overline{n}})_{\overline{n} \in Z \times Z}$ is a 1-amart and $(X_{\overline{n}}, F_{\overline{n}})_{\overline{n} \in Z \times Z}$ converges a.e..

<u>Proof</u> : Since $(X_{\overline{n}}, F_{\overline{n}})_{\overline{n} \in Z \times Z}$ is a uniformly integrable martingale, there is $X_{\infty} \in L^1$ such that

$$X_{\overline{n}} = E^{F_{\overline{n}}} X_{\infty} \tag{1}$$

for every $\overline{n} \in Z \times Z$. Indeed, the reasoning of theorem II.2.2.1 (i) ⇒ (ii) can immediately be adapted here.

Hence $(X_{\overline{n}})_{\overline{n} \in Z \times Z}$ converges to X_{∞} in the L^1-sense. This follows from the fact that $\|\cdot\|_1$ is complete - hence that we can restrict ourselves to subsequences $(X_{\overline{n}_k})_{k \in \mathbb{N}}$ where $(\overline{n}_k)_{k \in \mathbb{N}}$ increases - and by theorem II.1.3. We shall show now that $(X_{\tau})_{\tau \in T^1}$ converges to X_{∞} in the L^1-sense. For this, put $\Phi(x) = x \log^{+} x$. It follows that

$$\int_{\Omega} \Phi(|X_{\infty}|) < \infty \tag{2}$$

using Fatou's lemma on a subsequence of Z × Z. Jensen's inequality implies now that $(\Phi(|X_{\infty} - X_{\overline{n}}|))_{\overline{n} \in Z \times Z}$ is uniformly integrable by an easy calculation, using (1) and (2). Hence

$$\lim_{\overline{n} \in Z \times Z} \int_{\Omega} \Phi(|X_{\overline{n}} - X_{\infty}|) = 0$$

Let $\varepsilon > 0$ and choose $\overline{m} = (m_1, m_2) \in Z \times Z$ such that

$$\sup_{\bar{n} \in (Z \times Z)(\bar{m})} \int_\Omega |X_{\bar{m}} - X_{\bar{n}}| \leq \varepsilon \tag{3}$$

and

$$\sup_{\bar{n} \in (Z \times Z)(\bar{m})} \int_\Omega \Phi(|X_{\bar{m}} - X_{\bar{n}}|) \leq \varepsilon \tag{4}$$

For $\bar{n} \in (Z \times Z)(\bar{m})$, put $Y_{\bar{n}} = X_{\bar{n}} - X_{\bar{m}}$ and let $\tau \in T^1(\bar{m})$; so $\bar{m} \leq \tau \leq \bar{m}'$ where $\bar{m}' = (m_1', m_2') \in Z \times Z$. If $a \in \mathbb{N}$ with $m_1 \leq a \leq m_1'$, put

$$S_a = \sup_{m_2 \leq b \leq m_2'} |Y_{(a,b)}|$$

$$G_a = \sigma(\bigcup_b F_{(a,b)})$$

Consider $(Y_{(a,b)}, G_a)_{a \in \mathbb{N} \cap [m_1, m_1']}$ for every fixed $b \in \mathbb{N} \cap [m_2, m_2']$.
If $a \leq a'$, $a, a' \in \mathbb{N} \cap [m_1, m_1']$, then, if $A \in G_a$:

$$\int_A Y_{(a',b)} = \int_A (X_{(a',b)} - X_{(m_1,m_2)})$$

$$= \int_A (X_{(a,b)} - X_{(m_1,m_2)}) = \int_A Y_{(a,b)}$$

since $(X_{\bar{n}}, F_{\bar{n}})_{\bar{n} \in Z \times Z}$ is a 1-martingale. Consequently $(S_a, G_a)_{a \in \mathbb{N} \cap [m_1, m_1']}$ is a submartingale and

$$\left|\int_\Omega Y_\tau\right| \leq \int_\Omega |Y_\tau| \leq \int_\Omega S_{m_1'} .$$

Doob's submartingale inequality implies on the submartingale $(|Y_{(m_1',b)}|, F_{(m_1',b)})_{b \in \mathbb{N} \cap [m_2, m_2']}$, for each $\lambda > 0$ that

$$\lambda P(S_{m'_1} > \lambda) \leq \int_{\{S_{m_1'} > \lambda\}} |Y_{(m_1', m_2')}|$$

This inequality is in fact easily seen (cf. Neveu [1975], IV-2-9, p.69): By the equivalent definition in III.2.1, we have that

$$X_\sigma \leq E^{\mathcal{F}_\sigma} X_\tau \quad ,$$

a.e. from every $\sigma \in T$ and $\tau \in T(\sigma)$. Hence, for every $\sigma \in T$ and $n \in \mathbb{N}$ we have on $\{\sigma \leq n\}$ that

$$X_\sigma \leq E^{\mathcal{F}_\sigma} X_n$$

Define, for every $\lambda > 0$

$$\sigma_\lambda = \inf \{k \in \mathbb{N}(m_2) \,\|\, |Y_{(m_1',k)}| > \lambda\}$$

Then, on $\{\sigma_\lambda \leq m_2'\}$,

$$X_{\sigma_\lambda} \leq E^{\mathcal{F}_{\sigma_\lambda}} |Y_{(m_1',m_2')}|$$

Hence

$$\int_{\{\sigma_\lambda \leq m_2'\}} X_{\sigma_\lambda} \leq \int_{\{\sigma_\lambda \leq m_2'\}} |Y_{(m_1',m_2')}| \quad .$$

Since $\{\sigma_\lambda \leq m_2'\} = \{\sup_{m_2 \leq k \leq m_2'} |Y_{(m_1',k)}| > \lambda\} = \{S_{m_1'} > \lambda\}$ and since

$\int_{\{\sigma_\lambda \leq m_2'\}} X_{\sigma_\lambda} \geq \lambda P(\sigma_\lambda \leq m_2') = \lambda P(S_{m_1'} > \lambda)$, we have

$$\lambda P(S_{m_1'} > \lambda) \leq \int_{\{S_{m_1'} > \lambda\}} |Y_{(m_1',m_2')}| \quad ,$$

proving the above inequality.

So, for $\eta \in \,]0,1[$:

$$\int_\Omega S_{m_1'} = \int_0^\infty P(S_{m_1'} > \lambda)d\lambda \qquad \text{(using Fubini)}$$

$$\leq \eta + \int_\eta^\infty \frac{1}{\lambda}[\int_\Omega \chi_{\{S_{m_1'} > \lambda\}} |Y_{(m_1',m_2')}| dP]\, d\lambda$$

$$= \eta + \int_\Omega \frac{1}{\lambda}[\int_\eta^{S_{m_1'}} |Y_{(m_1',m_2')}| d\lambda]\, \chi_{\{S_{m_1'} > \lambda\}} dP$$

$$\leq \eta + \int_\Omega |Y_{(m_1',m_2')}| (\log S_{m_1'} - \log \eta)\, \chi_{\{S_{m_1'} > \lambda\}} dP$$

Now use that $a \log^+ b \leq a \log^+ a + \frac{b}{e}$. Then

$$\int_\Omega S_{m_1'} \leq \eta + \int_\Omega |Y_{(m_1',m_2')}| \log^+ |Y_{(m_1',m_2')}|$$

$$+ \frac{\int_\Omega S_{m_1'}}{e} + |\log \eta| \int_\Omega |Y_{(m_1',m_2')}|$$

$$\leq \eta + \varepsilon + \frac{\int_\Omega S_{m_1'}}{e} + \varepsilon\, |\log \eta| \;,$$

due to (3) and (4). Hence, given $\delta > 0$, if ε and η are chosen according to $0 < \varepsilon < \delta$, $0 < \eta < 1$ and $\eta + \varepsilon + \varepsilon\, |\log \eta| \leq \frac{\delta(e-1)}{e}$, then we see that

$$\int_\Omega |X_\tau - X_\infty| \leq \int_\Omega |X_\tau - X_{\bar{m}}| + \int_\Omega |X_\infty - X_{\bar{m}}|$$

$$= \int_\Omega |Y_\tau| + \int_\Omega |X_\infty - X_{\bar{m}}|$$

$$\leq \int_\Omega S_{m_1'} + \varepsilon$$

$$\leq \delta + \varepsilon \leq 2\delta \;,$$

if $\tau \in T^1$ is chosen to be larger than $\bar{m}$ (dependent on ε, hence on δ). So $(X_\tau)_{\tau \in T^1}$ converges in L^1. Now $(\mathcal{F}^1_\tau)_{\tau \in T^1}$ is totally ordered, hence it satisfies the Vitali condition V (see e.g. Neveu [1975] , p.100). There are now two ways to finish the above proof.

1° From the L^1-convergence of $(X_\tau)_{\tau \in T^1}$ the convergence in probability follows, and hence the essential convergence due to the Vitali condition V and the extension of I.3.5.5 remarked in VI.1.5. Hence $(X_{\overline{n}})_{\overline{n} \in Z \times Z}$ converges a.e..

2° Or we can conclude that $(X_{\overline{n}}, \mathcal{F}^1_{\overline{n}})_{\overline{n} \in Z \times Z}$ is an amart. So by the Vitali condition V and by theorem VI.1.6, applied for $E = \mathbb{R}$, the proof is finished.

The two methods depend on essentially the same stopping time - i.e. amart - arguments. □

We mention the following important special case, observed earlier by Cairoli [1970] .

Corollary VI.2.1.3 (Cairoli) : Let $(\mathcal{F}_{\overline{n}})_{\overline{n} \in Z \times Z}$ be a stochastic basis such that $\mathcal{F}^1_{\overline{n}}$ and $\mathcal{F}^2_{\overline{n}}$ are conditionally independent w.r.t. $\mathcal{F}_{\overline{n}}$. The every L log L-bounded martingale $(X_{\overline{n}}, \mathcal{F}_{\overline{n}})_{\overline{n} \in Z \times Z}$ converges a.e..

VI.2.2. Second application

Here we intend to prove a result of Gabriel [1977a] on L^1-bounded sums of independent measurable functions. The proof we present is also from Millet and Sucheston [1981b].

Theorem VI.2.2.1 (Gabriel) : Let $(Y_{\bar{n}})_{\bar{n} \in Z \times Z}$ be a family of real independent measurable functions and put

$$X_{\bar{n}} = \sum_{\substack{\bar{m} \leqslant \bar{n} \\ \bar{m} \in Z \times Z}} Y_{\bar{m}}$$

for every $\bar{n} \in Z \times Z$, and suppose that $(X_{\bar{n}})_{\bar{n} \in Z \times Z}$ is L^1-bounded. Then $(X_\tau)_{\tau \in T^1}$ converges in L^1 and so $(X_{\bar{n}}, F_{\bar{n}})_{\bar{n} \in Z \times Z}$ is a 1-amart, and $(X_{\bar{n}})_{\bar{n} \in Z \times Z}$ converges a.e..

Proof : We use the inequality proved by Gabriel [1977a] : there exists a constant K such that

$$\int_\Omega \sup_{\bar{n} \in Z \times Z} |X_{\bar{n}}| \leqslant K \sup_{\bar{n} \in Z \times Z} \int_\Omega |X_{\bar{n}}| ,$$

where K is independent of $(Y_{\bar{m}})_{\bar{m} \in Z \times Z}$. The interested reader is referred to Gabriel [1977a] for a proof - here we only show the amart-application. So $(X_{\bar{n}})_{\bar{n} \in Z \times Z}$ is uniformly integrable. It is easy to check that $(X_{\bar{n}}, F_{\bar{n}})_{\bar{n} \in Z \times Z}$ is a martingale, where $F_{\bar{n}}$ denotes $\sigma(\bigcup_{\bar{m} \leqslant \bar{n}} \sigma(Y_{\bar{m}}))$. The same argument as in the proof of theorem VI.2.1.2 now shows that there is a function $X_\infty \in L^1$ such that

$$\lim_{\bar{n} \in Z \times Z} \int_\Omega |X_{\bar{n}} - X_\infty| = 0$$

Hence, for $\varepsilon > 0$, let $\bar{m} = (m_1, m_2) \in Z \times Z$ be such that

$$\sup_{\bar{n} \in (Z \times Z)(\bar{m})} \int_\Omega |X_{\bar{n}} - X_{\bar{m}}| \leqslant \frac{\varepsilon}{3K}$$

Choose $\tau \in T^1(\overline{m})$. Then

$$\int_\Omega |X_\tau - X_{\overline{m}}| \leq \sum_{\overline{n}\in(Z\times Z)(\overline{m})} \int_{\{\tau=\overline{n}\}} \Big|\sum_{k_1\leq n_1}\ \sum_{m_2<k_2\leq n_2} Y_{\overline{k}}\Big|$$

$$+ \sum_{\overline{n}\in(Z\times Z)(\overline{m})} \int_{\{\tau=\overline{n}\}} \Big|\sum_{m_1<k_1\leq n_1}\ \sum_{k_2\leq m_2} Y_{\overline{k}}\Big|$$

$$\leq \int_\Omega \sup_{\overline{n}\in(Z\times Z)(\overline{m})} \Big|\sum_{k_1\leq n_1}\ \sum_{m_2<k_2\leq n_2} Y_{\overline{k}}\Big|$$

$$+ \int_\Omega \sup_{\overline{n}\in(Z\times Z)(\overline{m})} \Big|\sum_{m_1<k_1\leq n_1}\ \sum_{k_2\leq m_2} Y_{\overline{k}}\Big|$$

$$\leq K \sup_{\overline{n}\in(Z\times Z)(\overline{m})} \int_\Omega |X_{(n_1,n_2)} - X_{(n_1,m_2)}|$$

$$+ K \sup_{\overline{n}\in(Z\times Z)(\overline{m})} \int_\Omega |X_{(n_1,m_2)} - X_{(m_1,m_2)}|$$

$$\leq \varepsilon$$

So $(X_\tau)_{\tau\in T^1}$ converges (to X_∞) in the L^1-sense. The rest of the proof is of course identical with the end of the proof of theorem VI.2.1.2. □

An analogous result is proved in Edgar and Sucheston [1981], p.968-969, also using an amart method.

VI.2.3. Third application

This application is more theoretical in nature. Theorem VI.1.5.1 states - as a special case - that real L^1-bounded martingales $(X_i,F_i)_{i\in I}$ such that $(F_i)_{i\in I}$ satisfies V, converge essentially. Theorem VI.1.6 does the same for real amarts. The question is : Is V implied by these convergence properties? The answer is no for the first result : In Millet and Sucheston [1979b], Millet and Sucheston prove that L^1-bounded real martingales do already converge essentially under a condition "SV" which they prove to be strictly weaker than V.

For the second result, the answer is yes as is proved by Astbury [1978], yielding a theoretical application of amarts. The key is found in :

Theorem VI.2.3.1 (Astbury) : The stochastic basis $(F_i)_{i\in I}$ satisfies V if and only if every real potential $(Z_i,F_i)_{i\in I}$ essentially converges (to zero).

Here - as in the case I = $\mathbb{N}$ - $(Z_i,F_i)_{i\in I}$ is called a potential if $\lim_{\tau\in T}\int_\Omega Z_\tau = 0$. We see as in theorem V.1.4 with the explanation given in VI.1.5 that $\lim_{\tau\in T}\int_\Omega |Z_\tau| = 0$ if $(F_i)_{i\in I}$ satisfies V (in fact we do not have to suppose V - see the notes and remarks VI.3.1).

Proof : Sufficiency

Let $A \in F$ and for every $i \in I$, $A_i \in F_i$ such that

$$A \subset e\limsup_{i\in I} A_i$$

Define

$$\Phi = \{\{(i_j,B_j)\|\, j=1,\ldots,n\}\|\, n\in\mathbb{N};\ i_j\in I,\ B_j\in F_{i_j},\ B_j\subset A_{i_j} \text{ for every } j=1,\ldots,n;\ B_i\cap B_j=\phi \text{ for } i\neq j\}\ .$$

We adopt the notation : if $\varphi\in\Phi$,

$$\cup\varphi = \bigcup_{(i,B)\in\varphi} B$$

Define inductively sequences $\varphi_k \in \Phi$ and $r_k \in \mathbb{R}$:

$$\varphi_0 = \phi \quad \text{and} \quad r_0 = \sup_{\varphi\in\Phi,\varphi\supset\varphi_0} P(\cup(\varphi\setminus\varphi_0))$$

φ_k is any element of Φ such that $\varphi_k \supset \varphi_{k-1}$, $P(\cup(\varphi_k\setminus\varphi_{k-1})) \geq \frac{r_{k-1}}{2}$ and

$$r_k = \sup_{\varphi\in\Phi,\varphi\supset\varphi_k} P(\cup(\varphi\setminus\varphi_k))$$

Of course, φ_k exists; this follows from the definition of r_{k-1}. For $\varphi \supset \varphi_k$ we have

$$r_{k-1} \geq P(\cup(\varphi\setminus\varphi_{k-1}))$$

$$= P(\cup(\varphi\setminus\varphi_k)) + P(\cup(\varphi_k\setminus\varphi_{k-1}))$$

$$\geq \varphi(\cup(\varphi\setminus\varphi_k)) + \frac{r_{k-1}}{2}$$

So $\frac{r_{k-1}}{2} \geq r_k$ for every $k \in \mathbb{N}$. Hence $r_k \leq (\frac{1}{2})^k$, for every $k \in \mathbb{N}$. Put

$$\varphi_\infty = \bigcup_{k=0}^{\infty} \varphi_k$$

and write naturally $\cup\varphi_\infty$ for $\bigcup_{(i,B)\in\varphi_\infty} B$. For every $i \in I$, define

$$C_i = A_i \setminus \bigcup_{\substack{(j,B)\in\varphi_\infty \\ j\leq i}} B$$

$$Z_i = \chi_{C_i}$$

and $F_i = \sigma(\bigcup_{\substack{j\leq i \\ j\in I}} \sigma(Z_j))$. $(Z_i,F_i)_{i\in I}$ is a potential : Fix $k \in \mathbb{N}$ and choose $i \in I$ such that $i \geq j$ for every $(j,B) \in \varphi_k$. Choose $\tau \in T(i)$, and let $\tau(\Omega) = \{i_1,\ldots,i_n\}$. Define

$$\varphi = \varphi_k \cup \{(i_j, \{\tau = i_j\} \cap C_{i_j}) \| j = 1, \ldots, n\}$$

Obviously $\varphi_k \subset \varphi \in \Phi$. Now $Z_\tau = \chi_{\cup(\varphi \setminus \varphi_k)}$ as is easily seen. So

$$\int_\Omega |Z_\tau| = P(\cup(\varphi \setminus \varphi_k)) \leq (\tfrac{1}{2})^k$$

Now, since $(Z_i, F_i)_{i \in I}$ is a potential, the hypothesis ensures that $(Z_i)_{i \in I}$ essentially converges to zero. Hence

$$\phi = e \limsup_{i \in I} C_i$$

$$\supset e \limsup_{i \in I} (A_i \setminus \cup\varphi_\infty)$$

$$= e \limsup_{i \in I} A_i \setminus \cup\varphi_\infty$$

$$\supset A \setminus \cup\varphi_\infty \quad .$$

So, for every $\varepsilon > 0$, we can find a subclass of φ_∞ which satisfies the properties required in the definition of the Vitali condition V.

Necessity

Let $(Z_i, F_i)_{i \in I}$ be a potential. Let $\lambda > 0$ and $A = e \limsup_{i \in I} \{|Z_i| > \lambda\}$. For each $\varepsilon > 0$, choose $i \in I$ such that $\tau \in T(i)$ implies $\int_\Omega |Z_\tau| < \varepsilon$. Define

$$A_j \begin{cases} = \{|Z_j| > \lambda\} & \text{for } j \in I(i) \\ = \phi & \text{elsewhere} \end{cases}$$

So $A = e \limsup_{i \in I} A_i$. Hence, using the Vitali condition V by hypothesis, there are $i_1, \ldots, i_n \in I$ and $B_j \in F_j$ with $j = 1, \ldots, n$ such that

$$B_j \subset A_{i_j} \qquad \text{for } j = 1,\dots,n$$

$$B_i \cap B_j = 0 \qquad \text{if } i \neq j$$

and such that

$$P(A \setminus \bigcup_{j=1}^{n} B_j) \leqslant \varepsilon$$

Take $i_{n+1} \in I(i_j)$ for every $j = 1,\dots,n$ and define

$$\sigma \begin{cases} = i_j & \text{on } B_j, \text{ for } j = 1,\dots,n \\ = i_{n+1} & \text{elsewhere} \end{cases}$$

So $\sigma \in T(i)$ and hence

$$\varepsilon \geqslant \int_{\Omega} |Z_\tau| \geqslant \sum_{j=1}^{n} \int_{B_j} |Z_{i_j}|$$

$$\geqslant \lambda\, P(\bigcup_{j=1}^{n} B_j)$$

$$\geqslant \lambda\, (P(A) - \varepsilon)$$

For ε small, we so have $P(A) = 0$ for all $\lambda > 0$. So $(Z_i)_{i \in I}$ essentially converges to zero. □

Theorem VI.2.3.1 is already a characterization of V in terms of amarts : potentials are amarts. A characterization of V which involves the class of all L^1-bounded amarts can easily be deduced from it :

Theorem VI.2.3.2 (Astbury) : The stochastic basis satisfies V if and only if every L^1-bounded real amart essentially converges.

Proof : Sufficiency

Follows immediately from the previous theorem.

Necessity

Again by theorem V.1.4 for $E = \mathbb{R}$, extended to our case by the explanation given in VI.1.5, we see that the L^1-bounded amart $(X_i,F_i)_{i\in I}$ decomposes as

$$X_i = Y_i + Z_i$$

where $(Y_i,F_i)_{i\in I}$ is a martingale and $(Z_i,F_i)_{i\in I}$ is a potential. $(Z_i)_{i\in I}$ essentially converges by theorem VI.2.3.1. Now

$$\sup_{i\in I} \int_\Omega |Y_i| \leq \sup_{i\in I} \int |X_i|$$

is easily seen from the proof of this Riesz-decomposition theorem. Hence, from theorem VI.1.5.1 and by the Vitali condition V, $(Y_i)_{i\in I}$ essentially converges. Hence also $(X_i)_{i\in I}$. □

VI.3. Notes and remarks

VI.3.1. If we have a potential $(Z_i,F_i)_{i\in I}$ then we always have that

$$\lim_{\tau\in T} \int_\Omega |Z_\tau| = 0$$

even if $(F_i)_{i\in I}$ does not satisfy V. Indeed, theorem V.1.4 is extended to general index sets in a trivial way and without using V, since in corollary IV.1.10 we only use the integral form of inequality (2) in theorem IV.1.4 instead of (2'). Indeed, in theorem IV.1.4, inequality (2) is true - in the general index setting - without V as is easily seen. Inequality (2') follows from (2) by applying theorem VI.1.5.1 to the martingale; it is here that V is used.
Thus we see that this remark is also true for uniform potentials in general Banach spaces!

VI.3.2. As proved above, L^1-bounded martingales $(X_i,F_i)_{i\in I}$ where $(F_i)_{i\in I}$ satisfies the Vitali condition V essentially converge.
Although V is not necessary, by VI.2.3, it cannot be skipped

as we mentioned earlier. If we do not suppose V but if $I = \mathbb{N} \times \mathbb{N}$ and $(X_i)_{i \in I}$ is an L log L-bounded 1-amart then we also obtain essential convergence by theorem VI.2.1.2. This is f.i. the case in Cairoli's result VI.2.1.3. If we want to suppose neither V nor the L log L-boundedness condition, there is another possiblity of obtaining essential convergence due to Walsh [1979] : Let $I = Z \times Z$ and $(X_{\bar{n}}, F_{\bar{n}})_{\bar{n} \in Z \times Z}$ be a martingale which is L^1-bounded. Suppose in addition that for every $\bar{m} = (m_1, m_2) \leq \bar{n} = (n_1, n_2)$ in $Z \times Z$

$$E^{F^1_{\bar{m}} \wedge F^2_{\bar{m}}}(X_{\bar{n}} - X_{(n_1,m_2)} - X_{(m_1,n_2)} + X_{\bar{m}}) = 0$$

This is called a <u>strong martingale</u>. We have

<u>Theorem VI.3.2.1 (Walsh)</u> : Let $(X_{\bar{n}}, F_{\bar{n}})_{\bar{n} \in Z \times Z}$ be an L^1-bounded strong real martingale. Then $(X_{\bar{n}})_{\bar{n} \in Z \times Z}$ converges essentially (i.e., a.e. here).

For related results we refer the reader to Edgar [1982a] .

VI.3.3. In Millet and Sucheston [1980a] and other papers an "<u>ordered Vitali condition</u>" V' is studied. The definition of V' is derived from that of V by requiring $i_1 \leq \ldots \leq i_n$ in definition VI.1.2. In other words by requiring in VI.1.4 (i) τ to have a totally ordered range $\tau(\Omega)$.

Such elements of T are called <u>ordered stopping times</u> and the set of all finite ordered stopping times is denoted by T'. Another equivalent formulation of V' is obtained by replacing T by T' in VI.1.4.(ii). For instance V' is used to show that an adapted net $(X_i, F_i)_{i \in I}$ such that $(X_\tau)_{\tau \in T'}$ converges stochastically to X_∞ is essentially convergent, provided $(F_i)_{i \in I}$ satisfies V'. Other Vitali conditions are studied - see e.g. Millet and Sucheston [1980a] and [1979a] , ...

VI.3.4. An application of amarts in ergodic theory was apparently presented by Korzeniowski [1978b] with the intention of proving the Banach valued ergodic theorem of Mourier by using amart convergence results. This is a nice idea but - as L. Sucheston remarked in his review [1980] - the argument has an irreparable error.

VI.3.5. Another application of amarts is of a didactical nature, as was remarked by Edgar and Sucheston [1976a].
Amarts can be introduced very early in probability courses. For the real convergence theorem of amarts, we do not need conditional expectations, nor the Radon-Nikodym theorem. For the martingale convergence theorem, we do.
Moreover in our approach, we proved the real amart convergence theorem based on the uniform amart convergence theorem and hence on the martingale convergence theorem. A direct proof is seen as follows (Edgar and Sucheston [1976a]) : Let $(X_n, F_n)_{n\in\mathbb{N}}$ be a real L^1-bounded amart. Then it is of class (B); this is proved in V,2.8, using martingales, but it can be proved directly, using the lattice properties of amarts (cor. V.2.9 - proved directly : see Edgar and Sucheston [1976a], proposition 1.3 and cor.1.4; this is only a technical matter). From lemma II.1.5 we now have that :

$$\sup_{n\in\mathbb{N}} |X_n| < \infty ,$$

a.e.. So for $\lambda > 0$ large, $P(\sup_{n\in\mathbb{N}} |X_n| > \lambda)$ is arbitrarily small. So $X_n = -\lambda \vee X_n \wedge \lambda$ on a set with measure arbitrarily close to 1, for $\lambda > 0$ large. Furthermore, again using the lattice properties of amarts, $(-\lambda \vee X_n \wedge \lambda, F_n)_{n\in\mathbb{N}}$ is an amart such that, where $Y_n = -\lambda \vee X_n \wedge \lambda$

$$\int_\Omega \sup_{n\in\mathbb{N}} |Y_n| < \infty$$

For this type of amart the convergence a.e. is now seen as follows. By corollary I.3.5.2, let $(\sigma_n)_{n\in\mathbb{N}}$ and $(\tau_n)_{n\in\mathbb{N}}$ be two sequences in T, increasing to $+\infty$, such that $\lim_{n\to\infty} Y_{\tau_n} = \limsup_{n\in\mathbb{N}} Y_n$ and $\lim_{n\to\infty} Y_{\sigma_n} = \liminf_{n\in\mathbb{N}} Y_n$. By the dominated convergence theorem of Lebesgue, applicable on $(Y_n)_{n\in\mathbb{N}}$, we have :

$$\int_\Omega (\limsup_{n\in\mathbb{N}} Y_n - \liminf_{n\in\mathbb{N}} Y_n) = \lim_{n\to\infty}\int_\Omega (Y_{\tau_n} - Y_{\sigma_n}) = 0 ,$$

since $(Y_n,F_n)_{n\in\mathbb{N}}$ is an amart. So $\limsup_{n\in\mathbb{N}} Y_n = \liminf_{n\in\mathbb{N}} Y_n$, a.e., and hence $(Y_n)_{n\in\mathbb{N}}$ converges a.e.. □

From this theorem we can derive the Radon-Nikodym-theorem, using a method analogous to the one used in the proof of theorem II.2.2.1, (iii) ⇒ (i). As is well known, and as is in fact done in theorem I.2.2.1, we can prove the existence of conditional expectations using the Radon-Nikodym-theorem.

We so avoid the difficulty of deriving the Radon-Nikodym theorem from the martingale convergence theorem, in which conditional expectations must be defined, without using the Radon-Nikodym theorem.

Chapter VII : DISADVANTAGES OF AMARTS, CONVERGENCE OF GENERALIZED MARTINGALES IN BANACH SPACES - THE POINTWISE WAY

In chapter VI we have seen that amarts are good and fairly extensive generalizations of martingales. We have seen that there are nice applications of amart theory and also that some proofs are even simpler in the amart setting.

When working in Banach spaces E with (RNP), the most powerful results are strong a.e. convergence results. We have seen in section V.1 that L_E^1-bounded uniform amarts do converge strongly a.e.. But amarts did not do such a fine job! Even in separable duals of (RNP) spaces we have only weak convergence a.e. and this only when the amart is of class (B)! So to obtain strong convergence a.e. of L_E^1-bounded amarts must be very difficult. Indeed - as was first observed by A. Bellow in Bellow [1976] - this happens only when dim E < ∞, a very negative result indeed : it thus happens only in case the amart is a uniform amart. So we may say it never happens in case the amart is not uniform. Some other negative properties of amarts as well as the ones mentioned above will be proved in the first section. These negative results are mainly due to Bellow, Edgar and Sucheston, and Egghe.

Hence in trying to extend martingales beyond uniform amarts so that some strong convergence properties are valid we have to think in a different way. For uniform amarts, the differences

$$E^{F_\sigma} X_\tau - X_\sigma$$

$\sigma \in T$, $\tau \in T(\sigma)$ converge to zero in L_E^1-sense. For amarts, they do so in the $\|\cdot\|_{Pe}$-sense. Why not try generalizations in the pointwise way? In the next sections we shall develop the theory of such adapted sequences : pramarts, mils, games which become fairer with time, weak pramarts, weak mils, ... The main point in these sections is that

(i) When $E = \mathbb{R}$, they all generalize amarts but keep the convergence properties of real amarts without an additional assumption, except for games which become fairer with time which are in fact too general!

(ii) In general Banach spaces these new notions are amarts if and only if $\dim E < \infty$, which shows we are heading in the right direction. Indeed (ii) is a necessity in order to be able to expect good convergence properties (since amarts lack them!).

(iii) It will be checked that in Banach spaces pramarts and mils do have good strong convergence properties while weak pramarts and weak mils do have good weak convergence properties.

VII.1. Disadvantages of amarts

One group of negative results concerning vector-valued amarts are found in the next theorem. The results are from A. Bellow [1976]. However the proof is from Egghe [1980b] and Egghe [1982a] which gives a more unified proof of the theorem and which yields some sharper results; see also the notes and remarks concerning the case E is a Fréchet space.

Theorem VII.1.1 (A. Bellow) : The following assertions on a Banach space E are equivalent :

(i) $\dim E < \infty$.

(ii) Every Pettis uniformly integrable amart is L^1_E-convergent.

(iii) Every Pettis- convergent (and uniformly bounded) amart is L^1_E-convergent.

(iv) Every amart is a uniform amart.

(v) Every Pettis-convergent (and uniformly bounded) amart is a uniform amart.

(vi) Every L^1_E-bounded amart is of class (B).

(vii) For every uniformly integrable (or uniformly bounded) amart $(X_n,F_n)_{n\in\mathbb{N}}$, the real adapted sequence $(\|X_n\|,F_n)_{n\in\mathbb{N}}$ is an amart.

(viii) Every L^1_E-bounded amart is converging strongly a.e..

Proof : (i) implies all the other assertions, due to the fact that $\|\cdot\|_{Pe}$ is equivalent to $\|\cdot\|_1$, by the results of section V.1 and by the fact that projections of amarts are amarts. Now (ii) ⇒ (iii), (iv) ⇒ (v) are trivial and (v) ⇒ (iii) follows from section V.1. So we only show (iii) ⇒ (i), (vi) ⇒ (i), (vii) ⇒ (i), (viii) ⇒ (i).

(iii) ⇒ (i)

Suppose dim E = ∞ . Then by the well-known and deep theorem of Dvoretzky and Rogers [1950] there is a sequence $(x_n)_{n\in\mathbb{N}}$ in the unit ball of E such that $\sum_{n=1}^{\infty} \|x_n\| = \infty$, but such that $\Sigma\, x_n$ converges unconditionally.

1) Let $n_1 \in \mathbb{N}$ be the smallest number such that

$$\alpha_1 = \sum_{k=1}^{n_1} \|x_k\| \geq 1$$

Then $\sum_{k=1}^{n_1} \|y_k\| = 1$, with $y_k = \frac{x_k}{\alpha_1}$, $(k = 1,\ldots,n_1)$.

2) Let $n_2 \in \mathbb{N}(n_1)$ be the smallest number such that

$$\alpha_2 = \sum_{k=n_1+1}^{n_2} \|x_k\| \geq \|y_1\|$$

Then $\sum_{k=n_1+1}^{n_2} \|y_k\| = \|y_1\|$, with $y_k = \frac{\|y_1\|}{\alpha_2} x_k$, $(k = n_1+1,\ldots,n_2)$.

...

Let $n_{n_1+1} \in \mathbb{N}(n_{n_1})$ be the smallest number such that

$$\alpha_{n_1+1} = \sum_{k=n_{n_1}+1}^{n_{n_1+1}} \|x_k\| \geq \|y_{n_1}\|$$

Then $\sum_{k=n_{n_1}+1}^{n_{n_1+1}} \|y_k\| = \|y_{n_1}\|$, with

$$y_k = \frac{x_k}{\alpha_{n_1+1}} \|y_{n_1}\|, \ (k = n_{n_1}+1,\ldots,n_{n_1+1}) \quad .$$

3) Let $n_{n_1+2} \in \mathbb{N}(n_{n_1+1})$ be the smallest number such that

$$\alpha_{n_1+2} = \sum_{k=n_{n_1+1}+1}^{n_{n_1+2}} \|x_k\| \geqslant \|y_{n_1+1}\|$$

Then $\sum_{k=n_{n_1+1}+1}^{n_{n_1+2}} \|y_k\| = \|y_{n_1+1}\|$, with

$$y_k = \frac{x_k}{\alpha_{n_1+2}} \|y_{n_1+1}\|, \ (k = n_{n_1+1}+1,\ldots,n_{n_1+2})$$

and so on. The inductive step should be clear. Obviously $\sum_{k=1}^{\infty} \|y_k\| = \infty$. Also, since $\|y_n\| \leqslant \|x_n\|$ for every $n \in \mathbb{N}$ and since $\Sigma\, x_n$ converges unconditionally, the same is true for $\Sigma\, y_n$. This is an elementary result on unconditionalconvergent series; see e.g. Pietsch [1972], p.23-26. Now the sequence $(y_n)_{n \in \mathbb{N}}$ satisfies :

$$\sum_{k=1}^{n_1} \|y_k\| = 1 \qquad (\alpha_1)$$

$$\sum_{k=n_1+1}^{n_2} \|y_k\| = \|y_1\| \qquad (\beta_1)$$

$$\vdots \qquad\qquad \vdots$$

$$\sum_{k=n_{n_1}+1}^{n_{n_1+1}} \|y_k\| = \|y_{n_1}\| \qquad (\beta_{n_1})$$

$$\sum_{k=n_{n_1+1}+1}^{n_{n_1+2}} \|y_k\| = \|y_{n_1+1}\| \qquad (\gamma_1)$$

$$\vdots \qquad\qquad \vdots$$

and so on. The formulas (α_1); $(\beta_1),\ldots,(\beta_{n_1})$; $(\gamma_1),\ldots$ indicate also the way to divide $[0,1)$ into intervals of the same form. As a matter of fact, define

$$\pi_1 = \{A_1,\ldots,A_{n_1}\} \quad ,$$

where $A_1 = [0,\|y_1\|)$, $A_2 = [\|y_1\|,\|y_1\| + \|y_2\|),\ldots$

$$\pi_2 = \{A_{n_1+1},\ldots,A_{n_{n_1+1}}\} \geqslant \pi_1 \quad ,$$

where $A_{n_1+1} = [0,\|y_{n_1+1}\|)$,

$$A_{n_1+2} = [\|y_{n_1+1}\|,\|y_{n_1+1}\| + \|y_{n_1+2}\|), \ \ldots$$

In general : $\pi_n \geqslant \pi_m$ if and only if $n \in \mathbb{N}(m)$.
Define

$$X_1 = \sum_{k=1}^{n_1} \frac{y_k}{\|y_k\|}\chi_{A_k}$$

$$X_2 = \sum_{k=n_1+1}^{n_{n_1+1}} \frac{y_k}{\|y_k\|}\chi_{A_k}$$

$$X_3 = \sum_{k=n_{n_1+1}+1}^{n_{n_{n_1+1}+1}} \frac{y_k}{\|y_k\|} \chi_{A_k}$$

and so on. Put $F_n = \sigma(\pi_n)$ for every $n \in \mathbb{N}$. $(X_n, F_n)_{n\in\mathbb{N}}$ is now a uniformly bounded amart, $\|\cdot\|_{Pe}$-convergent to 0. We even prove that

$$\lim_{\tau \in T} \|X_\tau\|_{Pe} = 0$$

Indeed

$$\sup_{\substack{\|x'\|\leq 1 \\ x'\in E'}} \int_0^1 |x'(X_\tau)|$$

$$= \sup_{\substack{\|x'\|\leq 1 \\ x'\in E'}} \sum_{k=\min\tau}^{\max\tau} \int_{\{\tau=k\}} |x'(X_k)|$$

$$= \sup_{\substack{\|x'\|\leq 1 \\ x'\in E'}} \sum_{k=\min\tau}^{\max\tau} \sum_{j\in J_k} \frac{|x'(y_j)|}{\|y_j\|} P(A_j) , \qquad (1)$$

where J_k is the set of indices for which

$$\{\tau = k\} = \bigcup_{j\in J_k} A_j$$

Now $P(A_j) = \|y_j\|$, by construction and also $k \neq k'$ implies $J_k \cap J_{k'} = \phi$. Hence the sum in (1) is one where every index j appears not more than once. Furthermore the lowest j in this sum can be as high as we wish by taking τ large enough in T.

So we have that $\lim_{\tau\in T} \|X_\tau\|_{Pe} = 0$. Here we use the elementary fact that if Σz_n converges unconditionally then

$$\lim_{n\to\infty} \sup_{\|x'\|\leq 1} \sum_{k=n}^{\infty} |x'(z_k)| = 0 ,$$

(and conversely; see e.g. Pietsch [1972], p.25).
Now $(X_n)_{n\in \mathbb{N}}$ is not L^1_E-convergent. If it were convergent, its limit had to be zero also. But

$$\int_0^1 \|X_n\| = \sum_k \|y_k\| = 1$$

for every $n \in \mathbb{N}$, where in $\sum_k$, k varies over those indices appearing in the sum defining X_n. This ends this (main) part of the proof.

<u>(vi) ⇒ (i)</u>

Supposing dim E = ∞ we have again using the theorem of Dvoretzky and Rogers, the existence of a sequence $(x_n)_{n\in \mathbb{N}}$ such that $\sum x_n$ is unconditionally convergent but where $\sum_{n=1}^{\infty} \|x_n\| = \infty$. Choose any strictly decreasing sequence $(\gamma_n)_{n\in \mathbb{N}}$ in $[0,\frac{1}{2}]$, converging to zero and put $\delta_n = \gamma_n - \gamma_{n+1}$. Define, for every $n \in \mathbb{N}$:

$$A_o^n = [0,\gamma_{n+1}),\ B^n = [\gamma_{n+1},\gamma_n),\ A_1^n = [\gamma_n,1)$$

and

$$\pi_1 = \{A_0^1, B^1, A_1^1\}$$

$$\pi_2 = \pi_1 \vee \{A_0^2, B^2, A_1^2\}$$

$$\cdots$$

$$\pi_n = \pi_{n-1} \vee \{A_0^n, B^n, A_1^n\}$$

$$\cdots$$

Put

$$X_n = \frac{x_n}{\delta_n} \chi_{B^n}$$

and $F_n = \sigma(\pi_n)$, for every $n \in \mathbb{N}$. For every $\tau \in T$ we have :

$$\int_0^1 X_\tau = \sum_{k=\min\tau}^{\max\tau} \frac{x_k}{\delta_k} P(B^k \cap \{\tau = k\})$$

Since $P(B^k \cap \{\tau = k\}) \leq \delta_k$ and by the unconditional convergence of $\Sigma\, x_n$ we thus have

$$\lim_{\tau \in T} \int_0^1 X_\tau = 0$$

So $(X_n, F_n)_{n \in \mathbb{N}}$ is an amart. It is L^1_E-convergent, hence L^1_E-bounded, since $\int_0^1 \|X_n\| = \|x_n\|$ for every $n \in \mathbb{N}$. We now calculate :

$$\int_0^1 \|X_\tau\| = \sum_{k=\min\tau}^{\max\tau} \|x_k\| \frac{P(B^k \cap \{\tau = k\})}{\delta_k}$$

for $\tau = \tau_n$, where

$$\tau_n = \begin{cases} 1 & \text{on } B^1 \\ \vdots & \\ n & \text{on } B^n \\ n+1 & \text{on } [0,1) \setminus (\bigcup_{i=1}^{n} B^i) \end{cases} .$$

So $\tau_n \in T$ for every $n \in \mathbb{N}$ and it follows that

$$\int_0^1 \|X_{\tau_n}\| = \sum_{k=1}^{n} \|x_k\| + \int_{[0,1) \setminus (\bigcup_{i=1}^{n} B^i)} \|X_{n+1}\|$$

$$\geq \sum_{k=1}^{n} \|x_k\|$$

Hence

$$\sup_{\tau \in T} \int_0^1 \|X_\tau\| \geq \sup_{n \in \mathbb{N}} \int_0^1 \|X_{\tau_n}\| = \infty$$

Therefore $(X_n, F_n)_{n \in \mathbb{N}}$ is not of class (B). □

(vii) ⇒ (i)

We reuse the construction of $(y_n)_{n\in\mathbb{N}}$ made in the proof of (iii) ⇒ (i), when supposing that dim E = ∞. Thus (α_1); $(\beta_1),\ldots,(\beta_{n_1})$; $(\gamma_1),\ldots$ are fulfilled; see the proof of (iii) ⇒ (i). Now put

$$Y_n = \frac{y_n}{\|y_n\|}\chi_{A_n}$$

for every $n \in \mathbb{N}$. Then, for every $\tau \in T$

$$\int_0^1 Y_\tau = \sum_{k=\min\tau}^{\max\tau} \alpha_k y_k \quad ,$$

where $|\alpha_k| \leq 1$. So, since $\Sigma\, y_n$ converges unconditionally, $\lim_{\tau\in T}\int_0^1 Y_\tau = 0$; hence $(Y_n,F_n)_{n\in\mathbb{N}}$ is an amart. Also, $(Y_n)_{n\in\mathbb{N}}$ is uniformly bounded. But $(\|Y_n\|,F_n)_{n\in\mathbb{N}}$ is not an amart : define τ_k on A_j (for $A_j \in \pi_k$) as j. Then $\int_0^1 \|Y_{\tau_k}\| = 1$ for every $k \in \mathbb{N}$ and $(\tau_k)_{k\in\mathbb{N}}$ is cofinal in T. Now define for $A_j \in \pi_k$, τ'_k on A_j as $\rho_k(j)$, where ρ_k is a fixed permutation of the indices j appearing in the sets A_j of π_k, such that ρ_k has no fixed point. Then

$$\int_0^1 \|Y_{\tau'_k}\| = \sum_{j=\min\tau'_k}^{\max\tau'_k} \int_{\{\tau'_k=j\}} \|Y_j\|$$

Now $\{\tau'_k=j\} = A_{\rho_k^{-1}(j)}$. Since $\rho_k^{-1}(j) \neq j$, the intersection $A_{\rho_k^{-1}(j)} \cap A_j$ is empty. So

$$\int_0^1 \|Y_{\tau'_k}\| = 0$$

for every $k \in \mathbb{N}$ and also $(\tau'_k)_{k\in\mathbb{N}}$ is cofinal in T. So $(\|Y_n\|,F_n)_{n\in\mathbb{N}}$ cannot be an amart.

(viii) ⇒ (i)

Suppose dim E = ∞. The example constructed in the proof of (vii) ⇒ (i) is an L^1_E-convergent (to zero), hence L^1_E-bounded amart : indeed $\int_0^1 \|Y_n\| = \|y_n\|$ for every $n \in \mathbb{N}$. However, $(Y_n)_{n\in\mathbb{N}}$ does not converge strongly : for every $\omega \in [0,1)$, we have that $Y_n(\omega) = 0$ for a cofinal set of indices $n \in \mathbb{N}$ as well as $Y_n(\omega) = \frac{y_n}{\|y_n\|}$ for a cofinal set of indices $n \in \mathbb{N}$. □

Via another type of proof, (viii) ⇒ (i) above can be strengthened : this is a result of Edgar and Sucheston [1977b] which shows that even weak convergence a.e. cannot be expected for L^p_E-bounded amarts where $p \in [1,+\infty)$ is infinite dimensional Banach spaces. In fact, a stronger result is proved :

<u>Theorem VII.1.2 (Edgar and Sucheston)</u> : Let $\Phi : [0,+\infty) \to [0,+\infty)$ be a continuous increasing function with $\Phi(0) = \liminf_{t\to+\infty} \frac{\Phi(t)}{t} > 0$. The following assertions are equivalent for a Banach space E

(i) $\dim E < \infty$.

(ii) Every amart $(X_n,F_n)_{n\in\mathbb{N}}$ such that $\sup_{n\in\mathbb{N}} \int_\Omega \Phi(\ X_n\ ) < \infty$ converges weakly a.e..

(iii) For every L^1_E-bounded amart $(X_n,F_n)_{n\in\mathbb{N}}$ there is an $M > 0$ such that for every $\lambda > 0$

$$P\ (\sup_{n\in\mathbb{N}} \|X_n\| > \lambda) \leq \frac{M}{\lambda}$$

(iv) The same as in (ii) but now $(F_n)_{n\in\mathbb{N}}$ is a constant sequence.

(v) The same as in (iii) but now $(F_n)_{n\in\mathbb{N}}$ is a constant sequence.

<u>Proof</u> : <u>(i) ⇒ (ii)</u>

Since $\liminf_{t\to+\infty} \frac{\Phi(t)}{t} > 0$, $(X_n,F_n)_{n\in\mathbb{N}}$ is L^1_E-bounded; the result follows from theorem V.1.3.

<u>(i) ⇒ (iii)</u>

Since $(X_n,F_n)_{n\in\mathbb{N}}$ is now a uniform amart, the result follows from theorem V.1.6(iii).

(ii) ⇒ (iv) and (iii) ⇒ (v)
Are obvious.

(iv) ⇒ (i) and (v) ⇒ (i)
We now use Dvoretzky's theorem, Dvoretzky [1961], stating that ℓ^2 is finitely representable in any infinite dimensional Banach space. We repeat here the definition of finite representability : Let E and F be two Banach spaces. F is said to be finitely representable in E if for any finite dimensional subspace F_1 of F and for every $\varepsilon > 0$, there is an isomorphism V of F_1 into E such that

$$(1-\varepsilon)\|x\| \leq \|Vx\| \leq (1+\varepsilon)\|x\|$$

for every $x \in F_1$.
We return now to the proof of (iv) ⇒ (i). We suppose that dim E = ∞ so that Dvoretzky's result applies in the negative way. Let $F = \ell^2$. Denote by r_n the least integer larger than $2^n\ \Phi(2n)$. In the proof of (v) ⇒ (i) use $\Phi(t) = t$ for every $t \in \mathbb{R}^+$. Let $\{f_i^n \| n \in \mathbb{N},\ i = 1,\dots,r_n\}$ be a collection of orthonormal vectors in F. For each $n \in \mathbb{N}$, let V_n be an isomorphism of $F_n = \text{span}\ \{f_i^n \| n \in \mathbb{N},\ i = 1,\dots,r_n\}$ into E such that

$$\|V_n x\| \leq 2\|x\|$$

for every $x \in F_n$. Write $e_i^n = V_n f_i^n$. Let $\Omega = [0,1]$ and let $\pi_n = \{A_i^n \| i = 1,\dots,r_n\}$ be a partition of Ω such that $P(A_i^n) = \frac{1}{r_n}$. For every $n \in \mathbb{N}$ and $i = 1,\dots,r_n$, put

$$Y_i^n = n\ e_i^n\ \chi_{A_i^n}$$

Let $(X_m)_{m\in\mathbb{N}}$ be the sequence $(Y_i^n)_{n\in\mathbb{N},\, i=1,\dots,r_n}$ ordered so that $(i,n) \leq (i',n')$ if $n < n'$ or $n = n'$ and $i \leq i'$. So $X_m = Y_i^n$ implies $m = R_{n-1} + i$, where $R_{n-1} = \sum_{i=1}^{n-1} r_i$. Take F_n constantly the σ-algebra of the Lebesgue measurable sets. Let $\tau \in T$ and define

$$B_i^n = A_i^n \cap \{\tau = R_{n-1} + i\} \quad .$$

Then

$$X_\tau = \sum_n \sum_{i=1}^{r_n} n\, e_i^n \chi_{B_i^n}$$

where the sum is over the appropriate values of τ. $(X_m(\omega))_{m\in\mathbb{N}}$ is unbounded for every $\omega \in \Omega$, so it does not converge weakly and of course it is also seen that

$$\lim_{\lambda\to\infty} \lambda P\,(\sup_{n\in\mathbb{N}} \|X_n\| > \lambda) = \infty$$

Now, since for every $n \in \mathbb{N}$, $r_n \geq 2^n\,\Phi(2n) \geq 2^n\,\Phi(2)$, we see that $\sum_{n=1}^{\infty} \frac{n}{\sqrt{r_n}} < \infty$. Let $\tau \in T(R_{N-1})$, where $N \in \mathbb{N}$ is fixed. Then

$$\left\|\int_0^1 X_\tau\right\| \leq \sum_{n=N+1}^{\infty} \left\|\sum_{i=1}^{r_n} \int_0^1 n\, e_i^n \chi_{B_i^n}\right\|$$

$$\leq 2 \sum_{n=N+1}^{\infty} \left\|\sum_{i=1}^{r_n} \int_0^1 n\, f_i^n \chi_{B_i^n}\right\|$$

Since the f_i^n are orthonormal, we see that

$$\left\|\sum_{i=1}^{r_n} \int_0^1 n\, f_i^n \chi_{B_i^n}\right\|^2 = \sum_{i=1}^{r_n} n^2\, P(B_i^n)^2$$

$$\leq \frac{n^2}{r_n} \sum_{i=1}^{r_n} P(B_i^n) \leq \frac{n^2}{r_n}$$

Hence

$$\left\|\int_0^1 X_\tau\right\| \leq 2 \sum_{n=N+1}^{\infty} \frac{n}{\sqrt{r_n}} \,,$$

converging to zero for $N \to \infty$. To show that $\sup_{n\in\mathbb{N}} \int_0^1 \Phi(\|X_n\|) < \infty$, define

$Z_i^n = n\, f_i^n\, \chi_{A_i^n}$ for every $n \in \mathbb{N}$ and $i = 1,\dots,r_n$. Then of course

$$\|Y_i^n(\omega)\| \leq 2\|Z_i^n(\omega)\|$$

for every $\omega \in [0,1]$. So

$$\int_0^1 \Phi(\|Y_k^n\|) \leq \int_0^1 \Phi(2\|Z_k^n\|)$$

$$= \int_0^1 \Phi(2n\, \chi_{A_k^n})$$

$$= \frac{\Phi(2n)}{r_n} \leq \frac{1}{2^n} \quad .$$

Hence

$$\sup_{n \in \mathbb{N}} \int_0^1 \Phi(\|X_n\|) \leq 1 \qquad \square$$

Remark VII.1.3 : So theorem VII.1.2 is true in the special case of $\Phi(x) = x^p$, where $p \in [1,+\infty)$. If $(X_n)_{n \in \mathbb{N}}$ is uniformly bounded, the result is false due to corollary V.3.4.

VII.2. Pramarts, mils, GFT

In the previous section we have sufficiently motivated the need for an extension of the uniform amart concept in another direction than in the amart direction. So the new notions should be "very different" from amarts, at least in infinite dimensional Banach spaces. Of course since they will generalise uniform amarts, they will generalise amarts if E has finite dimension. Furthermore it will be seen that in finite dimensional Banach spaces the new notions have the same convergence properties as amarts, while in the infinite dimensional case they have almost the convergence properties of uniform amarts, thus have much better convergence properties than amarts.

However, it must be emphasized that one cannot extend the amart concept without losing the optional sampling theorem V.2.11 or the Riesz-decomposition theorem V.2.4 (even a very weak form thereof). Indeed we have the following result of Edgar and Sucheston [1976c] .

Theorem VII.2.1 (Edgar and Sucheston) : Let $\mathcal{A}$ be the class of sequences $(X_n)_{n\in\mathbb{N}}$ in L^1 such that $(X_n,\sigma(X_1,\ldots,X_n))_{n\in\mathbb{N}}$ is an amart. Let $\mathcal{B}$ be any class in L^1 such that

(i) $\mathcal{A}\subset\mathcal{B}$.

(ii) $(X_n)_{n\in\mathbb{N}}\in\mathcal{B}$ implies $(X_{\tau_n})_{n\in\mathbb{N}}\in\mathcal{B}$ for any increasing sequence (τ_n) such that $\tau_n\in T(n)$ for each $n\in\mathbb{N}$. Here T is w.r.t. $(\sigma(X_1,\ldots,X_n))_{n\in\mathbb{N}}$.

(iii) $(X_n)_{n\in\mathbb{N}}\in\mathcal{B}$ implies that X_n can be written as

$$X_n = Y_n + Z_n$$

where $(Y_n,\sigma(Y_1,\ldots,Y_n))_{n\in\mathbb{N}}$ is a martingale and where $(\int_\Omega Z_n)_{n\in\mathbb{N}}$ converges.

Then $\mathcal{A}=\mathcal{B}$.

Proof : Suppose that $(X_n)_{n\in\mathbb{N}}\in\mathcal{B}$. Then for any increasing sequence $(\tau_n)_{n\in\mathbb{N}}$ such that $\tau_n\in T(n)$ for every $n\in\mathbb{N}$ we have $(X_{\tau_n})_{n\in\mathbb{N}}\in\mathcal{B}$ due to (ii). So, due to (iii)

$$X_{\tau_n} = Y_n + Z_n$$

where $(Y_n,\sigma(Y_1,\ldots,Y_n))_{n\in\mathbb{N}}$ is a martingale and where $(\int_\Omega Z_n)_{n\in\mathbb{N}}$ converges. Hence $(\int_\Omega X_{\tau_n})_{n\in\mathbb{N}}$ converges. This of course implies that $(X_n)_{n\in\mathbb{N}}\in\mathcal{A}$, since $(\tau_n)_{n\in\mathbb{N}}$ is cofinal in T. □

The new notions we are discussing are pramarts and mils :

Definitions VII.2.2 : Let E be any Banach space and $(X_n,\mathcal{F}_n)_{n\in\mathbb{N}}$ an adapted sequence. We say that $(X_n, F_n)_{n \in \mathbb{N}}$ is an amart in probability (shortly pramart) if for every $\varepsilon > 0$, there is $\sigma_o \in T$ such that $\sigma \in T(\sigma_o)$ and $\tau \in T(\sigma)$ imply that

$$P(\{\| E^{F_\sigma} X_\tau - X_\sigma \| > \varepsilon\}) \leq \varepsilon$$

i.e. : $(\| E^{F_\sigma} X_\tau - X_\sigma \|)$ goes to zero in probability for $\sigma \in T$, uniformly in $\tau \in T(\sigma)$. This definition is due to Millet and Sucheston [1980a]. $(X_n, F_n)_{n \in \mathbb{N}}$ is called a martingale in the limit (shortly mil) if

$$\lim_{m \to \infty} \sup_{n \in \mathbb{N}(m)} \| E^{F_m} X_n - X_m \| = 0 \text{ , a.e. .}$$

Theorem VII.2.3 : Every pramart is a mil.

Proof : Using theorem I.3.5.5 (see also corollary I.3.5.6) we see that

$$(E^{F_\sigma} X_\tau - X_\sigma)_{\sigma \in T}$$

converges to zero a.e., uniformly in $\tau \in T(\sigma)$. Hence, since $\mathbb{N}$ is cofinal in T, $(X_n, F_n)_{n \in \mathbb{N}}$ is a mil. □

As will be seen on many occasions further on, the converse of theorem VII.2.3 is not true and in fact we shall see that mils are much more general than pramarts.

An abstract example of a case where mils are uniform amarts (hence pramarts) is seen as follows - see also Bellow [1981] : Let $(X_n)_{n \in \mathbb{N}}$ be a sequence of independent functions with values in a Banach space E. Put

$$S_n = \sum_{i=1}^{n} X_i$$

$$F_n = \sigma(X_1, \dots, S_n)$$

for every $n \in \mathbb{N}$. Then the following assertions are equivalent :

(i) $(S_n, F_n)_{n\in\mathbb{N}}$ is a mil.

(ii) $(S_n, F_n)_{n\in\mathbb{N}}$ is a uniform amart.

(iii) $\sum_{n=1}^{\infty} \int_\Omega X_n$ converges in E.

Indeed (i) ⇒ (iii) follows from

$$E^{F_m} S_n - S_m = \int_\Omega (S_n - S_m).$$

For (iii) ⇒ (ii) define

$$Y_n = S_n + \sum_{j=n+1}^{\infty} \int_\Omega X_j \quad .$$

$(Y_n, F_n)_{n\in\mathbb{N}}$ is a martingale and $(\sum_{j=n+1}^{\infty} \int_\Omega X_j)_{n\in\mathbb{N}}$ tends to zero. Hence $(S_n, F_n)_{n\in\mathbb{N}}$ is a uniform amart.

Since pointwise convergence occurs in the definition of pramart, the following result should be true :

Theorem VII.2.4 : Let $X \in L^1_E$ and let $(F_n)_{n\in\mathbb{N}}$ be a stochastic basis. Then for every $\sigma \in T$

$$(E^{F_\sigma} X)(\omega) = (E^{F_{\sigma(\omega)}} X)(\omega)$$

Proof : Fix $n \in \mathbb{N}$. We see that $\chi_{\{\sigma=n\}} E^{F_n} X$ is F_n-measurable and that for every $A \in F_n$, $A \cap \{\sigma = n\} \in F_\sigma$. So for every $A \in F_n$:

$$\int_\Omega \chi_A \chi_{\{\sigma=n\}} E^{F_n} X = \int_\Omega \chi_A \chi_{\{\sigma=n\}} E^{F_\sigma} X$$

Hence on $\{\sigma = n\}$, $E^{F_n} X = E^{F_\sigma} X$. □

That pramarts and mils are indeed very different from amarts is seen by the next theorem.

Theorem VII.2.5 : The following conditions on a Banach space E are equivalent :

(i) $\dim E < \infty$.

(ii) Every amart is a pramart.

(iii) Every amart is a mil.

Proof : (i) ⇒ (ii)

Now since the amart $(X_n,F_n)_{n\in\mathbb{N}}$ is in fact a uniform amart, we have

$$\lim_{\sigma\in T}\ \sup_{\tau\in T(\sigma)} \|E^{F_\sigma} X_\tau - X_\sigma\|_1 = 0$$

Hence $(\|E^{F_\sigma} X_\tau - X_\sigma\|)_{\sigma\in T}$ converges to zero in probability, uniformly in $\tau \in T(\sigma)$. So $(X_n,F_n)_{n\in\mathbb{N}}$ is a pramart.

(ii) ⇒ (iii)

Follows from theorem VII.2.3.

(iii) ⇒ (i)

We reuse most of the example given in theorem VII.1.2 (iii) ⇒ (i), also the notation. Put $r_n = 2^n$ and

$$X_n = \sum_{i=1}^{2^n} e_i^n \chi_{A_i^n}$$

where for each $n \in \mathbb{N}$, the sets

$$\pi_n = \{A_1^n,\ldots,A_{2^n}^n\}$$

are partitions of $[0,1]$ consisting of independent sets. Put $F_n = \sigma(X_1,\ldots,X_n)$. $(X_n,F_n)_{n\in\mathbb{N}}$ is an amart. This is seen by making the same calculation as in the proof of (iii) ⇒ (i) in theorem VII.1.2. Furthermore for every $m \in \mathbb{N}$ and $n \in \mathbb{N}(m)$ and for every $\omega \in [0,1]$, $(E^{F_m}X_n)(\omega) = \int_0^1 X_n \to 0$. So

$$\sup_{n\in\mathbb{N}(m)} \|E^{F_m} X_n - X_m\| \geq \|X_m\| - \sup_{n\in\mathbb{N}(m)} \|E^{F_m} X_n\| > \frac{1}{2} \qquad \square$$

Thus we have not only shown that $(X_n, F_n)_{n \in \mathbb{N}}$ is not a mil but it is not even a "game which becomes fairer with time" :

Definition VII.2.6 : An adapted sequence $(X_n, F_n)_{n \in \mathbb{N}}$ is called a game which becomes fairer with time (or game fairer with time, shortly GFT) if for every $\varepsilon > 0$, there is $m_o \in \mathbb{N}$ such that $m \in \mathbb{N}(m_o)$ and $n \in \mathbb{N}(m)$ imply

$$P(\|E^{F_m} X_n - X_m\| > \varepsilon) \leq \varepsilon \quad .$$

i.e. : $(\|E^{F_m} X_n - X_m\|)$ goes to zero in probability for $m \to \infty$, uniformly in $n \in \mathbb{N}(m)$. Obviously every mil is a GFT.
It will be seen that GFT are not so important since they do not have good convergence properties.

Remarks VII.2.7 :
1. We can exhibit the following chain of implications :

martingale $\Rightarrow$ uniform amart $\nearrow$ pramart $\Rightarrow$ mil $\Rightarrow$ GFT
martingale $\Rightarrow$ uniform amart $\searrow$ amart $\Rightarrow$ WS amart $\Rightarrow$ W amart

It is easy to see that there are pramarts which are not uniform amarts : take $E = \mathbb{R}$ and any a.e. convergent but not L^1-convergent sequence $(X_n)_{n \in \mathbb{N}}$ in L^1, and take $F_n = F = \sigma(X_1, X_2, \ldots)$, constantly. Then a pramart is just an a.e. convergent sequence while a uniform amart is an L^1-convergent sequence.

2. However, the following is valid : in any Banach space E any pramart $(X_n, F_n)_{n \in \mathbb{N}}$ such that $\sup_{n \in \mathbb{N}} \|X_n\| \in L^1$, is a uniform amart. Indeed since $\sup_{n \in \mathbb{N}} \|X_n\| \in L^1$, it follows that $(\sup_{\tau \in T(\sigma)} \|E^{F_\sigma} X_\tau - X_\sigma\|)_{\sigma \in T}$ is uniformly integrable, due to theorem I.2.2.1. Since $(X_n, F_n)_{n \in \mathbb{N}}$ is a pramart and by theorem I.3.5.5 and corollary I.3.5.6(iii), we see that

$$\lim_{\sigma \in T} \sup_{\tau \in T(\sigma)} \|E^{F_\sigma} X_\tau - X_\sigma\| = 0 \text{ , a.e.,}$$

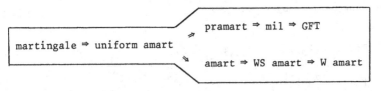

Consequently $\lim_{\sigma \in T} \int_\Omega \sup_{\tau \in T(\sigma)} \| E^{F_\sigma} X_\tau - X_\sigma \| = 0$ since $\mathbb{N}$ is cofinal in T. Hence, $(X_n, F_n)_{n \in \mathbb{N}}$ is certainly a uniform amart.

3. Supposing that a pramart $(X_n, F_n)_{n \in \mathbb{N}}$ is of class (B), is not enough to obtain a uniform amart as the next example of Millet and Sucheston [1980a], in $\mathbb{R}$ shows. It is based upon the general example of V.5.13 of Krengel and Sucheston. We also reuse the notation of V.5.13. From example V.5.13.3 and theorem V.5.3 we see that $(X_n, F_n)_{n \in \mathbb{N}}$ is of class (B) and is not an amart. We now show that it is a pramart. Let $\tau \in T(\sigma)$, $\sigma \in T(g_n)$, for $n \in \mathbb{N}$ fixed. Now the supports of X_σ and X_τ are included in

$$\bigcup_{i_1=1}^{p_1} \cdots \bigcup_{i_{n-1}=1}^{p_{n-1}} A_{n-1}^{i_1,\ldots,i_{n-1}}$$

and this belongs to $F_{g_n} \subset F_\sigma$. Hence $E^{F_\sigma} X_\tau - X_\sigma = 0$ on the complement of this set. Furthermore by V.5.13.3

$$P\Big(\bigcup_{i_1=1}^{p_1} \cdots \bigcup_{i_{n-1}=1}^{p_{n-1}} A_{n-1}^{i_1,\ldots,i_{n-1}}\Big) = \prod_{i=1}^{n-1} \alpha_i = \frac{1}{2^{n-1}} \to 0$$

Hence $(X_n, F_n)_{n \in \mathbb{N}}$ is a pramart.

4. Compare theorem VII.2.5 also with theorem 1 in Edgar and Sucheston [1977a], p.316 (and its lengthy proof!). Indeed the key of our proof is the powerful theorem I.3.5.5.

The rest of this chapter deals mainly with a.e. convergence, optional sampling and Riesz-decomposition theorems for the new notions mentioned above. We shall show that a.e. convergence in the real case holds for mils. However optional sampling is only true for pramarts and so they cannot satisfy a Riesz-decomposition condition, due to theorem VII.2.1. We start with the optional sampling.

Theorem VII.2.8 (Millet and Sucheston [1980a]) : Let E be any Banach space. Let $(X_n,F_n)_{n\in\mathbb{N}}$ be a pramart. Suppose that $(\tau_k)_{k\in\mathbb{N}}$ is an increasing sequence in T. Define $Y_k = X_{\tau_k}$ and $G_k = F_{\tau_k}$ for each $k \in \mathbb{N}$. Then $(Y_k,F_k)_{k\in\mathbb{N}}$ is a pramart.

Proof : Put $\tau_\infty = \lim\limits_{k\to\infty} \tau_k$. It is easily seen that if $\sigma \in T'$ (the finite stopping times w.r.t. $(G_k)_{k\in\mathbb{N}}$) then $\tau_\sigma \in T$ and $G_\sigma = F_{\tau_\sigma}$. Let $\sigma,\sigma' \in T'$. So

$$E^{G_\sigma} Y_{\sigma'} = E^{F_{\tau_\sigma}} X_{\tau_{\sigma'}}$$

Let $\varepsilon > 0$. Choose $n_0 \in \mathbb{N}$ such that $\tau \in T(n_0)$ and $\tau' \in T(\tau)$ implies

$$P(\{\|E^{F_\tau} X_{\tau'} - X_\tau\| > \varepsilon\}) \leq \varepsilon \tag{1}$$

and choose $K_0 \in \mathbb{N}$ such that

$$P(\{\tau_\infty > n_0\} \setminus \{\tau_{K_0} > n_0\}) \leq \varepsilon \tag{2}$$

If $\sigma \in T'(K_0)$ and $\sigma' \in T'(\sigma)$ then

$$P(\{\|E^{G_\sigma} Y_{\sigma'} - Y_\sigma\| > \varepsilon\})$$

$$\leq \varepsilon + P(\{\|E^{F_{\tau_\sigma}} X_{\tau_{\sigma'}} - X_{\tau_\sigma}\| > \varepsilon\} \cap \{\tau_\sigma > n_0\})$$

$$+ P(\{\|E^{F_{\tau_\sigma}} X_{\tau_{\sigma'}} - X_{\tau_\sigma}\| > \varepsilon\} \cap \{\tau_\infty \leq n_0\}) ,$$

due to (1) and (2). So

$$P(\{\|E^{G_\sigma} Y_{\sigma'} - Y_\sigma\| > \varepsilon\}) \leq \varepsilon +$$

$$P(\{\|E^{F_{\tau_\sigma \vee n_0}} X_{\tau_{\sigma'} \vee n_0} - X_{\tau_\sigma \vee n_0}\| > \varepsilon\}) +$$

$$\sum_{i=1}^{n_0} P(\{\|E^{F_{\tau_\sigma}} X_{\tau_{\sigma'}} - X_{\tau_\sigma}\| > \varepsilon\} \cap \{\tau_\infty = i\}) .$$

For every $i \leq n_0$, choose $K_i \in \mathbb{N}$ such that

$$P(\{\tau_\infty = i\} \,\Delta \bigcap_{k \geq K_i} \{\tau_k = i\}) \leq \frac{\varepsilon}{n_0} \tag{3}$$

Choose $\sigma \in T'(K)$ where $K = \max\{K_0, K_1, \ldots, K_{n_0}\}$ and choose $\sigma' \in T'(\sigma)$. Hence, on the set $\bigcap_{k \geq K_i} \{\tau_k = i\} \in F_i$ we have $\tau_\sigma = \tau_{\sigma'} = i$ and so, using theorem VII.2.4, on this set

$$E^{F_{\tau_\sigma}} X_{\tau_{\sigma'}} - X_{\tau_\sigma} = 0$$

Consequently

$$P(\{\|E^{G_\sigma} Y_{\sigma'} - Y_\sigma\| > \varepsilon\}) \leq 3\varepsilon . \qquad \square$$

So we also have the optional stopping theorem for pramarts (cf. remark V.1.9).

We now show that the above theorem is false for mils : we show that even the optional stopping theorem is false for mils and also for GFT, also that a weak form of the optional sampling theorem, not implying the optional stopping theorem, is also false for mils and for GFT :

Theorem VII.2.9 (Edgar and Sucheston [1977a]) :

(i) The optional stopping theorems fails for mils and GFT : there exists a real mil $(X_n, F_n)_{n \in \mathbb{N}}$ and a stopping time σ with $P(\{\sigma = \infty\}) > 0$ such that $(X_{n \wedge \sigma}, F_{n \wedge \sigma})_{n \in \mathbb{N}}$ is not even a GFT.

(ii) The optional sampling theorem in a weak form fails for mils and GFT : there exists a real mil $(X_n, F_n)_{n \in \mathbb{N}}$ and an increasing sequence $(\tau_n)_{n \in \mathbb{N}}$ in T such that $\lim_{n \to \infty} \tau_n = \infty$, uniformly and such that $(X_{\tau_n}, F_{\tau_n})_{n \in \mathbb{N}}$ is not a GFT.

Proof : (i) Let $(A_n)_{n\in\mathbb{N}}$ be a sequence of independent measurable sets in a probability space (Ω,F,P), such that $P(A_1) = 0$, $P(A_n) = \frac{1}{n^2}$ for $n \in \mathbb{N}$ (2). Put $X_n = n\,\chi_{A_n}$ and $F_n = \sigma(X_1,\dots,X_n)$. Using the Borel-Cantelli lemma we see that $\lim_{n\to\infty} X_n = 0$ a.e.. So due to independence we have

$$E^{F_m} X_n - X_m = \int_\Omega X_n - X_m \to 0, \text{ a.e. },$$

hence $(X_n,F_n)_{n\in\mathbb{N}}$ is a mil. Put

$$\sigma = \inf\{n\in\mathbb{N} \,\|\, X_n \neq 0\} \quad .$$

It follows that $P(\sigma=\infty) = \prod_{n=2}^{\infty} (1 - \frac{1}{n^2}) = \frac{1}{2}$ since $(A_n)_{n\in\mathbb{N}}$ consists of independent sets. So, if we put $Y_n = X_{\sigma\wedge n}$, $G_n = F_{\sigma\wedge n}$, then we can show that for $\omega\in\{\sigma=\infty\}$

$$\sup_{n\in\mathbb{N}(m)} |E^{G_m} Y_n - Y_m|(\omega) = +\infty \tag{1}$$

proving that $(Y_n,G_n)_{n\in\mathbb{N}}$ is not a GFT. To obtain this, let $M > 0$ be fixed and choose $m,n \in \mathbb{N}$ so that $n \in \mathbb{N}(m)$ and so that $\sum_{k=m+1}^{n} \frac{1}{k} \geq 2M$. Now $\omega\in\{\sigma=\infty\} \subset (\Omega\setminus A_2) \cap \dots \cap (\Omega\setminus A_m)$ for each $m \in \mathbb{N}$ and $\bigcap_{k=2}^{m} (\Omega\setminus A_k)$ is an atom in G_m. So

$$(E^{G_m} Y_n)(\omega) = \sum_{k=m+1}^{n} E^{G_m} (X_k\,\chi_{\{\sigma=k\}})$$

since $X_k(\omega) = 0$ for $k = 2,\dots,m$. Hence

$$(E^{G_m} Y_n)(\omega) = \sum_{k=m+1}^{n} k\,E^{G_m} (\chi_{A_2^c\cap\dots\cap A_{k-1}^c\cap A_k}(\omega))$$

$$= \sum_{k=m+1}^{n} k \frac{\prod_{j=2}^{k-1} (1 - \frac{1}{j^2}) \frac{1}{k^2}}{\prod_{j=2}^{m} (1 - \frac{1}{j^2})}$$

due to independence. Hence

$$(E^{G_m} Y_n)(\omega) \geq \frac{1}{2} \sum_{k=m+1}^{n} \frac{1}{k} \geq M \quad ,$$

for each $M > 0$, $m \in \mathbb{N}$, $n \in \mathbb{N}(m)$. Therefore (1) is satisfied, since $Y_m(\omega) = 0$.

(ii) Let $(X_n, F_n)_{n \in \mathbb{N}}$ be as in (i). Put $N_1 = 1$, and given N_k and $M > 0$ arbitrary, let $N_{k+1} > N_k$ be the smallest element in $\mathbb{N}$ such that

$$\sum_{n=N_{k+1}}^{N_{k+1}} \frac{1}{n} \geq M$$

Define $\tau_k \in T$ by

$$\tau_k = \begin{cases} \inf \{n \| N_k + 1 \leq n \leq N_{k+1}, X_n \neq 0\}, \text{ if this set is non-empty} \\ \\ N_{k+1} \text{ otherwise.} \end{cases}$$

Put $Y_k = X_{\tau_k}$, $G_k = F_{\tau_k}$ for each $k \in \mathbb{N}$. Then

$$E^{G_{k-1}} Y_k = \sum_{n=N_k+1}^{N_{k+1}} \int_{\{\tau_k = n\}} X_n$$

$$= \sum_{n=N_k+1}^{N_{k+1}-1} n \cdot \prod_{j=N_{k+1}}^{n-1} (1 - \frac{1}{j^2}) \frac{1}{n^2} + \int_{\{\tau_k = N_{k+1}\}} X_{N_{k+1}}$$

Now

$$\int_{\{\tau_k = N_{k+1}\}} X_{N_{k+1}} = N_{k+1} \left(\prod_{j=N_k+1}^{N_{k+1}-1} (1 - \frac{1}{j^2}) \frac{1}{N_{k+1}^2} + \prod_{j=N_k+1}^{N_{k+1}} (1 - \frac{1}{j^2}) \right)$$

$$\geqslant N_{k+1} \prod_{j=N_k+1}^{N_{k+1}-1} (1 - \frac{1}{j^2}) \quad .$$

Hence, for each $k \in \mathbb{N}$

$$E^{G_{k-1}} Y_k \geqslant \sum_{n=N_k+1}^{N_{k+1}} n \prod_{j=N_k+1}^{n-1} (1 - \frac{1}{j^2}) \frac{1}{n^2}$$

$$\geqslant \frac{1}{2} \sum_{n=N_k+1}^{N_{k+1}} \frac{1}{n} \geqslant M$$

Furthermore $\lim_{n \to \infty} Y_n = 0$, a.e. obviously. So $(Y_n, F_n)_{n \in \mathbb{N}}$ is not a GFT. □

Corollary VII.2.10 : There exists a mil which is not a pramart.

Proof : This is immediate from theorem VII.2.9 and theorem VII.2.8. □

Corollary VII.2.11 : The Riesz-decomposition theorem, in the weak form of theorem VII.2.1 (iii), is false for pramarts hence also for mils.

Proof : This follows immediately from theorem VII.2.8, VII.2.1 and remark VII.2.7 which says that the class of pramarts is strictly larger than the class of amarts. □

A very good result true for mils, hence also for pramarts, is that L^1-bounded real mils do converge a.e.. It is a result of Mucci [1976] , using a downcrossing argument.

Theorem VII.2.12 (Mucci) : Let $(X_n, F_n)_{n\in\mathbb{N}}$ be an L^1-bounded real mil. Then $(X_n)_{n\in\mathbb{N}}$ converges a.e. to an integrable function.

Proof : The argument is a classic one. Let $a < b$ be two arbitrary real numbers. Define $(\tau_n)_{n\in\mathbb{N}\cup\{0\}}$ inductively in T as follows : let $\tau_0 = 0$. Let $(\alpha_n)_{n\in\mathbb{N}}$ be a decreasing sequence of positive numbers such that $\sum_{n=1}^{\infty} \alpha_n < \infty$. Fix $N \in \mathbb{N}$. Define τ_{2n-1} as the first $m \in \mathbb{N}$ such that $m \leq N$, $m > \tau_{2n-2}$, $X_m > b$ and such that

$$\sup_{m'\in\mathbb{N}(m)} |E^{F_m} X_{m'} - X_m| < \alpha_n \tag{1}$$

If such an m does not exist, put $\tau_{2n-1} = N$. Define τ_{2n} as the first $m \in \mathbb{N}$ such that $m \leq N$, $m > \tau_{2n-1}$, $X_m < a$ and such that

$$\sup_{m'\in\mathbb{N}(m)} |E^{F_m} X_{m'} - X_m| < \alpha_n \tag{2}$$

again. If such an m does not exist, put $\tau_{2n} = N$. We have, for every $n \in \mathbb{N}$

$$\int_\Omega X_{\tau_{2n-1}} - \int_\Omega X_{\tau_{2n}}$$

$$= \sum_{k=1}^{N} \int_{\{\tau_{2n-1}=k\}} (X_k - E^{F_k} X_N) + \sum_{k=1}^{N} \int_{\{\tau_{2n}=k\}} (E^{F_k} X_N - X_k) < 2\alpha_n$$

So

$$\sum_{n=1}^{\infty} \int_\Omega (X_{\tau_{2n-1}} - X_{\tau_{2n}}) < 2 \sum_{n=1}^{\infty} \alpha_n < \infty . \tag{3}$$

Define

$$\overline{\varphi}(N,a,b) =$$

$$\sum_{n=1}^{\infty} \chi_{\{X_{\tau_{2n-1}} \geqslant b\}} \chi_{\{X_{\tau_{2n}} \leqslant a\}} \chi_{\{\sup_{m \in \mathbb{N} \cup \{0\}} |E^{F_{\tau_{2n}}} X_{\tau_{2n}+m} - X_{\tau_{2n}}| < \alpha_n\}}$$

This function is the "number of downcrossing" subject to conditions (1) and (2). We have :

$$\sum_{n=1}^{\infty} (X_{\tau_{2n-1}} - X_{\tau_{2n}}) \geqslant (b-a)\,\overline{\varphi}(N,a,b) - |b| - |X_N|$$

due to the definition of τ_{2n-1} and τ_{2n} : once τ_{2n-1} or τ_{2n} is N, all the further τ_i are also; in this case it might be that $X_{\tau_{2n-1}} \leqslant b$ or $X_N = X_{\tau_{2n}} \geqslant a$.

So

$$\overline{\varphi}(N,a,b) \leqslant \frac{1}{b-a} \left(\sum_{n=1}^{\infty} (X_{\tau_{2n-1}} - X_{\tau_{2n}}) + |b| + |X_N| \right) .$$

Put $\overline{\varphi}(a,b) = \liminf_{N \in \mathbb{N}} \overline{\varphi}(N,a,b)$. Then, integrating, using the lemma of Fatou and (3) :

$$\int_{\Omega} \overline{\varphi}(a,b) \leqslant \frac{1}{b-a} (2\alpha + |b| + \sup_{n \in \mathbb{N}} \int_{\Omega} |X_n|) < \infty$$

So $P(\overline{\varphi}(a,b) < \infty) = 1$. Let $\varphi(a,b)$ be the usual downcrossing function (see e.g. Chow and Teicher [1978] or Chung [1974]). Since we are working with mils it is clear that

$$\varphi(a,b) < \infty \text{ if and only if } \overline{\varphi}(a,b) < \infty$$

Hence $(X_n)_{n \in \mathbb{N}}$ converges a.e.. □

From this the L^p-convergence ($p \in \,]1,+\infty[$) of course follows if and only if $(X_n)_{n \in \mathbb{N}}$ is L^p-bounded, using standard arguments.

We said already that theorem VII.2.12 is the only positive result we can mention concerning mils. Indeed, to the negative list (theorem VII.2.9, corollary VII.2.11) we can add the following two negative results, the first being negative for pramarts as well :

Theorem VII.2.13 (Edgar and Sucheston [1977a]) : There exists a real L^1-bounded pramart $(X_n, F_n)_{n \in \mathbb{N}}$ such that

$$\sup_{\lambda > 0} \lambda P(\sup_{n \in \mathbb{N}} |X_n| > \lambda) = \infty$$

Proof : Let $(A_n)_{n \in \mathbb{N}}$ be a measurable partition of a probability space (Ω, F, P) such that $P(A_n) = \frac{1}{n} - \frac{1}{n+1}$ for every $n \in \mathbb{N}$. Put $X_n = n(n+1)\chi_{A_n}$ and $F_n = \sigma(X_1, \ldots, X_n)$ for every $n \in \mathbb{N}$. Of course, $\int_\Omega X_n = 1$ for every $n \in \mathbb{N}$ but

$$i(i+1)\ P(\sup_{n \in \mathbb{N}} |X_n| \geq i(i+1)) = i(i+1) \sum_{n \in \mathbb{N}(i)} P(A_n) = n+1 \to \infty$$

Furthermore $(X_n, F_n)_{n \in \mathbb{N}}$ is a pramart : for every $\sigma \in T$, and $\tau \in T(\sigma)$

$$E^{F_\sigma} X_\tau - X_\sigma$$

$$= \sum_{k=\min\sigma}^{\max\sigma} [(E^{F_k} X_\tau - X_k)\chi_{\{\sigma = k\}}]$$

$$= \sum_{k=\min\sigma}^{\max\sigma} \chi_{\{\sigma=k\}} [\sum_{\substack{j=\min\tau \\ j \geq k}}^{\max\tau} E^{F_k}(X_j \chi_{\{\tau = j\}}) - X_k]$$

$$= \sum_{k=\min\sigma}^{\max\sigma} \chi_{\{\sigma=k\}} [\sum_{\substack{j=\min\tau \\ j \geq k}}^{\max\tau} \frac{j(j+1)P(A_j \cap \{\tau = j\})}{P(\Omega \setminus \bigcup_{i=1}^{k} A_j)} \chi_{\Omega \setminus \bigcup_{i=1}^{k} A_i} - k(k+1)\chi_{A_k}]$$

Let $\omega \in \Omega$ be arbitrary. Hence $\omega \in A_{m_o}$ for a certain $m_o \in \mathbb{N}$. For every $\sigma > m_o$, $\sigma \in T$ we thus have that $\omega \notin A_k$ and $\omega \notin \Omega \setminus \bigcup_{i=1}^{k} A_i$ for every $k \in \sigma(\Omega)$. Hence, uniformly in $\tau \in T(\sigma)$, if $\omega \in \Omega$, there is an $M \in \mathbb{N}$ such that $\sigma \in T(M)$ implies

$$\sup_{\tau \in T(\sigma)} \| E^{F_\sigma} X_\tau(\omega) - X_\sigma(\omega) \| = 0$$

proving the pramart property.

This was not the case for uniform amarts : see V.1.6 and also VII.1.2.

It takes more time to elaborate the following negative result of Bellow and Dvoretzky [1980a] concerning mils :

<u>Theorem VII.2.14 (Bellow and Dvoretzky)</u> : There exists a real mil $(X_n, F_n)_{n \in \mathbb{N}}$ such that

a) $(X_n)_{n \in \mathbb{N}}$ converges to zero a.e..

b) $(X_n)_{n \in \mathbb{N}}$ converges to zero in L^1.

c) $(|X_n|, F_n)_{n \in \mathbb{N}}$ is not a mil.

So the set of L^1-bounded mils is not a vector lattice.

<u>Proof</u> : Let (Ω, F, P) be an arbitrary probability space. Divide Ω into a partition of four sets

$$\pi_1 = \{A_1', B_1', A_1'', B_1''\}$$

where $P(A_1') = P(A_1'') = \frac{1}{4}$ and $P(B_1') = P(B_1'')$. Let

$$G_1 = \{A_1', A_1''\}$$

$$c_1 = 4$$

We have the fixed numbers

$$k_1 = \# \pi_1 = 4$$

$$\Delta(\pi_1) = \sup_{C \in \pi_1} P(C) < \frac{1}{2}$$

$$\ell(G_1) = P(A_1') + P(A_1'') = \frac{1}{2} .$$

Rewrite

$$\pi_1 = \{C_{1,1}; C_{1,2}; C_{1,3}; C_{1,4}\} \quad .$$

Each $C_{1,i}$ (i = 1,2,3,4) is divided into four disjoint sets :

$$C_{1,i} = A'_{2,i} \cup B'_{2,i} \cup A''_{2,i} \cup B''_{2,i}$$

where

$$P(A'_{2,i}) = P(A''_{2,i}) = \frac{1}{2^3} P(C_{1,i})$$

$$P(B'_{2,i}) = P(B''_{2,i})$$

Let

$$\pi_2 = \{A'_{2,i}; B'_{2,i}; A''_{2,i}; B''_{2,i} \| 1 \leq i \leq k_1\}$$

$$G_2 = \{A'_{2,i}; A''_{2,i} \| 1 \leq i \leq k_1\}$$

$$c_2 = 2^3$$

and we have the fixed numbers

$$k_2 = \# \pi_2 = 4 k_1 = 4^2$$

$$\Delta(\pi_2) = \sup_{C \in \pi_2} P(C) \leq \frac{1}{2} \Delta(\pi_1) < \frac{1}{2^2}$$

$$\ell(G_2) = \sum_{i=1}^{k_1} (P(A'_{2,i}) + P(A''_{2,i}))$$

$$= \frac{2}{2^3} \sum_{i=1}^{k_1} P(C_{1,i}) = \frac{1}{2^2}$$

These are the first two steps of our construction (the second one spelled out for clarity). Inductively, assume now that π_j, G_j, c_j have been constructed for $1 \leqslant j \leqslant p$. Rewrite

$$\pi_p = \{C_{p,i} \| 1 \leqslant i \leqslant k_p\}$$

Each $C_{p,i}$ is divided into four disjoint sets

$$C_{p,i} = A'_{p+1,i} \cup B'_{p+1,i} \cup A''_{p+1,i} \cup B''_{p+1,i}$$

where

$$P(A'_{p+1,i}) = P(A''_{p+1,i}) = \frac{1}{2^{p+2}} P(C_{p,i})$$

$$P(B'_{p+1,i}) = P(B''_{p+1,i})$$

Put

$$\pi_{p+1} = \{A'_{p+1,i};\ B'_{p+1,i};\ A''_{p+1,i};\ B''_{p+1,i} \| 1 \leqslant i \leqslant k_p\}$$

$$G_{p+1} = \{A'_{p+1,i};\ A''_{p+1,i} \| 1 \leqslant i \leqslant k_p\}$$

$$c_{p+1} = 2^{p+2}$$

and we have the fixed numbers

$$k_{p+1} = \# \pi_{p+1} = 4\, k_p = 4^{p+1}$$

$$\Delta(\pi_{p+1}) = \frac{1}{2} \Delta(\pi_p) < \frac{1}{2^{p+1}}$$

$$\ell(G_{p+1}) = \sum_{i=1}^{k_p} (P(A'_{p+1,i}) + P(A''_{p+1,i}))$$

$$= \frac{2}{2^{p+2}} \sum_{i=1}^{k_p} P(C_{p,i}) = \frac{1}{2^{p+1}}$$

Consequently $\lim_{p\to\infty} \Delta(\pi_p) = 0$ and $\sum_{p=1}^{\infty} \ell(G_p) < +\infty$. So much for the partitions. We now define the following functions

$$Y_1 = c_1 \chi_{A_1'} - c_1 \chi_{A_1''}$$

and generally, if $p > 1$:

$$Y_{p,i} = c_p \chi_{A'_{p,i}} - c_p \chi_{A''_{p,i}} \quad (1 \leq i \leq k_{p-1})$$

Define the following σ-algebras

$$\mathcal{G}_1 = \sigma(\pi_1)$$

and generally, if $p > 1$

$$\mathcal{G}_{p,i} = \sigma(\pi_{p-1} \cup \{A'_{p,j}; B'_{p,j}; A''_{p,j}; B''_{p,j} \| 1 \leq j \leq i\})$$

where $1 \leq i \leq k_{p-1}$.

Clearly $\mathcal{G}_1 \subset \mathcal{G}_{2,i}$ where $1 \leq i \leq k_1$ and $\mathcal{G}_{p,i} \subset \mathcal{G}_{p,j}$ for every $p > 1$ and $1 \leq i \leq j \leq k_{p-1}$ and $\mathcal{G}_{p,i} \subset \mathcal{G}_{p+1,j}$ for any $1 \leq i \leq k_{p-1}$, $1 \leq j \leq k_p$. Thus, if we arrange $(\mathcal{G}_{p,i})$ in this lexicographical way, this sequence becomes increasing and clearly, every $Y_{p,i}$ is $\mathcal{G}_{p,i}$-measurable. Obviously, $\lim_{(p,i)} Y_{p,i} = 0$, a.e. since $\ell(G_p) = P(\bigcup_{1 \leq i \leq k_{p-1}} \text{support}(Y_{p,i}))$ goes to zero for $p \to \infty$. Also

$$\int_\Omega |Y_{p,i}| = 2\, c_p\, P(A'_{p,i}) = 2\, P(C_{p-1,i}) \leq 2\, \Delta(\pi_{p-1})$$

and hence $\lim_{(p,i)} \int_\Omega |Y_{p,i}| = 0$ also. For $p > 1$

$$E^{\mathcal{G}_{p,i-1}} Y_{p,i} = 0 \qquad \text{for} \quad 1 \leq i \leq k_{p-1}$$

$$E^{\mathcal{G}_{p-1,k_{p-2}}} Y_{p,1} = 0$$

having examined all the possible cases! So $(Y_{p,i}, \mathcal{G}_{p,i})_{(p,i)}$ is a mil. But for $(|Y_{p,i}|, \mathcal{G}_{p,i})_{(p,i)}$ we have, if $p > 1$ and $1 \leq i \leq k_{p-1}$

$$E^{G_{p,i-1}}|Y_{p,i}|(\omega) = \frac{2\ c_p\ P(A'_{p,i})}{P(C_{p-1,i})} = 2$$

for $\omega \in C_{p-1,i}$ since $G_{p,i-1}$ refines π_{p-1} to the pth-step only until the $(i-1)^{th}$-place. Also

$$E^{G_{p-1,k_{p-2}}}|Y_{p,1}|(\omega) = \frac{2\ c_p\ P(A'_{p,i})}{P(C_{p-1,i})} = 2$$

for $\omega \in C_{p-1,i}$ obviously. So on $\Omega = \bigcup_{i=1}^{k_{p-1}} C_{p-1,i}$ and for every $(p,i) \leq (p',i')$ we have

$$E^{G_{p,i}}|Y_{p',i'}| = 2$$

Since $\lim_{(p,i)} |Y_{p,i}| = 0$, a.e., $(|Y_{p,i}|, G_{p,i})_{(p,i)}$ cannot be a mil. □

Compare this result with theorem V.1.6.

Whether the lattice property is true for pramarts is not known at the present time, so far as I am aware.

Theorem VII.2.12 is the most general a.e. convergence theorem for real adapted sequences we know so far. Indeed, we can easily show that this theorem fails for GFT. Indeed, w.r.t. constant σ-algebras $(F_n)_{n\in\mathbb{N}}$ a GFT is just a sequence $(X_n)_{n\in\mathbb{N}}$ converging in probability. So just take a sequence $(X_n)_{n\in\mathbb{N}}$ in L^1, converging in probability but not converging a.e., which is L^1-bounded.

However, the L^1-convergence theorem ("Uniformly integrable mils converge in L^1-sense", which follows immediately from theorem VII.2.12) extends to GFT as shown by Subramanian [1973] ; see also Bru and Heinich [1979a] . We can give the proof here but we prefer to wait until the next chapter; there a much more general result (vector-valued), will be proved in an easier way. So we have :

Theorem VII.2.15 (Subramanian) : Uniformly integrable real GFT converge in the L^1-sense.

Now we come to the vector-valued convergence properties of pramarts and mils, which was one of the main reasons for studying these notions. It will be seen that strong convergence a.e. obtainsfor pramarts and mils under fairly reasonable conditions although the main problem remains open.

Problem VII.2.16 : Let E have (RNP). Do L^1_E-bounded pramarts (or mils) converge strongly a.e.?

This problem was formulated by L. Sucheston in 1979. We present two results, one for mils (hence for pramarts also), and one for pramarts only. For another important and very new result on pramarts, see VII.3.6. The mil result was first proved by Bellow and Dvoretzky [1980b] . The proof given here is that of Bellow and Egghe [1982] . We first need a lemma :

Lemma VII.2.17 (Bellow and Egghe) : Let E have (RNP) and let $(X_n, \mathcal{F}_n)_{n \in \mathbb{N}}$ be an adapted sequence, such that there is a subsequence $(X_{n_k})_{k \in \mathbb{N}}$ with the following properties :

(1) $(X_{n_k})_{k \in \mathbb{N}}$ is uniformly integrable.

(2) $\limsup_{k \in \mathbb{N}} \left(\sup_{\ell \in \mathbb{N}(k)} \| E^{\mathcal{F}_{n_k}} X_{n_\ell}(\omega) - X_{n_k}(\omega) \| \right) = 0$, a.e..

Then a.e.

(3) $\limsup_{m,n \in \mathbb{N}} \|X_n(\omega) - X_m(\omega)\| \leq 2 \limsup_{m \in \mathbb{N}} \left(\sup_{n \in \mathbb{N}(m)} \| E^{\mathcal{F}_m} X_n(\omega) - X_m(\omega) \| \right)$.

Proof : By (1) and theorem I.2.2.1 it follows that

$$\left(E^{\mathcal{F}_{n_k}} X_{n_\ell} - X_{n_k}\right)_{\substack{k \in \mathbb{N} \\ \ell \in \mathbb{N}(k)}}$$

is uniformly integrable. Hence, using (2)

$$\limsup_{k \in \mathbb{N}} \left(\sup_{\ell \in \mathbb{N}(k)} \| E^{\mathcal{F}_{n_k}} X_{n_\ell} - X_{n_k} \|_1 \right) = 0$$

It is now standard to see that (A_T), with T = norm-topology, in theorem IV.1.4 is satisfied; hence also inequality (3). □

Theorem VII.2.18 (Bellow and Dvoretzky) : Let E have (RNP). Every mil with a uniformly integrable subsequence converges strongly a.e..

Proof : This is trivial from lemma VII.2.17 and the definition of mil. □

It is also trivial that a mil can have a uniformly integrable subsequence without making the whole sequence even L_E^1-bounded. Indeed, take the constant σ-algebra and any a.e. convergent sequence in L^1 ($E = \mathbb{R}$) which is not L^1-bounded, but has a uniformly integrable subsequence. Another example is given by $X_n = x_n \chi_{[\gamma_{n+1},\gamma_n]}$ where $(x_n)_{n\in\mathbb{N}}$ is any sequence in an arbitrary Banach space E and where $(\gamma_n)_{n\in\mathbb{N}}$ is a strictly decreasing sequence in $[0,1) = \Omega$. So $(X_n,\mathcal{F}_n)_{n\in\mathbb{N}}$ becomes a pramart where $\mathcal{F}_n = \sigma(X_1,\dots,X_n)$ for every $n \in \mathbb{N}$. It is now easy to choose x_n so that $(X_n)_{n\in\mathbb{N}}$ has a uniformly integrable subsequence without $(X_n)_{n\in\mathbb{N}}$ being L_E^1-bounded. For uniform amarts, the existence of a uniformly integrable subsequence implies that the whole sequence is uniformly integrable, due to theorem V.1.4.

The theorem above is valid for mils, hence also for pramarts. In addition, only for pramarts, the following result can be proved.

Theorem VII.2.19 (Millet and Sucheston [1980a]) : Let E have (RNP). Then every pramart of class (B) converges strongly a.e..

Proof : The main point in this proof is that pramarts do have the optional stopping property by theorem VII.2.8. Indeed, proceed exactly as in II.2.4.8 : now class (B) is needed to obtain, with the notation of II.2.4.8,

$$\sup_{n\in\mathbb{N}} \|X_{n\wedge\sigma}\| < \infty$$

since inequality (*) there fails now. Hence we may and do suppose that our pramart $(X_n,\mathcal{F}_n)_{n\in\mathbb{N}}$ satisfies $\sup_{n\in\mathbb{N}} \|X_n\| \in L^1$. But then theorem VII.2.18 finishes the proof. □

Remark VII.2.20 : Since theorem VII.2.8 is not valid for mils, by VII.2.9, the above proof cannot be extended to mils. So for mils we have the problem.

Problem VII.2.21 : Let E have (RNP). Do mils of class (B) converge strongly a.e.? We do have a partial result :

Theorem VII.2.22 (Edgar [1979]) : Let E be a subspace of a separable dual. Then every mil of class (B) converges strongly a.e..

Proof : Indeed, take $\mathcal{T}$ = w*-topology of the separable dual Banach space E. Then we see that, due to metrizability of $\mathcal{T}$ on bounded sets, closed bounded sets are $\mathcal{T}$-sequentially compact. Thus we see that in theorem IV.1.9, case I, (a) and (b') are satisfied. So $(A_{\mathcal{T}})$ follows and hence strong convergence a.e., due to theorem IV.1.4. □

We remark that the assumption "subspace of a separable dual" is strictly stronger than "(RNP)" as was proved in detail in section IV.2.

Also from the inequalities proved in section IV.1, a result of Peligrad [1976] for mils follows trivially.

Theorem VII.2.23 (Peligrad) : Suppose E has (RNP) and suppose that $(X_n, \mathcal{F}_n)_{n\in\mathbb{N}}$ is an L^1_E-bounded mil such that

$$\limsup_{m\in\mathbb{N}} \left(\sup_{n\in\mathbb{N}(m)} \| E^{\mathcal{F}_m} X_n - X_m \|_1 \right) = 0 \qquad (1)$$

Then $(X_n)_{n\in\mathbb{N}}$ converges strongly a.e..

Proof : (1) trivially implies $(A_{\mathcal{T}})$ with $\mathcal{T}$ = norm topology on E in theorem IV.1.4. Hence, convergence follows. □

We proceed with the results on finitely generated mils ($\mathcal{T}$ is as in section IV.1) :

Theorem VII.2.24 (Bellow and Egghe [1982]) : Assume that E has (RNP) and that $(X_n, \mathcal{F}_n)_{n\in\mathbb{N}}$ is a finitely generated adapted sequence. Suppose there exists a subsequence $(X_{n_k})_{k\in\mathbb{N}}$ which is L^1_E-bounded and such that for every $m \in \mathbb{N}$ and $h \in L^\infty(\Omega,F_m,P)$, the T-closure of the set

$$o(h) = \{\int_\Omega h X_{n_k} \| k \in \mathbb{N}(m)\}$$

is T-sequentially compact. Then, a.e.,

$$\limsup_{m,n\in\mathbb{N}} \|X_m(\omega) - X_n(\omega)\| \leq 2 \limsup_{m\in\mathbb{N}} \sup_{n\in\mathbb{N}(m)} \|E^{F_m} X_n(\omega) - X_m(\omega)\| .$$

Proof : Due to the L^1_E-boundedness and the finiteness of each F_m we see that $(E^{F_m} X_{n_k})_{k\in\mathbb{N}(m)}$ is uniformly bounded, hence uniformly integrable for every $m \in \mathbb{N}$. So theorem IV.1.9 applies yielding the claimed inequality. □

Corollary VII.2.25 (Bellow and Egghe [1982]) : Let E have (RNP) and let $(X_n,F_n)_{n\in\mathbb{N}}$ be a finitely generated mil such that there is a subsequence $(X_{n_k})_{k\in\mathbb{N}}$ which is L^1_E-bounded and such that for every $m \in \mathbb{N}$ and $h \in L^\infty(\Omega,F_m,P)$, the T-closure of the set

$$o(h) = \{\int_\Omega h X_{n_k} \| k \in \mathbb{N}(m)\}$$

is T-sequentially compact. Then $(X_n)_{n\in\mathbb{N}}$ converges strongly a.e.. This is in particular the case for every finitely generated mil with an L^1_E-bounded subsequence, if E is a subspace of a separable dual Banach space.

Proof : This follows readily from theorem VII.2.24. □

We close this section by introducing a new type of Banach space valued adapted sequence : the weak mil, and we show that with this notion we can extend the weakly a.e. convergence theorem of Brunel and Sucheston on W amarts (theorem V.4.5).

Definition VII.2.26 (Egghe [1983]) : We say that an adapted sequence $(X_n,F_n)_{n\in\mathbb{N}}$ is a weak mil, W mil shortly, if the double sequence

$$(E^{F_m} X_n - X_m)_{\substack{m \in \mathbb{N} \\ n \in \mathbb{N}(m)}}$$

converges weakly to zero a.e. for $m \to \infty$, uniformly in $n \in \mathbb{N}(m)$.

The relationship with W amarts is not quite clear, but due to the weak a.e. convergence theorem V.4.5 of Brunel and Sucheston, the only important case is that the adapted sequence is of class (B) and that E' is separable. In this case we can prove :

Theorem VII.2.27 (Egghe [1983]) : Let E be a Banach space with separable dual E'. Let $(X_n, F_n)_{n \in \mathbb{N}}$ be a W amart of class (B). Then $(X_n, F_n)_{n \in \mathbb{N}}$ is a W mil.

Proof : Since for every $x' \in E'$, $(x'(X_n), F_n)_{n \in \mathbb{N}}$ is a scalar amart, it is a uniform amart, hence a pramart and hence a mil, due to theorem VII.2.3. Put

$$G_m(\omega) = \sup_{n \in \mathbb{N}(m)} \| E^{F_m} X_n(\omega) - X_m(\omega) \|$$

Since

$$\sup_{\tau \in T} \int_\Omega G_\tau = \sup_{\tau \in T} \int_\Omega \sup_{n \in \mathbb{N}(\tau)} \| E^{F_\tau} X_n - X_\tau \|$$

$$\leq \sup_{\tau \in T} \sup_{\tau' \in T(\tau)} \int_\Omega \| E^{F_\tau} X_{\tau'} - X_\tau \| \quad \text{(by lemma I.3.5.7)}$$

$$\leq 2 \sup_{\tau \in T} \int_\Omega \| X_\tau \| < \infty$$

we see, using lemma II.1.5, that

$$\{ E^{F_m} X_n(\omega) - X_m(\omega) \| m \in \mathbb{N},\ n \in \mathbb{N}(m) \}$$

is a.e. bounded. This together with the fact that $(x'(X_n), F_n)_{n \in \mathbb{N}}$ is a mil for every $x' \in E'$, implies that $(X_n, F_n)_{n \in \mathbb{N}}$ is a W mil. □

For this type of adapted sequences it is now easy to extend theorem V.4.5 :

Theorem VII.2.28 (Egghe [1983]) : Let E be a reflexive Banach space and let $(X_n, F_n)_{n\in\mathbb{N}}$ be a W mil of class (B). Then $(X_n)_{n\in\mathbb{N}}$ converges weakly a.e..

Proof : We present two proofs :

First proof (L. Sucheston) : This follows readily from theorem V.4.6, together with lemma II.1.5 (cf. remark V.4.7).

Second proof (L. Egghe) : This proof is in fact essentially the same as the first but is based upon theorem IV.1.9. Apply case I in IV.1.9 for T = weak*-topology on E. Condition (a) there is satisfied since $(X_n)_{n\in\mathbb{N}}$ is certainly L_E^1-bounded and since bounded sets in E are weakly relatively compact, and hence their closure is weakly sequentially compact, due to Eberlein's theorem. So there exists a function $X_\infty \in L_E^1$ such that $\lim_{n\to\infty} x'(X_n) = x'(X_\infty)$, a.e. for every $x' \in E'$, since $(x'(X_n), F_n)_{n\in\mathbb{N}}$ is a mil for every $x' \in E'$. Now it follows from lemma II.1.5 that $\sup_{n\in\mathbb{N}} \|X_n(\omega)\| < \infty$, a.e.. Hence $(X_n)_{n\in\mathbb{N}}$ converges weakly a.e. to X_∞. □

VII.3. Notes and remarks

VII.3.1. Theorem VII.1.1 has been extended by Egghe [1980b] and [1982a] to Fréchet spaces. In this case Egghe obtains a characterization of nuclearity in Fréchet spaces. For the definition and properties of nuclear spaces, see Pietsch [1972] and Schaefer [1971]. As mentioned in VII.1 the proof given in theorem VII.1.1 is that of Egghe and in fact all the proofs implying (i) carry over to Fréchet spaces by changing the norm into p(.) where p is a continuous semi-norm on the Fréchet space. In theorem VII.1.1, the proofs assuming (i) are trivial from the scalar case. This is not so in case E is a Fréchet space; in this case, (i) reads : E is nuclear. But the result is nevertheless true. It uses several characterizations of nuclearity found in Pietsch [1972]. It must be emphasized that the properties in brackets in (ii), (v), (vii) must be deleted in the Fréchet case.

VII.3.2. An operatorversion of (i) ⇒ (iv) in theorem VII.1.1 appears in Ghoussoub [1979b], thus characterizing absolutely summing operators. These are operators from a Banach space E into a Banach space F which transform unconditional convergent series into absolutely convergent ones. The essence of the proof remains the same.

VII.3.3. In Egghe [1981], before theorem VII.2.18 was known, Egghe proved this result for pramarts in an entirely different way than in VII.2.18. The key result in the reasoning in Egghe [1981] has some independent interest :

Theorem VII.3.3.1 (Egghe) : Let E be any Banach space and $(X_n, \mathcal{F}_n)_{n \in \mathbb{N}}$ a pramart. If there is a subsequence $(X_{n_k})_{k \in \mathbb{N}}$ which is Cesàro-mean convergent, then $(X_n)_{n \in \mathbb{N}}$ itself converges strongly a.e..

Proof : Let $(\tau_n)_{n \in \mathbb{N}}$ be any increasing sequence in T. Put

$$X_\infty = \|\cdot\|_1 - \lim_{k \to \infty} U_k$$

where

$$U_k = \frac{1}{k} \sum_{i=1}^{k} X_{n_i}$$

For any $m, n \in \mathbb{N}$ and $\omega \in \Omega$ we have

$$\|X_{\tau_m}(\omega) - X_{\tau_n}(\omega)\|$$

$$\leq \|X_{\tau_m}(\omega) - E^{\mathcal{F}_{\tau_m}} U_k(\omega)\| + \|E^{\mathcal{F}_{\tau_m}} U_k(\omega) - E^{\mathcal{F}_{\tau_m}} X_\infty(\omega)\| +$$

$$+ \| E^{F_{\tau_m}} X_\infty(\omega) - E^{F_{\tau_n}} X_\infty(\omega)\| + \|E^{F_{\tau_n}} X_\infty(\omega) - E^{F_{\tau_n}} U_k(\omega)\|$$

$$+ \|E^{F_{\tau_n}} U_k(\omega) - X_{\tau_n}(\omega)\| .$$

Now :

$$\|X_{\tau_m} - E^{F_{\tau_m}} U_k\| \leq \frac{1}{k} \sum_{i=1}^{k} \|X_{\tau_m} - E^{F_{\tau_m}} X_{n_i}\|$$

$$=: \frac{1}{k} \Sigma_1^{(m)} + \frac{1}{k} \Sigma_2^{(m)} ,$$

where $\Sigma_1^{(m)}$ is summation over those indices i such that $n_i \not\geq \tau_m$. Since $(n_i)_{n \in \mathbb{N}}$ is cofinal in T, we only have a finite number of n_i such that $n_i \not\geq \tau_m$. $\Sigma_2^{(m)}$ contains the other terms. So

$$\frac{1}{k} \Sigma_2^{(m)} \leq \sup_{n_i \in \mathbb{N}(\tau_m)} \|X_{\tau_m} - E^{F_{\tau_m}} X_{n_i}\|$$

Fix $\varepsilon > 0$. Since, $(E^{F_{\tau_m}} X_\infty, F_{\tau_m})_{m \in \mathbb{N}}$ is L_E^1-convergent to X_∞ by theorem II.1.3 it is convergent in probability. So, choose $m_o \in \mathbb{N}$ such that for $m,n \in \mathbb{N}(m_o)$

$$P(\{\|E^{F_{\tau_m}} X_\infty - E^{F_{\tau_n}} X_\infty\| > \frac{\varepsilon}{5}\}) \leq \frac{\varepsilon}{5} .$$

Since $(X_n, F_n)_{n \in \mathbb{N}}$ is a pramart it follows from theorem I.3.5.5 that

$$\lim_{\sigma \in T} \sup_{\tau \in T(\sigma)} \|E^{F_\sigma} X_\tau - X_\sigma\| = 0$$

where lim denotes the limit in probability. So choose $m_1 \in \mathbb{N}$ such that for $m \in \mathbb{N}(m_1)$

$$P(\{\sup_{n_i \geq \tau_m} \|X_{\tau_m} - E^{F_{\tau_m}} X_{n_i}\| > \frac{\varepsilon}{10}\}) \leq \frac{\varepsilon}{10}$$

Fix $m,n \in \mathbb{N}(m_o \vee m_1)$. Choose a $k \in \mathbb{N}$ such that

(i) $\frac{1}{k} \Sigma_1^{(m)} < \frac{\varepsilon}{10}$

(ii) $\frac{1}{k} \Sigma_1^{(n)} < \frac{\varepsilon}{10}$

(iii) $P(\{\|E^{F_{\tau_m}} U_k - E^{F_{\tau_m}} X_\infty\| > \frac{\varepsilon}{5}\}) \leq \frac{\varepsilon}{5}$

(iv) $P(\{\|E^{F_{\tau_n}} U_k - E^{F_{\tau_n}} X_\infty\| > \frac{\varepsilon}{5}\}) \leq \frac{\varepsilon}{5}$

Now it is clear that if $m,n \in \mathbb{N}(m_o \vee m_1)$ then

$$P(\{\|X_{\tau_m} - X_{\tau_n}\| > \varepsilon\}) < \varepsilon$$

Since convergence in probability is determined by a complete metric by I.3.4, we see that $(X_\tau)_{\tau \in T}$ converges in probability. Now apply, theorem I.3.5.5 yielding that $(X_n)_{n \in \mathbb{N}}$ converges strongly a.e. □

From this theorem, theorem VII.2.18 - but only for pramarts-has been deduced, using an earlier result of Millet and Sucheston [1980a] proving that in (RNP) spaces, uniformly integrable pramarts converge strongly a.e.. At present however theorems VII.2.18 and VII.2.19 are the best results on strong convergence a.e. of mils resp. pramarts, in Banach spaces with (RNP), and with fairly easy proofs.

VII.3.4. For a discussion of pramarts and mils $(X_i, F_i)_{i \in I}$, where I is a directed index set as in chapter VI, see Millet and Sucheston [1980a] and [1979a] and other articles. In chapter VII we have often used theorem I.3.5.5. As proved in chapter VI this result remains true in the case of a directed index set I if we suppose that $(F_i)_{i \in I}$ satisfies the Vitali-condition V. So, in most cases, the results of chapter VII carry over to $(X_i,F_i)_{i\in I}$ if we assume the Vitali condition V.

VII.3.5. In section VII.2 we gave an example of a class of adapted sequences in which every mil is a uniform amart. Now let $E = \mathbb{R}$. In Millet and Sucheston [1980a] some classes of adapted sequences are indicated such that in these classes the notion of pramart and amart coincide.
A first example is :

Example VII.3.5.1 (Millet and Sucheston) : Suppose that $(X_n)_{n\in\mathbb{N}}$ is a sequence of independent functions in L^1 such that $\liminf_{n\in\mathbb{N}} \int_\Omega X_n^- < \infty$ or $\liminf_{n\in\mathbb{N}} \int_\Omega X_n^+ < \infty$. Then $(X_n,F_n)_{n\in\mathbb{N}}$, where $F_n = \sigma(X_1,\dots,X_n)$ for every $n \in \mathbb{N}$, is an amart if and only if it is a pramart.

Another result is :

Example VII.3.5.2 (Millet and Sucheston) : Let $(Y_n,F_n)_{n\in\mathbb{N}}$ be an adapted sequence of positive functions and let $(c_n)_{n\in\mathbb{N}}$ be an increasing sequence with $c_1 > 0$. Suppose that for every $n \in \mathbb{N}$, Y_{n+1} is independent of F_n. Put

$$X_n = \frac{1}{c_n}\sum_{i=1}^{n} Y_i$$

then $(X_n,F_n)_{n\in\mathbb{N}}$ is an amart if and only if it is a pramart.

For the proofs we refer to Millet and Sucheston [1980a], p.109-110.

VII.3.6. Recently M. Słaby proved the following nice result.

Theorem VII.3.6.1 (M. Słaby [1983b]) : Let E be a weakly sequentially complete Banach space with (RNP). Then every pramart with an L_E^1-bounded subsequence converges strongly a.e., to an integrable function.

For the proof we refer to chapter VIII where the subpramart notion is studied. This is needed in the proof of theorem VII.3.6.1.

This almost solves L. Sucheston's problem. It is not known if the weak sequential completeness of E can be deleted. However theorem VII.3.6.1 above solves the problem completely for Banach lattices since a Banach lattice with (RNP) is weakly sequentially complete since c_o cannot be embedded in E - see theorem III.1.3.

Even more recently, N.E. Frangos proved.

Theorem VII.3.6.2 (N.E. Frangos [1983]) : Let E be a subspace of a separable dual Banach space. Let $(X_n, F_n)_{n \in \mathbb{N}}$ be a pramart with an L^1_E-bounded subsequence. Then $(X_n)_{n \in \mathbb{N}}$ converges strongly a.e. to an integrable function.

Also for this proof we refer to chapter VIII again because we need subpramart convergence results. Also this result almost solves L. Sucheston's problem. It is not known if we can change "subspace of a separable dual" into "(RNP)".

VII.3.7. One can also introduce the notion of weak pramart :

Definition VII.3.7.1 (Egghe [1983]) : An adapted sequence $(X_n, F_n)_{n \in \mathbb{N}}$ is called a weak pramart (W pramart shortly) if

$$(E^{F_\sigma} X_\tau - X_\sigma)_{\substack{\sigma \in T \\ \tau \in T(\sigma)}}$$

converges weakly to zero a.e. for $\sigma \in T$, uniformly in $\tau \in T(\sigma)$.

Obviously, every W pramart is a W mil. However in chapter VIII we shall see that theorem VII.2.28, which is also true for W pramarts can be refined for this class : we do not have to suppose class (B); L^1_E-boundedness is enough, as remarked to the author by L. Sucheston. For the proof, see chapter VIII where the subpramart notion is used.

VII.3.8. This seems to be a good place to mention a general strong a.e. convergence result, valid for general adapted sequences and in general Banach spaces. The result is due to Bellow and Dvoretzky [1979] and requires some introduction. Let $(X_n, F_n)_{n\in \mathbb{N}}$ be an arbitrary adapted sequence with values in an arbitrary Banach space E. Let T_f denote the set of all <u>finite stopping</u> times, i.e. $\tau \in T_f$ if and only if

$$P(\{\tau < +\infty) = 1$$

Let $S \subset T_f$. For $\sigma \in T_f$, write $S(\sigma) = \{\tau \in S \| \tau \geqslant \sigma\}$. We say that S is <u>dense</u> if for any $\varepsilon \in]0,1[$ there is $n \in \mathbb{N}$ such that for any $\tau \in T_f(n)$ there is $\tau' \in S$ with

$$P(\tau' \neq \tau) \leqslant \varepsilon$$

For $\tau \in T_f$, define, as usual

$$F_\tau = \{A \in \sigma(\bigcup_n F_n) \| A \cap \{\tau = n\} \in F_n, \text{ for every } n \in \mathbb{N}\}$$

and $X_\tau = \sum_{k=1}^{\infty} X_k \chi_{\{\tau=k\}}$.

We say that S satisfies the <u>localization condition</u> if for every finite family $(\tau_i)_{i\in I}$ in S and for every finite partition $(A_i)_{i\in I}$ of Ω with $A_i \in F_{\tau_i}$ for every $i \in I$ we have that $\tau \in S$ where τ is defined as

$$\tau(\omega) = \tau_i(\omega) \quad \text{for} \quad \omega \in A_i$$

for every $i \in I$.

We say that S is <u>abundant</u> if S contains a set S' such that S' satisfies :

(i) S' is dense.

(ii) S' satisfies the localization condition.

(iii) For every $n \in \mathbb{N}$, $S'(n) \neq \phi$.

If S itself satisfies (i), (ii) and (iii), S is called abundant in the strict sense.

These definitions can be found in Bellow and Dvoretzky [1979] where the following general result is proved :

Theorem VII.3.8.1 (Bellow and Dvoretzky) : Let $(X_n, \mathcal{F}_n)_{n \in \mathbb{N}}$ be an adapted sequence. Suppose that $(X_n)_{n \in \mathbb{N}}$ is L^1_E-bounded. Consider the following three properties :

(i) $(X_n)_{n \in \mathbb{N}}$ converges strongly a.e..

(ii) There is a set $S \subset T_f$ which is abundant in the strict sense, such that the set $\{X_\tau \| \tau \in S\}$ is L^1_E-bounded and such that

$$\lim_{\sigma \in S} \sup_{\tau \in S(\sigma)} \| E^{\mathcal{F}_\sigma} X_\tau - X_\sigma \|_1 = 0$$

(iii) There exists a decreasing sequence $(S_n)_{n \in \mathbb{N}}$, where every S_n is dense, and every set $\{X_\tau \| \tau \in S_n\}$ is L^1_E-bounded such that

$$\lim_{n \to \infty} \sup_{\sigma, \tau \in S_n} \| E^{\mathcal{F}_\sigma} (X_\tau - X_\sigma) \chi_{\{\sigma \leq \tau\}} \|_1 = 0$$

Then (i) ⇒ (ii) and if E has (RNP), then (i) ⇔ (ii) ⇔ (iii)

Earlier, finite stopping times were used by Chow in Chow [1963] where the following result is proved :

Theorem VII.3.8.2 (Chow) : Let $(X_n, \mathcal{F}_n)_{n \in \mathbb{N}}$ be a real sub-martingale such that

$$\int_\Omega X_\tau^+ < \infty$$

for every $\tau \in T_f$. Then $\lim_{n \to \infty} X_n$ exists a.e. and is $> -\infty$.

The same result for mils was proved recently by Yamasaki in Yamasaki [1981] :

Theorem VII .3.8.3 (Yamasaki) : Let $(X_n, F_n)_{n\in\mathbb{N}}$ be a real mil such that

$$\int_\Omega X_\tau^+ < \infty$$

for every $\tau \in T_f$. Then $\lim_{n\to\infty} X_n$ exists a.e., and is $> -\infty$.

For a last extension of this type of theorem : see VIII.3.5.

VII.3.9. Theorem IV.3.3.3 extends to mils, with exactly the same proof :

Theorem VII.3.9.1 (Egghe) : Let $A \in F$ and let $(X_n, F_n)_{n\in\mathbb{N}}$ be a mil on A, with values in a separable dual Banach space. Suppose that

$$\sup_{\tau\in T} \int_\Omega \|X_\tau\| < \infty .$$

Then there is a function $X_\infty \in L^1_E$ such that

$$\lim_{n\to\infty} X_n\Big|_A = X_\infty\Big|_A ,$$

a.e. on A.

As indicated in IV.3.3, we say that $(X_n, F_n)_{n\in\mathbb{N}}$ is a mil on A if

$$\lim_{m\to\infty} \sup_{n\in\mathbb{N}(m)} \|E^{F_m} X_n(\omega) - X_m(\omega)\| = 0 ,$$

a.e. on A.

Chapter VIII : CONVERGENCE OF GENERALIZED SUB- AND SUPER-MARTINGALES IN BANACH LATTICES

This chapter will study extensions of results and notions described in chapter III. These results are not only important in themselves, but they can be used to prove some results concerning pramarts, mils and so on.

VIII.1. Subpramarts, superpramarts and related notions

In view of the definitions of sub- (super-) martingale, pramart, mil and GFT, the following definitions are natural. They are taken from Millet and Sucheston [1980a] in the real case and from Egghe [1984a] or Słaby [1983b] in the general case.

Definition VIII.1.1 : Let E be any Banach lattice and $(X_n, F_n)_{n \in \mathbb{N}}$ an adapted sequence. We say that $(X_n, F_n)_{n \in \mathbb{N}}$ is a subpramart (resp. superpramart) if for every $\varepsilon > 0$ there is a $\sigma_o \in T$ such that for every $\sigma \in T(\sigma_o)$ and $\tau \in T(\sigma)$ we have

$$\overline{P}(\{\omega \in \Omega \| E^{F_\sigma} X_\tau(\omega) - X_\sigma(\omega) \geqslant -\varepsilon e \ (\leqslant \varepsilon e) \text{ for a certain } e \in E^+, \|e\| = 1\}) \geqslant 1 - \varepsilon$$

where $\overline{P}$ denotes the outer measure.

The following easy exercise is left to the reader : $(X_n, F_n)_{n \in \mathbb{N}}$ is a sub-(super-)pramart if and only if for every $\varepsilon > 0$, there is a $\sigma_o \in T$ such that $\sigma \in T(\sigma_o)$ implies

$$\sup_{\tau \in T(\sigma)} P(\{\omega \in \Omega \| \|(X_\sigma - E^{F_\sigma} X_\tau)^+(\omega)\| \geq \varepsilon\}) \leq \varepsilon$$

(resp. $$\sup_{\tau \in T(\sigma)} P(\{\omega \in \Omega \| \|(E^{F_\sigma} X_\tau - X_\sigma)^+(\omega)\| \geq \varepsilon\}) \leq \varepsilon) \quad .$$

Definition VIII.1.2 : $(X_n, F_n)_{n \in \mathbb{N}}$ is said to be a submil (resp. supermil) if there is a null set N in Ω such that for every $\omega \in \Omega \setminus N$ and for every $\varepsilon > 0$, there is $m_o \in \mathbb{N}$ such that $m \in \mathbb{N}(m_o)$ implies that $E^{F_m} X_n(\omega) - X_m(\omega) \geq -\varepsilon e$ ($\leq \varepsilon e$) for a certain $e \in E^+$, $\|e\| = 1$, uniformly in $n \in \mathbb{N}(m)$. Alternatively $(X_n, F_n)_{n \in \mathbb{N}}$ is a sub-(super-)mil if and only if

$$\lim_{m \in \mathbb{N}} \sup_{n \in \mathbb{N}(m)} \|(X_m - E^{F_m} X_n)^+(\omega)\| = 0, \text{ a.e.}$$

(resp. $$\lim_{m \in \mathbb{N}} \sup_{n \in \mathbb{N}(m)} \|(E^{F_m} X_n - X_m)^+(\omega)\| = 0, \text{ a.e.}) \ .$$

Definition VIII.1.3 : $(X_n, F_n)_{n \in \mathbb{N}}$ is said to be a game which becomes better (resp. worse) with time (abbreviated GBT resp. GWT) if for every $\varepsilon > 0$ there is a $m_o \in \mathbb{N}$ such that for every $m \in \mathbb{N}(m_o)$ and $n \in \mathbb{N}(m)$ we have

$$\overline{P}(\{\omega \in \Omega \| E^{F_m} X_n(\omega) - X_m(\omega) \geq -\varepsilon e \ (\leq \varepsilon e) \quad \text{for a certain}$$

$$e \in E^+, \ \|e\| = 1\}) \geq 1 - \varepsilon$$

Alternatively, $(X_n, F_n)_{n \in \mathbb{N}}$ is a GBT (resp. GWT) if and only if for every $\varepsilon > 0$, there is a $m_o \in \mathbb{N}$ such that $m \in \mathbb{N}(m_o)$ implies

$$\sup_{n \in \mathbb{N}(m)} P(\{\omega \in \Omega \| \|(X_m - E^{F_m} X_n)^+(\omega)\| \geq \varepsilon\}) \leq \varepsilon$$

(resp. $$\sup_{n \in \mathbb{N}(m)} P(\{\omega \in \Omega \| \|(E^{F_m} X_n - X_m)^+(\omega)\| \geq \varepsilon\}) \leq \varepsilon) \quad .$$

Properties and examples VIII.1.4

We have :

Submartingale ⇒ subpramart ⇒ submil ⇒ GBT.

Supermartingale ⇒ superpramart ⇒ supermil ⇒ GWT.

This follows from theorem I.3.5.5.

The converse of these implications is not true. This is trivial for GBT $\not\Rightarrow$ submil or GWT $\not\Rightarrow$ supermil. Indeed, take a sequence $(X_n)_{n\in\mathbb{N}}$, converging in probability but not a.e. and take $F_n = F = \sigma((X_n)_{n\in\mathbb{N}})$ for every $n \in \mathbb{N}$. It is also trivial that subpramart $\not\Rightarrow$ submartingale and superpramart $\not\Rightarrow$ supermartingale. Also submil $\not\Rightarrow$ subpramart or equivalently supermil $\not\Rightarrow$ superpramart. Indeed we can even construct a real mil which is not a subpramart : take $(X_n,F_n)_{n\in\mathbb{N}}$ from theorem VII.2.9(ii) and define

$$X'_n = -X_n$$

Then $(X'_n,F_n)_{n\in\mathbb{N}}$ is a mil and not a subpramart as follows immediately from the proof. Of course $(X_n,F_n)_{n\in\mathbb{N}}$ is a mil which is not a superpramart.

As we have seen in chapter VII, pramarts are mils. However, subpramarts are not mils as the following example of Millet and Sucheston [1980a] shows.

Let $(A_n)_{n\in\mathbb{N}}$ be independent sets such that $P(A_n) = \frac{1}{n^2}$, $F_n = \sigma(A_1,\ldots,A_n)$ and $X_n = n^2\,\chi_{A_n}$, for every $n \in \mathbb{N}$. For $\sigma \in T(n)$ and $\tau \in T(\sigma)$ we have

$$P(X_\sigma - E^{F_\sigma} X_\tau > \varepsilon)$$

$$\leq P(X_\sigma > 0)$$

$$\leq \sum_{k\in\mathbb{N}(n)} P(A_k) \to 0 \quad \text{for} \quad n \to \infty .$$

So $(X_n,F_n)_{n\in\mathbb{N}}$ is a subpramart. However

$$\lim_{n\to\infty} |X_n - E^{F_n} X_{n+1}| = 1, \quad \text{a.e.} ,$$

due to independence. Hence $(X_n, F_n)_{n \in \mathbb{N}}$ is not a mil.

A.e. convergence of submils cannot be expected as the following result of Millet and Sucheston [1980a] shows :

Example VIII.1.5 : There is a real L^1-bounded submil which does not converge a.e..

Proof : Let $(A_n)_{n\in\mathbb{N}}$ be independent sets such that $P(A_n) = \frac{1}{n^2}$ for every $n \in \mathbb{N}$. Define

$$\begin{cases} X_{2n+1} = 1 \\ X_{2n} = n^2 \chi_{A_n} \end{cases}$$

and $F_n = \sigma(X_1,\ldots,X_n)$, for every $n \in \mathbb{N}$. Then

$$\limsup_{m\in\mathbb{N}} \left(\sup_{n\in\mathbb{N}(m)} (X_m - E^{F_m} X_n)\right) = \limsup_{m\in\mathbb{N}} (X_m - 1) = 0$$

However, $P(X_{2n} = 0) \to 1$. So

$$\limsup_{n\in\mathbb{N}} X_n = 1, \text{ a.e.}$$

$$\liminf_{n\in\mathbb{N}} X_n = 0, \text{ a.e..} \qquad \square$$

Compare this with the positive result : theorem VII.2.12 for mils. However subpramarts and superpramarts behave much better. We continue now to establish these good results concerning real sub-(super-) pramarts. The method of proving the subpramart a.e. convergence theorem is especially interesting since it is applied later on in other important results.

First of all we establish the optional sampling theorem for sub-(super-)

pramarts. This fails for submils, due to theorem VII.2.9(ii) and its proof.

Theorem VIII.1.6 (Millet and Sucheston [1980a]) : Let E be any Banach lattice and $(X_n, F_n)_{n\in\mathbb{N}}$ an arbitrary sub-(super-)pramart. Then $(X_{\tau_k}, F_{\tau_k})_{k\in\mathbb{N}}$ is a sub-(super-)pramart for every increasing sequence $(\tau_k)_{k\in\mathbb{N}}$ in T.

Proof : Just follow the lines of the proof of theorem VII.2.8, now using the subpramart notion and remark that if $(X_n, F_n)_{n\in\mathbb{N}}$ is a superpramart, then $(-X_n, F_n)_{n\in\mathbb{N}}$ is a subpramart. □

The real supermartingale convergence theorem extends for real superpramarts without any additional condition.

Theorem VIII.1.7 (Millet and Sucheston [1980a]) : Positive real superpramarts converge a.e..

Proof :

a) We first show that $P\{\liminf_{n\in\mathbb{N}} X_n = \infty\} = 0$.

Let $\varepsilon > 0$. Choose $n \in \mathbb{N}$ and $M \in \mathbb{R}^+$ such that $P(A_k) < \varepsilon$ for every $k \in \mathbb{N}$, where

$$A_k = \{E^{F_n} X_{n+k} - X_n > \varepsilon\} \cup \{X_n > M\}$$

Given $K \in \mathbb{R}^+$, choose $k \in \mathbb{N}$ such that

$$P(\{X_{n+k} > K\}) > P(\liminf_{n\in\mathbb{N}} X_n = \infty) - \varepsilon$$

Now

$$K\,\chi_{\Omega\setminus A_k} E^{F_n} \chi_{\{X_{n+k} \geq K\}} \leq \chi_{\Omega\setminus A_k} E^{F_n} X_{n+k} \leq M + \varepsilon$$

by the definition of A_k. Hence, taking $\omega \in \Omega \setminus A_k$,

$$M + \varepsilon \geq \int_\Omega K \chi_{\Omega \setminus A_k} E^{F_n} \chi_{\{X_{n+k} \geq K\}}$$

$$= K \int_\Omega E^{F_n} (\chi_{(\Omega \setminus A_k) \cap \{X_{n+k} \geq K\}}) \quad ,$$

since A_k is F_n-measurable. Hence

$$M + \varepsilon \geq K\, P((\Omega \setminus A_k) \cap \{X_{n+k} \geq K\})$$

$$\geq K\, (P(X_{n+k} \geq K) - \varepsilon)$$

$$\geq K\, (P(\liminf_{n \in \mathbb{N}} X_n = \infty) - 2\varepsilon) \ ,$$

for any $K \in \mathbb{R}^+$ and $\varepsilon > 0$. So $P(\liminf_{n \in \mathbb{N}} X_n = \infty) = 0$.

b) From (a) and the positivity of $(X_n)_{n \in \mathbb{N}}$, if $(X_n)_{n \in \mathbb{N}}$ does not converge a.e. then there exist $\alpha, \beta \in \mathbb{R}$ such that $P(A) > 0$ where

$$A = \{\liminf_{n \in \mathbb{N}} X_n < \alpha < \beta < \limsup_{n \in \mathbb{N}} X_n\}$$

Fix $\varepsilon > 0$ and $M \in \mathbb{N}$. Choose a set B and $M_1 \in \mathbb{N}(M)$ such that $B \in F_{M_1}$ and $P(A \,\Delta\, B) \leq \delta$ where $\delta = \frac{\varepsilon^2}{8\delta}$. Choose $M_2 \in \mathbb{N}(M_1)$ and $M_3 \in \mathbb{N}(M_2)$ such that

$$P(A \setminus \{\inf_{M_1 \leq n \leq M_2} X_n < \alpha < \beta < \sup_{M_2 \leq n \leq M_3} X_n\}) \leq \delta$$

Define

$$C_1 = \{\inf_{M_1 \leq n \leq M_2} X_n < \alpha\} \cap B \qquad \text{and}$$

$$C_2 = \{\sup_{M_2 \leq n \leq M_3} X_n > \beta\} \cap C_1$$

Define $\tau_1, \tau_2 \in T$ as follows

$$\tau_1(\omega) = \begin{cases} M_2 & \text{if } \omega \notin C_1 \\ \inf \{n \in \mathbb{N} \| M_1 \leq n \leq M_2 \text{ and } X_n(\omega) < \alpha\}, & \text{if } \omega \in C_1 \end{cases}$$

$$\tau_2(\omega) = \begin{cases} M_2 & \text{if } \omega \notin C_1 \\ M_3 & \text{if } \omega \in C_1 \setminus C_2 \\ \inf \{n \in \mathbb{N} \| M_2 \leq n \leq M_3 \text{ and } X_n(\omega) > \beta\}, & \text{if } \omega \in C_2 \end{cases}$$

Then

$$X_{\tau_2} - X_{\tau_1} \geq (\beta - \alpha)\chi_{C_2} + X_{\tau_2}\chi_{C_1 \setminus C_2} - \alpha\, \chi_{C_1 \setminus C_2}$$

since $X_{\tau_1} = X_{\tau_2}$ on $\Omega \setminus C_1$. Due to the positivity of $(X_n)_{n \in \mathbb{N}}$ we may delete $X_{\tau_2}\, \chi_{C_1 \setminus C_2}$.

Now apply $E^{F_{\tau_1}}$ to the above inequality.

$$E^{F_{\tau_1}}\, X_{\tau_2} - X_{\tau_1} \geq (\beta - \alpha)\, E^{F_{\tau_1}}(\chi_{C_2}) - \alpha\, E^{F_{\tau_1}}(\chi_{C_1 \setminus C_2})$$

$$\geq (\beta - \alpha)\, E^{F_{\tau_1}}(\chi_A)$$

$$- (\beta - \alpha)\, E^{F_{\tau_1}}(\chi_{A \Delta C_2}) - \alpha\, E^{F_{\tau_1}}(\chi_{C_1 \setminus C_2})$$

Next we shall show that

$$(\beta - \alpha)\, E^{F_{\tau_1}}(\chi_{A \Delta C_2}) + \alpha\, E^{F_{\tau_1}}(\chi_{C_1 \setminus C_2}) < \varepsilon$$

outside a set of measure $\leq \varepsilon$. Indeed, first remark that for every measurable set D and every $\eta > 0$

$$P(E^{F_{\tau_1}}(\chi_D) > \eta) \leq \frac{P(D)}{\eta} \qquad (1)$$

Indeed

$$\eta P(E^{F_{\tau_1}}(\chi_D) > \eta) \leqslant \int_\Omega E^{F_{\tau_1}}(\chi_D) = \int_\Omega \chi_D = P(D) \ .$$

Now take $\eta = \frac{\varepsilon}{4\beta}$. Then

$$(\beta - \alpha)\ E^{F_{\tau_1}}(\chi_{A \,\Delta\, C_2})$$

$$= (\beta - \alpha)\ E^{F_{\tau_1}}(\chi_{A \,\Delta\, C_2}) \cdot \chi_{\{E^{F_{\tau_1}}(\chi_{A \,\Delta\, C_2}) > \eta\}}$$

$$+ (\beta - \alpha)\ E^{F_{\tau_1}}(\chi_{A \,\Delta\, C_2}) \cdot \chi_{\{E^{F_{\tau_1}}(\chi_{A \,\Delta\, C_2}) \leqslant \eta\}}$$

Here $P(E^{F_{\tau_1}}(\chi_{A \,\Delta\, C_2}) > \eta) \leqslant \frac{P(A \,\Delta\, C_2)}{\eta}$ due to (1)

$$\leqslant \frac{P(A \,\Delta\, B)}{\eta}$$

$$\leqslant \frac{\varepsilon^2}{8\beta} \cdot \frac{4\beta}{\varepsilon} = \frac{\varepsilon}{2}$$

while $(\beta - \alpha)\ E^{F_{\tau_1}}(\chi_{A \,\Delta\, C_2})\chi_{\{E^{F_{\tau_1}}(\chi_{A \,\Delta\, C_2}) \leqslant \eta\}} \leqslant (\beta - \alpha)\eta$

We see analogously that

$$\alpha\ E^{F_{\tau_1}}(\chi_{C_1 \setminus C_2}) \leqslant \alpha\ \eta$$

except on a set of measure $\leqslant \frac{\varepsilon}{2}$ (take the same η). So, except on a set of measure $\leqslant \varepsilon$, we have

$$(\beta-\alpha)\, E^{F_{\tau_1}}(\chi_{A\,\Delta\, C_2}) + \alpha\, E^{F_{\tau_1}}(\chi_{C_1 \setminus C_2})$$

$$\leqslant (\beta-\alpha)\eta + \alpha\eta = \beta\eta = \beta\,\frac{\varepsilon}{4\beta} = \frac{\varepsilon}{4} < \varepsilon .$$

Hence, except on a set of measure $\leqslant\varepsilon$ one has :

$$E^{F_{\tau_1}} X_{\tau_2} - X_{\tau_1} \geqslant (\beta-\alpha)\, E^{F_{\tau_1}}(\chi_A) - \varepsilon .$$

Using the superpramart property, choose $M \in \mathbb{N}$ so large that, if $\tau_1 \in T(M)$ and $\tau_2 \in T(\tau_1)$, then $E^{F_{\tau_1}} X_{\tau_2} - X_{\tau_1} \leqslant \varepsilon$, outside of a set of measure $\leqslant \varepsilon$.

We can now conclude that we can construct an increasing sequence $(\tau_n)_{n\in\mathbb{N}}$ in T such that

$$(\beta-\alpha)\, E^{F_{\tau_n}}(\chi_A) \leqslant \frac{1}{n}$$

outside a set of measure $\leqslant\frac{1}{n}$; this means that $(E^{F_{\tau_n}}(\chi_A))_{n\in\mathbb{N}}$ converges to zero in probability. Furthermore the sequence is bounded. Hence we have :

$$\lim_{n\to\infty} \int_\Omega E^{F_{\tau_n}}(\chi_A) = 0$$

Hence $P(A) = 0$, a contradiction. □

We remark that this proof does not use the supermartingale result. On the other hand, the "upcrossing" method that is used is well-known and in fact goes back to J.L. Doob. So we see that, without adding an additional assumption, we could extend the supermartingale convergence theorem. We are now going to do the same for the submartingale convergence theorem (cf. also theorem III.2.2). First some lemmas are required.

Lemma VIII.1.8 (Millet and Sucheston [1980a]) : Let $(X_n, F_n)_{n \in \mathbb{N}}$ be a real positive adapted sequence. For every $m \in \mathbb{N}$ put

$$R_m = \inf_{\tau \in T(m)} E^{F_m} X_\tau$$

Then, for every $\sigma \in T$,

$$R_\sigma = \inf_{\tau \in T(\sigma)} E^{F_\sigma} X_\tau$$

Proof : Let $\sigma \in T$ and $m \in \sigma(\Omega)$ be fixed. If $\tau \in T(\sigma)$, then of course $\tau \geq m$ on $\{\sigma = m\}$. Define

$$\tau' \begin{cases} = \tau & \text{on } \{\sigma = m\} \\ = \max \tau & \text{on } \{\sigma \neq m\} \end{cases}$$

Then certainly $\tau' \in T(m)$ and on $\{\sigma = m\}$ we have

$$E^{F_\sigma} X_\tau = E^{F_m} X_{\tau'}$$

So

$$\inf_{\tau' \in T(m)} E^{F_m} X_{\tau'} \leq \inf_{\tau \in T(\sigma)} E^{F_\sigma} X_\tau \tag{1}$$

on $\{\sigma = m\}$. Conversely if $\tau \in T(m)$, define

$$\tau' \begin{cases} = \tau & \text{on } \{\sigma = m\} \\ = \max \sigma & \text{on } \{\sigma \neq m\} \end{cases}$$

Then $\tau' \in T(\sigma)$ and on $\{\sigma = m\}$ we have

$$E^{F_\sigma} X_{\tau'} = E^{F_m} X_\tau$$

Hence now, on $\{\sigma = m\}$

$$\inf_{\tau' \in T(\sigma)} E^{F_\sigma} X_{\tau'} \leqslant \inf_{\tau \in T(m)} E^{F_m} X_\tau \qquad (2)$$

From (1) and (2), equality on $\{\sigma = m\}$ follows for every m. Hence

$$R_\sigma = \inf_{\tau \in T(\sigma)} E^{F_\sigma} X_\tau \qquad \square$$

The following lemma and its proof will also be used in the vector valued subpramart convergence problem, namely in lemma VIII.1.15 :

Lemma VIII.1.9 (Millet and Sucheston [1980a]) : Let $(X_n, F_n)_{n \in \mathbb{N}}$ be a real positive adapted sequence. Then $(X_n, F_n)_{n \in \mathbb{N}}$ is a subpramart if and only if there exists a positive submartingale $(R_n, F_n)_{n \in \mathbb{N}}$ such that $R_n \leqslant X_n$, a.e. for every $n \in \mathbb{N}$ and such that $(X_n - R_n)_{n \in \mathbb{N}}$ converges to zero, a.e..

Proof : For every $n \in \mathbb{N}$ put

$$R_n = \inf_{\tau \in T(n)} E^{F_n} X_\tau$$

Hence, using lemma VIII.1.8, for every $\sigma \in T$

$$R_\sigma = \inf_{\tau \in T(\sigma)} E^{F_\sigma} X_\tau$$

Now $(R_n, F_n)_{n \in \mathbb{N}}$ is the required submartingale : for every $n \in \mathbb{N}$, $0 \leqslant R_n \leqslant X_n$, a.e. so that $R_n \in L^1$, for every $n \in \mathbb{N}$. Also the same argument as in lemma I.3.5.4 shows that there exists a sequence $\tau_n \in T(\sigma)$ such that

$$R_\sigma = \inf_{n \in \mathbb{N}} E^{F_\sigma} X_{\tau_n}$$

By a similar argument as f.i. in lemma I.3.5.4, we can assume that $(E^{F_\sigma} X_{\tau_n})_{n \in \mathbb{N}}$ decreases to R_σ. If $\sigma \in T(\sigma')$ where $\sigma' \in T$ arbitrary we so have

$$E^{F_{\sigma'}} R_\sigma = E^{F_{\sigma'}} (\lim_{n\to\infty} E^{F_\sigma} X_{\tau_n}) = \lim_{n\to\infty} E^{F_{\sigma'}} X_{\tau_n} \geq \inf_{\tau\in T(\sigma')} E^{F_{\sigma'}} X_\tau$$

Hence, $(R_n, F_n)_{n\in\mathbb{N}}$ is a submartingale.

If $(R'_n, F_n)_{n\in\mathbb{N}}$ is another submartingale such that $X_n \geq R'_n$, a.e. for every $n \in \mathbb{N}$, then $R_n \geq R'_n$, a.e. : indeed, for every $m \in \mathbb{N}$ and $n \in \mathbb{N}(m)$

$$E^{F_m} X_n \geq E^{F_m} R'_n \geq R'_m$$

Hence

$$\inf_{n\in\mathbb{N}(m)} E^{F_m} X_n = R_m \geq R'_m$$

Suppose that $(X_n - R_n)_{n\in\mathbb{N}}$ does not converge to zero a.e.. It follows from theorem I.3.5.5 that $(X_\sigma - R_\sigma)_{\sigma\in T}$ does not converge to zero in probability. Hence there is $\varepsilon > 0$ and a sequence $(\sigma_n)_{n\in\mathbb{N}}$ in T such that

$$P(\{X_{\sigma_n} - R_{\sigma_n} > \varepsilon\}) > \varepsilon$$

From the fact that $(E^{F_\sigma} X_{\tau_n})_{n\in\mathbb{N}}$ converges pointwise, hence in probability, to R_σ for every $\sigma \in T$ we deduce that we can choose $\tau_n \in T(\sigma_n)$ such that

$$P(\{X_{\sigma_n} - E^{F_{\sigma_n}} X_{\tau_n} > \tfrac{\varepsilon}{2}\}) > \tfrac{\varepsilon}{2}$$

contradicting the subpramart assumption.

Conversely, suppose there is a submartingale $(R'_n, F_n)_{n\in\mathbb{N}}$ such that $R'_n \leq X_n$, a.e. and such that $(X_n - R'_n)_{n\in\mathbb{N}}$ converges to zero a.e.. As remarked before, $X_n \geq R_n \geq R'_n$, a.e.. Hence $(X_n - R_n)_{n\in\mathbb{N}}$ converges to zero a.e.. Hence obviously $(X_\sigma - R_\sigma)_{\sigma\in T}$ also. But for every $\sigma \in T$ and $\tau \in T(\sigma)$ we have that

$$X_\sigma - E^{F_\sigma} X_\tau \leq X_\sigma - R_\sigma$$

So $(X_n, F_n)_{n\in\mathbb{N}}$ is a subpramart. □

Lemma VIII.1.10 (Millet and Sucheston [1980a]) : Let $(X_n, F_n)_{n\in\mathbb{N}}$ be a real subpramart. Then for every $\lambda \in \mathbb{R}$, $(X_n \vee \lambda, F_n)_{n\in\mathbb{N}}$ is a subpramart. So $(X_n^+, F_n)_{n\in\mathbb{N}}$ is a subpramart.

Proof : For every $\varepsilon > 0$, choose $m \in \mathbb{N}$ such that $\sigma \in T(m)$ and $\tau \in T(\sigma)$ imply

$$P(\{X_\sigma - E^{F_\sigma} X_\tau > \varepsilon\}) \leq \varepsilon$$

Define

$$\tau' \begin{cases} = \sigma & \text{on } \{X_\sigma < 0\} \\ = \tau & \text{on } \{X_\sigma \geq 0\} \end{cases}$$

Then

$$X_\sigma^+ - X_\tau^+ \leq X_\sigma - X_{\tau'}$$

So

$$X_\sigma^+ - E^{F_\sigma}(X_\tau^+) \leq X_\sigma - E^{F_\sigma} X_{\tau'}$$

Hence $(X_n^+, F_n)_{n\in\mathbb{N}}$ is a subpramart. More generally, if λ is given, obviously $(\lambda + X_n, F_n)_{n\in\mathbb{N}}$ is a subpramart. The general property now follows from the above and from the identity

$$X_n \vee \lambda = \lambda + (X_n - \lambda)^+$$

for every $n \in \mathbb{N}$. □

We are now in a position to state and prove the a.e. convergence theorem for real subpramarts, extending completely the submartingale convergence theorem.

Theorem VIII.1.11 (Millet and Sucheston [1980a]) : Let $(X_n,F_n)_{n\in\mathbb{N}}$ be a L^1-bounded real subpramart. Then $(X_n)_{n\in\mathbb{N}}$ converges a.e.. In fact we do not need the full L^1-boundedness : we only need

$$\liminf_{n\in\mathbb{N}} \int_\Omega X_n^+ + \liminf_{n\in\mathbb{N}} \int_\Omega X_n^- < \infty$$

which is satisfied f.i. in case there is only a subsequence of $(X_n)_{n\in\mathbb{N}}$ which is L^1-bounded.

Proof : a) Suppose first that $(X_n,F_n)_{n\in\mathbb{N}}$ is a positive subpramart such that $\liminf_{n\in\mathbb{N}} \int_\Omega X_n < \infty$. Let $(R_n,F_n)_{n\in\mathbb{N}}$ denote the approximating submartingale, constructed in lemma VIII.1.9. Obviously $(R_n)_{n\in\mathbb{N}}$ is L^1-bounded. From the positive submartingale convergence theorem III.2.2 we deduce that $(R_n)_{n\in\mathbb{N}}$ converges a.e.. Hence, from lemma VIII.1.9, it now follows that $(X_n)_{n\in\mathbb{N}}$ converges a.e..

b) From (a), the convergence follows for subpramarts $(X_n,F_n)_{n\in\mathbb{N}}$ for which there is an $\alpha \in \mathbb{R}$ such that $\alpha \leq X_n$, a.e. for every $n \in \mathbb{N}$.

c) Now consider the general case. Suppose first that $(X_n)_{n\in\mathbb{N}}$ oscillates : there exist $\alpha,\beta \in \mathbb{R}$ such that, on a set of positive measure,

$$\liminf_{n\in\mathbb{N}} X_n < \alpha < \beta < \limsup_{n\in\mathbb{N}} X_n \qquad (1)$$

Now from lemma VIII.1.10, $(X_n \vee \alpha, F_n)_{n\in\mathbb{N}}$ is a subpramart and $\liminf_{n\in\mathbb{N}} \int_\Omega X_n \vee \alpha < \infty$. So, from (b), $(X_n)_{n\in\mathbb{N}}$ converges. However (1) implies the contrary. So $(X_n)_{n\in\mathbb{N}}$ cannot oscillate. Now since

$$\liminf_{n\in\mathbb{N}} X_n = \limsup_{n\in\mathbb{N}} X_n \in \overline{\mathbb{R}} = \mathbb{R} \cup \{-\infty,+\infty\},$$

we also have

$$\liminf_{n\in\mathbb{N}} X_n^+ = \limsup_{n\in\mathbb{N}} X_n^+ \quad \text{and} \quad \liminf_{n\in\mathbb{N}} X_n^- = \limsup_{n\in\mathbb{N}} X_n^-$$

as is readily seen. Hence

$$\int_\Omega \left|\lim_{n\to\infty} X_n\right|$$

$$= \int_\Omega \lim_{n\to\infty} |X_n|$$

$$= \int_\Omega (\lim_{n\to\infty} X_n^+ + \lim_{n\to\infty} X_n^-) \leq \liminf_{n\in\mathbb{N}} \int_\Omega X_n^+ + \liminf_{n\in\mathbb{N}} \int_\Omega X_n^-$$

So $\lim_{n\to\infty} X_n \in L^1$. □

Thus as we have shown, the condition

$$\liminf_{n\in\mathbb{N}} \int_\Omega X_n^+ + \liminf_{n\in\mathbb{N}} \int_\Omega X_n^- < \infty$$

suffices to guarantee the a.e. convergence of real subpramarts. That this condition is strictly weaker than L^1-boundedness, i.e. Doob's condition, is seen in the next example : Take $\Omega = [0,1]$, P = Lebesgue-measure, $F_n = F$, the σ-algebra of all Borelsets in $[0,1]$. Let

$$X_n = (-1)^n n^2 \chi_{[0,\frac{1}{n}]}$$

Then $(X_n,F_n)_{n\in\mathbb{N}}$ is even a pramart as is obviously following from the fact that $(F_n)_{n\in\mathbb{N}}$ is constant. But

$$\liminf_{n\in\mathbb{N}} \int_\Omega X_n^+ + \liminf_{n\in\mathbb{N}} \int_\Omega X_n^- = 0$$

and

$$\int_\Omega |X_n| = n$$

for every $n \in \mathbb{N}$.

We now come to vector valued convergence theorems and we shall show that sub-(super-)pramarts behave quite nicely when strong a.e. convergence is concerned. Especially subpramarts do : very recently, Słaby [1983b] has solved the problem of Egghe (Egghe [1984]) completely, proving that Heinich's theorem III.2.2 extends to subpramarts without any additional hypothesis. His proof is based upon the proof of Egghe [1984] of a partial result concerning this problem and an additional abstract but important remark of Słaby. A bit later, but independently, Frangos did the same, based on a result of Talagrand. We shall give the proofs in complete detail. In the next section we shall give an important application of this result, cf. the above mentioned theorems VII.3.6.1 and VII.3.6.2. After this, the theorem V.3.20 of Ghoussoub and Talagrand is extended to superpramarts, yielding the same characterization. The proof is of Egghe [1984]. This section ends with a proof of theorem VII.2.15, but, as promised there, in a much more general, i.e. vector-valued, setting and valid for GWT (or GBT).

We start with the subpramart convergence theorem in Banach lattices. A number of lemmas are required.

Lemma VIII.1.12 (Egghe [1984a]) : Let $(X_n, F_n)_{n \in \mathbb{N}}$ be a positive sub-pramart and let $x' \in (E')^+$ be arbitrary. Then $(x'(X_n), F_n)_{n \in \mathbb{N}}$ is a positive subpramart and $(\|X_n\|, F_n)_{n \in \mathbb{N}}$ is a positive subpramart.

Proof : The proof of the second assertion is very similar to that of the first. So we only prove the latter. Fix $x' \in (E')^+ \setminus \{0\}$. For every $\varepsilon > 0$, choose $\sigma_o \in T$ such that for every $\sigma \in T(\sigma_o)$ and $\tau \in T(\sigma)$

$$\overline{P}(\{X_\sigma - E^{F_\sigma} X_\tau \leq \frac{\varepsilon}{\|x'\|} \cdot e \| \exists e \in E^+, \|e\| = 1\}) \geq 1 - \varepsilon$$

Fix $e \in E^+$ with $\|e\| = 1$ and put

$$A_e = \{X_\sigma - E^{F_\sigma} X_\tau \leq \frac{\varepsilon}{\|x'\|} e\}.$$

For every $\omega \in A_e$ we have

$$0 \leq X_\sigma(\omega) \leq \frac{\varepsilon}{\|x'\|} e + E^{F_\sigma} X_\tau(\omega)$$

Thus

$$0 \leq x'(X_\sigma)(\omega) \leq \varepsilon + E^{F_\sigma} x'(X_\tau)(\omega)$$

Hence

$$\{x'(X_\sigma) - E^{F_\sigma} x'(X_\tau) \leq \varepsilon\} \supset A_e$$

for every $e \in E^+$ with $\|e\| = 1$. So

$$\{x'(X_\sigma) - E^{F_\sigma} x'(X_\tau) \leq \varepsilon\} \supset \bigcup_{\substack{e \in E^+ \\ \|e\| = 1}} A_e$$

Hence

$$P(\{x'(X_\sigma) - E^{F_\sigma} x'(X_\tau) \leq \varepsilon\}) \geq 1 - \varepsilon$$

proving that $(x'(X_n), F_n)_{n \in \mathbb{N}}$ is a subpramart. □

As mentioned in section II.2, theorem II.2.4.5 is going to be generalized for subpramarts. We shall extend it here since it is one of the lemmas needed in proving the subpramart convergence theorem. However, theorem II.2.4.5 is false when the term "real submartingale" is replaced by "real subpramart" as the following trivial example of Egghe shows :

Example VIII.1.13 : Take constant functions

$$X_n^m = \frac{m^2 a_n}{m^2 + n^2}$$

where $(a_n)_{n \in \mathbb{N}}$ is bounded, positive and non-convergent in $\mathbb{R}$ and where all σ-algebras F_n concerned are constant : $F_n = \{\Omega\}$, $\Omega = [0,1]$. We have $\lim_{n \to \infty} X_n^m = 0$ for every $m \in \mathbb{N}$. Since the σ-algebras are constant, this means that $(X_n^m, F_n)_{n \in \mathbb{N}}$ is a pramart for every $m \in \mathbb{N}$, hence a subpramart. We also have

$$\sup_{n \in \mathbb{N}} \int_\Omega \sup_{m \in \mathbb{N}} X_n^m = \sup_{n \in \mathbb{N}} a_n < \infty$$

but since $(X_n^m)_{m \in \mathbb{N}}$ is increasing

$$\sup_{m \in \mathbb{N}} X_n^m = \lim_{m \to \infty} X_n^m = a_n$$

which is bounded and nonconvergent. So $(\sup_{m \in \mathbb{N}} X_n^m, F_n)_{n \in \mathbb{N}}$ is not a subpramart; this follows from the following elementary lemma :

Lemma VIII.1.13.1 : Let $(a_n)_{n \in \mathbb{N}}$ be a sequence within $\mathbb{R}$ such that for each $\varepsilon > 0$, there is an $n_0 \in \mathbb{N}$ such that, if $n \in \mathbb{N}(n_0)$ and $n' \in \mathbb{N}(n)$, then

$$a_n \leqslant \varepsilon + a_{n'}$$

and such that $(a_n)_{n \in \mathbb{N}}$ is bounded above. Then $(a_n)_{n \in \mathbb{N}}$ converges. □

So in theorem II.2.4.5 we must add an additional assumption in order to obtain the validity of this theorem for subpramarts, but so that the result is still natural and applicable. In order to do so let us define :

Definition VIII.1.14 (Egghe [1984a]) : Let $(X_n^m, F_n)_{n \in \mathbb{N}}$ be a sequence of real (we only need the real case here) subpramarts. It is called a uniform sequence of subpramarts if for every $\varepsilon > 0$, there is $\sigma_0 \in T$ such that if $\sigma \in T(\sigma_0)$ and $\tau \in T(\sigma)$, then

$$P(\{\sup_{m \in \mathbb{N}} (X_\sigma^m - E^{F_\sigma} X_\tau^m) \leqslant \varepsilon\}) \geqslant 1 - \varepsilon$$

This is a very natural notion as the following examples show :

1. Every sequence of submartingales is obviously an example of a uniform sequence of subpramarts.

2. Let $(X_n, F_n)_{n \in \mathbb{N}}$ be a pramart in a Banach space E and let F be another Banach space. Denote by $\mathcal{L}(E,F)$ the space of all continuous linear operators on E into F. Suppose $T_m \in \mathcal{L}(E,F)$ for every $m \in \mathbb{N}$ such that

$\sup_{m \in \mathbb{N}} \|T_m\| < \infty$. Then $(\|T_m(X_n)\|, F_n)_{n \in \mathbb{N}}$ is a uniform sequence of subpramarts.

3. Let $(X_n, F_n)_{n \in \mathbb{N}}$ be a positive subpramart in a Banach lattice E and for every $m \in \mathbb{N}$, let $x'_m \in (E')^+$ with $\sup_{m \in \mathbb{N}} \|x'_m\| < \infty$. Then $(x'_m(X_n), F_n)_{n \in \mathbb{N}}$ is a uniform sequence of subpramarts.

Replacement of "sequences of submartingales" in theorem II.2.4.5 by "uniform sequence of subpramarts" gives a true result as the next proof shows, and is really an extension of II.2.4.5, due to example 1 in VIII.1.14 :

Lemma VIII.1.15 (Egghe [1984a]) : Let $(X_n^m, F_n)_{n \in \mathbb{N}}$ be a uniform sequence of positive real subpramarts. Suppose that there is a subsequence $(n_k)_{n \in \mathbb{N}}$ such that

$$\sup_{k \in \mathbb{N}} \int_\Omega \sup_{m \in \mathbb{N}} X_{n_k}^m < \infty \quad . \tag{1}$$

Then each subpramart $(X_n^m, F_n)_{n \in \mathbb{N}}$ converges a.e. to an integrable function X^m and we have

$$\lim_{n \to \infty} (\sup_{m \in \mathbb{N}} X_n^m) = \sup_{m \in \mathbb{N}} X_\infty^m \text{ , a.e..}$$

Proof : As we have observed, only (1) is needed instead of the requirement

$$\sup_{n \in \mathbb{N}} \int_\Omega \sup_{m \in \mathbb{N}} X_n^m < \infty \quad , \tag{1'}$$

For submartingales however, (1) is the same as (1'), but for subpramarts there is even a difference for one sequence : see the remarks after theorem VII.2.18.

Note also that (1) is equivalent with

$$\liminf_{n \in \mathbb{N}} \int_\Omega \sup_{m \in \mathbb{N}} X_n^m < \infty \quad .$$

Now to the proof, which is an elaboration of lemma VIII.1.9. For every $m \in \mathbb{N}$, define

$$R_n^m = \inf_{\tau \in T(n)} E^{F_n} X_\tau^m$$

Then, as in the proof of lemma VIII.1.9, $(R_n^m, F_n)_{n \in \mathbb{N}}$ is a submartingale and $0 \leqslant R_n^m \leqslant X_n^m$, for every $n,m \in \mathbb{N}$. Suppose

$$\lim_{\tau \in T} (\sup_{n \in \mathbb{N}} (X_\tau^m - R_\tau^m)) \neq 0 \tag{2}$$

Then, due to theorem I.3.5.5, $(\sup_{m \in \mathbb{N}} (X_\tau^m - R_\tau^m))_{\tau \in T}$ does not converge to zero in probability. So there is an $\varepsilon > 0$ and an increasing sequence $(\sigma_n)_{n \in \mathbb{N}}$ in T such that

$$P(\sup_{m \in \mathbb{N}} (X_{\sigma_n}^m - R_{\sigma_n}^m) > \varepsilon) > \varepsilon$$

For each $n \in \mathbb{N}$, choose $m_n \in \mathbb{N}$ such that

$$P(\sup_{m \in \{1,\dots,m_n\}} (X_{\sigma_n}^m - R_{\sigma_n}^m) > \tfrac{\varepsilon}{2}) > \tfrac{\varepsilon}{2} . \tag{3}$$

Define, for each $j \in \{1,\dots,m_n\}$

$$A_j = \{\omega \in \Omega \| j \text{ is the first index in } \mathbb{N} \text{ for which}$$

$$\sup_{m \in \{1,\dots,m_n\}} (X_{\sigma_n}^m(\omega) - R_{\sigma_n}^m(\omega)) = X_{\sigma_n}^j(\omega) - R_{\sigma_n}^j(\omega)\}$$

Hence $(A_j)_{j \in \{1,\dots,m_n\}}$ is a disjoint family of F_{σ_n}-measurable sets. (4)
Using the classical argument in lemma I.3.5.4 we can arrange for a sequence $\gamma_{k,j}^n \in T(\sigma_n)$ for each $n \in \mathbb{N}$ such that

$$(E^{F_{\sigma_n}} X_{\gamma_{k,j}^n}^j)_{k \in \mathbb{N}}$$

decreases to $R^j_{\sigma_n}$, a.e.. So there exists $\tau^j_n \in T(\sigma_n)$ such that

$$P(E^{F_{\sigma_n}} X^j_{\tau^j_n} - R^j_{\sigma_n} > \frac{\varepsilon}{4} P(A_j)) \leq \frac{\varepsilon}{4} P(A_j) \qquad (5)$$

Define $\tau_n = \tau^j_n$ on A_j. So $\tau_n \in T(\sigma_n)$ for every $n \in \mathbb{N}$. (6)
Now, by (3)

$$\frac{\varepsilon}{2} < P(\sup_{m \in \{1,\ldots,m_n\}} (X^m_{\sigma_n} - R^m_{\sigma_n}) > \frac{\varepsilon}{2})$$

$$= \sum_{j=1}^{m_n} P(\{X^j_{\sigma_n} - R^j_{\sigma_n} > \frac{\varepsilon}{2}\} \cap A_j) \qquad \text{(by (4))}$$

$$\leq \sum_{j=1}^{m_n} P(\{X^j_{\sigma_n} - E^{F_{\sigma_n}} X^j_{\tau^j_n} > \frac{\varepsilon}{2} (1 - \frac{P(A_j)}{2})\} \cap A_j)$$

$$+ \sum_{j=1}^{m_n} P(\{E^{F_{\sigma_n}} X^j_{\tau^j_n} - R^j_{\sigma_n} > \frac{\varepsilon}{2} \frac{P(A_j)}{2}\} \cap A_j)$$

$$\leq \sum_{j=1}^{m_n} P(\{X^j_{\sigma_n} - E^{F_{\sigma_n}} X^j_{\tau_n} > \frac{\varepsilon}{4}\} \cap A_j) + \frac{\varepsilon}{4}, \text{ by (5) and (6)}$$

$$= P(\bigcup_{j=1}^{m_n} \{X^j_{\sigma_n} - E^{F_{\sigma_n}} X^j_{\tau_n} > \frac{\varepsilon}{4}\} \cap A_j) + \frac{\varepsilon}{4}$$

$$\leq P(\sup_{j \in \mathbb{N}} (X^j_{\sigma_n} - E^{F_{\sigma_n}} X^j_{\tau_n}) > \frac{\varepsilon}{4}) + \frac{\varepsilon}{4}$$

Hence

$$P(\sup_{j \in \mathbb{N}} (X^j_{\sigma_n} - E^{F_{\sigma_n}} X^j_{\tau_n}) > \frac{\varepsilon}{4}) > \frac{\varepsilon}{4} ,$$

a contradiction to the fact that $(X^m_n, F_n)_{n \in \mathbb{N}}$ is a uniform sequence of subpramarts. So (a) cannot be true. Hence

$$\lim_{n\to\infty} (\sup_{m\in\mathbb{N}} (X_n^m - R_n^m)) = 0, \text{ a.e..} \tag{7}$$

Since $0 \leq R_n^m \leq X_n^m$ for every $m,n \in \mathbb{N}$ we also have

$$\sup_{k\in\mathbb{N}} \int_\Omega \sup_{m\in\mathbb{N}} R_{n_k}^m < \infty \quad .$$

Since $(\sup_{m\in\mathbb{N}} R_n^m, F_n)_{n\in\mathbb{N}}$ is a submartingale we so have also

$$\sup_{n\in\mathbb{N}} \int_\Omega \sup_{m\in\mathbb{N}} R_n^m < \infty$$

Now apply II.2.4.5, yielding

$$\lim_{n\to\infty} R_n^m =: R_\infty^m \text{ exists, a.e.} \tag{8}$$

and

$$\lim_{n\to\infty} (\sup_{m\in\mathbb{N}} R_n^m) = \sup_{m\in\mathbb{N}} R_\infty^m \text{ , a.e..} \tag{9}$$

Since

$$\sup_{m\in\mathbb{N}} X_n^m \leq \sup_{m\in\mathbb{N}} (X_n^m - R_n^m) + \sup_{m\in\mathbb{N}} R_n^m$$

we also have

$$0 \leq \sup_{m\in\mathbb{N}} X_n^m - \sup_{m\in\mathbb{N}} R_n^m \leq \sup_{m\in\mathbb{N}} (X_n^m - R_n^m) \ . \tag{10}$$

From (7), (9) and (10) it now follows that

$$\lim_{n\to\infty} (\sup_{m\in\mathbb{N}} X_n^m) = \sup_{m\in\mathbb{N}} R_\infty^m \text{ , a.e..}$$

From (7) however, for each $m \in \mathbb{N}$:

$$\lim_{n\to\infty} R_n^m = \lim_{n\to\infty} X_n^m \text{ , a.e..}$$

Hence

$$\lim_{n\to\infty} (\sup_{m\in\mathbb{N}} X_n^m) = \sup_{m\in\mathbb{N}} X_\infty^m \text{ , a.e..} \qquad \square$$

Remarks VIII.1.16 :

1. It is easy to see from the definition that $(\sup_{m\in\mathbb{N}} X_n^m, F_n)_{n\in\mathbb{N}}$ is a subpramart if $((X_n^m, F_n)_{n\in\mathbb{N}})_{m\in\mathbb{N}}$ is a uniform sequence of subpramarts. Of course, as follows from example VIII.1.13 also, this assertion is false if we delete the assumption "uniform".

2. From these lemmas it is now relatively easy, see e.g. Egghe [1984a] or see the argument in VIII.1.18 on $(Y_{n_k})_{k\in\mathbb{N}}$, to prove a partial subpramart a.e. convergence theorem : Let E be a Banach lattice with (RNP) and let $(X_n, F_n)_{n\in\mathbb{N}}$ be a positive subpramart with a subsequence $(n_k)_{n\in\mathbb{N}}$ in $\mathbb{N}$ such that

   (a) $(\int_A X_{n_k})_{k\in\mathbb{N}}$ converges weakly for every $A \in \bigcup_n F_n$

   (b) $\sup_{k\in\mathbb{N}} \int_\Omega \|X_{n_k}\| < \infty$

   Then $(X_n)_{n\in\mathbb{N}}$ converges strongly a.e..

   Condition (a) and (b) can be replaced by (a') $(X_{n_k})_{k\in\mathbb{N}}$ is uniformly integrable or (a")

   $$\sup_{\tau\in T} \int_\Omega \|X_\tau\| < \infty$$

   as is shown in Egghe [1984a] . However the problem of Egghe is solved if one can delete condition (a). This is what Słaby did in Słaby [1983b] recently (*), and also Frangos (*) in Frangos [1983] independently

(*) I thank M. Słaby and N. Frangos for sending me so promptly their manuscript containing this result.

In addition to the previous lemmas Słaby proved the following non-trivial

Lemma VIII.1.17 (Słaby) : Let E be any Banach space and let $(X_n, F_n)_{n\in \mathbb{N}}$ be an L_E^1-bounded adapted sequence. Then there exists a subsequence $(n_k)_{k\in \mathbb{N}}$ in $\mathbb{N}$ such that for every $k \in \mathbb{N}$

$$X_{n_k} = Y_{n_k} + Z_{n_k}$$

where Y_{n_k} and Z_{n_k} are F_{n_k}-measurable, $(Y_{n_k})_{k\in \mathbb{N}}$ is uniformly integrable and where $\lim_{k\to\infty} Z_{n_k} = 0$, a.e..

Proof : For every $m \in \mathbb{N}$, put

$$g_m(t) = \sup_{n\in \mathbb{N}(m)} \int_{\{\|X_n\| > t\}} \|X_n\| dP$$

Since g_1 decreases with t and is positive, $\lim_{t\to\infty} g_1(t) = \alpha$ exists. Let $(t_k)_{k\in \mathbb{N}}$ be an increasing sequence such that $\lim_{k\to\infty} t_k = \infty$ and

$$\alpha \leq g_1(t_k) < \alpha + \frac{1}{k} \tag{1}$$

Now obviously

$$g_m \leq g_1 \tag{2}$$

and

$$g_1 \leq \sup_{m>n\geq 1} \int_{\{\|X_n\| > t\}} \|X_n\| dP + \sup_{n\in \mathbb{N}(m)} \int_{\{\|X_n\| > t\}} \|X_n\| dP$$

So

$$\lim_{t\to\infty} g_1(t) \leq \lim_{t\to\infty} g_m(t) \tag{3}$$

2) and (3) now imply

$$\lim_{t\to\infty} g_m(t) = \alpha \tag{4}$$

Hence for every m and $k \in \mathbb{N}$

$$g_m(t_k) \geq \alpha$$

From (1), choose an increasing sequence $(n_k)_{k\in\mathbb{N}}$ in $\mathbb{N}$ such that

$$\int_{\{\|X_{n_k}\| > t_k\}} \|X_{n_k}\| dP > \alpha - \frac{1}{k} \tag{5}$$

for every $k \in \mathbb{N}$. Put $Y_{n_k} = X_{n_k} \chi_{\{\|X_{n_k}\| \leq t_k\}}$ and $Z_{n_k} = X_{n_k} \chi_{\{\|X_{n_k}\| > t_k\}}$ for every $k \in \mathbb{N}$. So $(Y_{n_k}, F_{n_k})_{k\in\mathbb{N}}$ and $(Z_{n_k}, F_{n_k})_{k\in\mathbb{N}}$ are adapted sequences.

Furthermore

$$P(\|Z_{n_k}\| \geq \varepsilon) \leq P(\|X_{n_k}\| > t_k) \leq \frac{1}{t_k} \sup_{n\in\mathbb{N}} \int_\Omega \|X_n\|$$

So $(Z_{n_k})_{k\in\mathbb{N}}$ goes to 0 in probability. By passing to a subsequence of $(n_k)_{k\in\mathbb{N}}$, which we still denote by $(n_k)_{k\in\mathbb{N}}$, we see that $(Z_{n_k})_{k\in\mathbb{N}}$ goes to 0 a.e.. Put

$$g(t) = \sup_{k\in\mathbb{N}} \int_{\{\|Y_{n_k}\| > t\}} \|Y_{n_k}\| dP$$

Then

$$\lim_{t\to\infty} g(t) = \lim_{i\to\infty} g(t_i)$$

$$= \lim_{i\to\infty} \sup_{k\in\mathbb{N}(i)} \int_{\{t_i < \|X_{n_k}\| \leq t_k\}} \|X_{n_k}\| dP$$

$$= \lim_{i\to\infty} \sup_{k\in\mathbb{N}(i)} \left(\int_{\{\|X_{n_k}\| > t_i\}} \|X_{n_k}\| dP - \int_{\{\|X_{n_k}\| > t_k\}} \|X_{n_k}\| dP \right)$$

$$\leq \lim_{i\to\infty} \sup_{k\in\mathbb{N}(i)} (\alpha + \frac{1}{i} - \alpha + \frac{1}{k}) \leq \frac{2}{i}$$

So $\lim_{t\to\infty} g(t) = 0$. Hence $(Y_{n_k})_{k\in\mathbb{N}}$ is uniformly integrable. □

Now we have enough material to prove the following interesting theorem, being the complete solution of the problem of Egghe, see remark VIII.1.16(2) :

Theorem VIII.1.18 (Sľaby [1983b]) : Let E be a Banach lattice with (RNP) and let $(X_n,F_n)_{n\in\mathbb{N}}$ be a positive subpramart with an L^1_E-bounded subsequence. Then $(X_n)_{n\in\mathbb{N}}$ converges strongly a.e..

Proof : Using lemma VIII.1.17 and the fact that a subsequence of $(X_n)_{n\in\mathbb{N}}$ is L^1_E-bounded we can assume that

$$X_{n_k} = Y_{n_k} + Z_{n_k}$$

where $(X_{n_k})_{k\in\mathbb{N}}$ is L^1_E-bounded, $(Y_{n_k})_{k\in\mathbb{N}}$ is uniformly integrable and $\lim_{k\to\infty} Z_{n_k} = 0$, a.e.. By lemma VIII.1.12, for every $x' \in (E')^+$ $(x'(X_n),F_n)_{n\in\mathbb{N}}$ is a real subpramart satisfying the boundedness condition in theorem VIII.1.11. Thus there is a function $f_{x'} \in L^1$ such that

$$\lim_{n\to\infty} x'(X_n) = f_{x'}\ , \text{ a.e..} \qquad (1)$$

So $\lim_{k\to\infty} x'(Y_{n_k}) = f_{x'}$, a.e. and hence in the L^1-norm. So, for $A \in \sigma(\bigcup_n F_n)$

$$\lim_{k\to\infty} \int_A x'(Y_{n_k})$$

exists. Hence, $(\int_A Y_{n_k})_{k\in\mathbb{N}}$ is weakly Cauchy. Since the Banach lattice has (RNP), it does not contain c_o as an isomorphic copy. Hence, by theorem III.1.3, E is weakly sequentially complete.
Then let

$$\mu(A) = w\text{-}\lim_{k\to\infty} \int_A Y_{n_k}$$

for every $A \in \sigma(\bigcup_n F_n)$. It is classical, see e.g. lemma IV.1.3, to prove that μ has bounded variation, that μ is countably additive on $\sigma(\bigcup_n F_n)$ and that $\mu \ll P$. Applying (RNP), let $X_\infty \in L^1_E$ be such that

$$\mu(A) = \int_A X_\infty$$

for every $A \in \sigma(\bigcup_n F_n)$. Obviously

$$\int_A x'(X_\infty) = \int_A f_{x'}$$

for every $A \in \sigma(\bigcup_n F_n)$. Hence $x'(X_\infty) = f_{x'}$, a.e. for every $x' \in (E')^+$. Hence (1) implies

$$\lim_{n\to\infty} x'(X_n) = x'(X_\infty), \text{ a.e.} \tag{2}$$

for every $x' \in (E')^+$. Now invoke theorem III.2.3 and let $|||\cdot|||$ denote the new Kadec-norm equivalent with $\|\cdot\|$, as described in theorem III.2.3. Also let $D \in (E')^+$ be as in theorem III.2.3. From the fact that $((x'(X_n),F_n)_{n\in\mathbb{N}})_{x'\in D}$ is a uniform sequence of positive subpramarts, from (2) and from lemma VIII.1.15 it now follows that

$$\lim_{n\to\infty} |||X_n||| = |||X_\infty||| \text{ , a.e..} \tag{3}$$

Theorem III.2.3 concludes the proof since it implies that $\lim_{n\to\infty} X_n = X_\infty$ strongly a.e.. □

Very recently another proof of theorem VIII.1.18 became possible using the following new result of Talagrand; see Talagrand [1981] :

<u>Theorem VIII.1.19 (Talagrand)</u> : If E is a Banach lattice with (RNP), then E is isomorphic with (a subspace of) a separable dual Banach lattice.

This result was unavailable at the time the results VIII.1.15 - VIII.1.18 were proved. Use of VIII.1.19 gives an easy proof of Sľaby's result. This second proof was given by Frangos [1983], independently of Sľaby's proof.

VIII.1.20 - Second proof of theorem VIII.1.18 : By theorem VIII.1.19, let E be isomorphic with the separable Banach lattice F' and let D be a countable dense subset of F^+. By lemma VIII.1.12 and theorem VIII.1.11, we have that

$$(x'(X_n))_{n\in\mathbb{N}} \text{ converges a.e., for every } x' \in D \qquad (1)$$

$$(\|X_n\|)_{n\in\mathbb{N}} \text{ converges a.e..} \qquad (2)$$

Now since D is countable, let $N \in \mathcal{F}$ be such that $P(N) = 0$ and such that $(x'(X_n(\omega)))_{n\in\mathbb{N}}$ and $(\|X_n(\omega)\|)_{n\in\mathbb{N}}$ converges on $\Omega \setminus N$ for every $x' \in D$. Then it is obvious that, for every fixed $\omega \in \Omega \setminus N$

$$\{x' \in F \| (x'(X_n(\omega)))_{n\in\mathbb{N}} \text{ converges}\}$$

is closed in F. Hence $(x'(X_n(\omega)))_{n\in\mathbb{N}}$ converges for every $x' \in F$. This limit is then a linear functional on F, which is bounded since

$$|\lim_{n\to\infty} x'(X_n(\omega))| \leq \sup_{n\in\mathbb{N}} \|X_n(\omega)\| < \infty \quad ,$$

if $\|x'\| \leq 1$, by (2). So, for every $\omega \in \Omega\setminus N$, there exists an $X_\infty(\omega) \in F' = E$ such that

$$\lim_{n\to\infty} x'(X_n(\omega)) = x'(X_\infty(\omega)) \ , \qquad (3)$$

for every $x' \in F$. This argument so far is taken from Neveu [1975], p.108. Now invoke theorem III.2.3, which is applicable since a Banach lattice with (RNP) is ordercontinuous, by III.1.2. Then let C be a countable norming subset of F^+ for $|\!|\!|\cdot|\!|\!|$, an equivalent norm on E, such that $|\!|\!|\cdot|\!|\!|$ satisfies the properties described in III.2.3 w.r.t. C. By (3) we have, for every $\omega \in \Omega\setminus N$

$$\lim_{n\to\infty} x'(X_n(\omega)) = x'(X_\infty(\omega)) \tag{4}$$

for every $x' \in C$. From lemma VIII.1.15 we now have also

$$\lim_{n\to\infty} |\|X_n(\omega)\|| = |\|X_\infty(\omega)\|| \tag{5}$$

on $\Omega \setminus N$, since C is countable and norming for $|\|\cdot\||$. The properties of $|\|\cdot\||$ in III.2.3 now imply that $\lim_{n\to\infty} X_n = X_\infty$, strongly a.e. □

Another application of the previous results will appear in the next section. It will solve almost completely the problem of Sucheston on strong pramart convergence.

We now come to the superpramart convergence theorem in Banach lattices, extending a result of Ghoussoub and Talagrand (theorem V.3.20). As follows from this result, all we can expect is weak convergence a.e..

Theorem VIII.1.21 (Egghe [1982c]) : The following assertions for a Banach lattice are equivalent :

(i) E has (RNP) and every separable sublattice F of E has a quasi-interior point in F' (i.e. condition (i) in theorem V.3.20).

(ii) E has (RNP) and for every sublattice F of E there is a subset A of F' such that $\# A \leq \text{dens } F$ for which $H_A = F'$.

(iii) Every positive superpramart of class (B) converges weakly a.e..

Proof : (i) ⇒ (ii) :
See theorem V.3.19.

(i) ⇒ (iii) :
Let $(X_n, F_n)_{n\in\mathbb{N}}$ be a positive superpramart of class (B). A similar device as in the proof of theorem VII.2.19 or II.2.4.8, now using theorem VIII.1.6 shows that we can suppose $(X_n)_{n\in\mathbb{N}}$ to be uniformly integrable. So, for every $x' \in (E')^+$ $(x'(X_n), F_n)_{n\in\mathbb{N}}$ is a uniformly integrable superpramart. Hence, for every $\varepsilon > 0$, there exists $\delta > 0$ such that $P(A) \leq \delta$ implies that $\sup_{n\in\mathbb{N}} \int_A x'(X_n) < \frac{\varepsilon}{4}$. For $\eta = \min(\frac{\varepsilon}{2}, \delta)$, there is $\sigma_o \in T$ such that for every $\sigma \in T(\sigma_o)$ and $\tau \in T(\sigma)$

$$P(\{E^{F_\sigma} x'(X_\tau) - x'(X_\sigma) \leq \eta\}) \geq 1 - \eta \quad .$$

Put

$$A_{\varepsilon,x',\sigma,\tau} =: \{E^{F_\sigma} x'(X_\tau) - x'(X_\sigma) \leq \eta\} \in F_\sigma$$

Then for every $m \in \mathbb{N}$, every $A \in F_m$ and every $n \in \mathbb{N}(m \vee \sigma_0)$ and $n' \in \mathbb{N}(n)$ we have

$$\int_{A \setminus A_{\varepsilon,x',n,n'}} (x'(X_{n'}) - x'(X_n)) \leq \frac{\varepsilon}{2} \, .$$

Also

$$\int_{A \cap A_{\varepsilon,x',n,n'}} (x'(X_{n'}) - x'(X_n))$$

$$= \int_{A \cap A_{\varepsilon,x',n,n'}} (E^{F_n} x'(X_{n'}) - x'(X_n))$$

$$\leq \eta \leq \frac{\varepsilon}{2} \text{ , since } A \cap A_{\varepsilon,x',n,n'} \in F_n \quad .$$

Hence the sequence $(\int_A x'(X_n))_{n \in \mathbb{N}}$ satisfies : for each $\varepsilon > 0$, there is a $n_0 \in \mathbb{N}$ such that if $n \in \mathbb{N}(n_0)$ and $n' \in \mathbb{N}(n)$ then

$$\int_A x'(X_{n'}) - \int_A x'(X_n) \leq \varepsilon$$

while $(\int_A x'(X_n))_{n \in \mathbb{N}}$ is also bounded. An appeal to the elementary lemma VIII.1.13 shows now that

$$(\int_A x'(X_n))_{n \in \mathbb{N}}$$

converges for every $A \in \bigcup_{m \in \mathbb{N}} F_m$. Using uniform integrability of $(X_n)_{n \in \mathbb{N}}$ once more we see that

$$(\int_A x'(X_n))_{n\in\mathbb{N}}$$

converges for every $A \in \sigma(\bigcup_n F_n)$. Hence, for every $A \in \sigma(\bigcup_n F_n)$ the sequence $(\int_A X_n)_{n\in\mathbb{N}}$ is weakly Cauchy, hence weakly convergent since E has (RNP). Call this limitmeasure μ. Obviously (RNP) is applicable on μ, yielding $X_\infty \in L^1_E$ such that

$$\int_A X_\infty = \mu(A) = \text{w-}\lim_{n\to\infty} \int_A X_n \quad .$$

Now since, for each $x' \in (E')^+$, the positive superpramart converges a.e., by theorem VIII.1.7 , we have necessarily

$$\lim_{n\to\infty} x'(X_n) = x'(X_\infty), \text{ a.e.}$$

for each $x' \in (E')^+$. From lemma II.1.5 one obtains

$$\sup_{n\in\mathbb{N}} \|X_n\| < \infty \text{ , a.e. .}$$

Since F' has a quasi-interior point, where

$$F = \overline{\text{span}} \bigcup_{n\in\mathbb{N}} X_n(\Omega) \quad ,$$

being a separable sublattice of E, it now follows that $(X_n)_{n\in\mathbb{N}}$ converges to X_∞, weakly a.e..

<u>(iii) ⇒ (i)</u>

This is implied by theorem V.3.20 (iv) ⇒ (i). □

We end this section with a proof of theorem VII.2.15, but immediately in a much more general setting : we work in Banach lattices and with GWT (or GBT). It is noted already that GWT (or GBT) do not converge a.e.; this is even the case for positive submils. The only thing we can hope for is L^1_E-convergence of uniformly integrable GWT (or GBT). Compare the next result also with theorem III.4.1.

Theorem VIII.1.22 (Egghe [1982c]) : The following assertions for a Banach lattice E are equivalent :

(i) E is isomorphic to a subspace of $\ell^1(\Gamma)$ for a certain Γ.

(ii) Every uniformly integrable GBT (or GWT) is L^1_E-convergent.

Proof : (ii) ⇒ (i)
This follows already from theorem III.4.1.

(i) ⇒ (ii)
Exactly as in the proof of theorem VIII.1.21 we can prove, using uniform integrability of the GBT $(X_n,\mathcal{F}_n)_{n\in\mathbb{N}}$, that $\text{w-}\lim_{n\to\infty}\int_A X_n$ exists for every $A\in\sigma(\bigcup_n \mathcal{F}_n)$. But, since $\ell^1(\Gamma)$ is a Schur space, we then have that

$$\lim_{n\to\infty}\int_A X_n \tag{1}$$

exists, for every $A\in\sigma(\bigcup_n \mathcal{F}_n)$. From the definition of GBT it follows that the sequence

$$(|E^{\mathcal{F}_m}X_n - X_m| - (E^{\mathcal{F}_m}X_n - X_m))_{\substack{n\in\mathbb{N}(m)\\ m\in\mathbb{N}}}$$

converges to zero in probability and hence, since this sequence is uniformly integrable

$$\lim_{m\in\mathbb{N}}\ \sup_{n\in\mathbb{N}(m)}\ \int_\Omega \| |E^{\mathcal{F}_m}X_n - X_m| - (E^{\mathcal{F}_m}X_n - X_m)\| = 0\ .$$

So

$$\lim_{m\in\mathbb{N}}\ \sup_{n\in\mathbb{N}(m)}\ (\int_\Omega |E^{\mathcal{F}_m}X_n - X_m| - \int_\Omega (E^{\mathcal{F}_m}X_n - X_m)) = 0\ . \tag{2}$$

From (1) and (2) it now follows that

$$\lim_{m\in\mathbb{N}}\ \sup_{n\in\mathbb{N}(m)}\ \int_\Omega \|E^{\mathcal{F}_m}X_n - X_m\| = 0\ .$$

We use now lemma III.3.6, yielding

$$\lim_{m,n\to\infty} \int_\Omega \|X_n - X_m\| = 0 .$$

So, $(X_n)_{n\in\mathbb{N}}$ is L^1_E-convergent. □

VIII.2. Applications to pramart-convergence

In this short section we shall complete the results of chapter VII, as promised in VII.3.6 and VII.3.7. The first deals with two results on pramart convergence :

Theorem VIII.2.1 (Słaby [1983b]) : Let E be a weakly sequentially complete Banach space with (RNP). Then every pramart with an L^1_E-bounded subsequence converges strongly a.e. to an integrable function.

Proof : The proof is essentially the same as that one of theorem VIII.1.18. We include it for the sake of completeness and since it is short. We apply lemma VIII.1.17 once more. So we can assume that

$$X_{n_k} = Y_{n_k} + Z_{n_k}$$

where $(n_k)_{k\in\mathbb{N}}$ is a subsequence of $\mathbb{N}$, where $(Y_{n_k})_{k\in\mathbb{N}}$ is uniformly integrable and where $\lim_{k\to\infty} Z_{n_k} = 0$, a.e.. We can of course assume that E is separable. Then, let $|||\cdot|||$ be the new Kadec-norm equivalent with $\|\cdot\|$, as described in theorem II.2.4.4. Also let $D \subset E'$ be as in theorem II.2.4.4. Since for every $x' \in E'$, $(x'(X_n),\mathcal{F}_n)_{n\in\mathbb{N}}$ is a real pramart with an L^1-bounded subsequence, theorem VIII.1.11 implies the existence of a function $f_{x'} \in L^1$ such that

$$\lim_{n\to\infty} x'(X_n) = f_{x'} \text{ , a.e..}$$

Also, as in the proof of theorem VIII.1.18, the sequence

$$\left(\int_A Y_{n_k}\right)_{k\in\mathbb{N}}$$

is weakly convergent for every $A \in \sigma(\bigcup_n F_n)$, now using the weak sequential completeness of E. Let X_∞ be the Radon-Nikodym derivative of the limit-measure $\text{w-}\lim_{k\to\infty} \int_A Y_{n_k}$, as in theorem VIII.1.18. Again we have

$$\lim_{n\to\infty} x'(X_n) = x'(X_\infty), \text{ a.e.} \tag{1}$$

for every $x' \in E'$. So also

$$\lim_{n\to\infty} |x'(X_n)| = |x'(X_\infty)|, \text{ a.e..}$$

Now $((x'(X_n),F_n)_{n\in\mathbb{N}})_{x'\in D}$ is a uniform sequence of positive sub-pramarts, satisfying the boundedness condition in lemma VIII.1.15. So this lemma guarantees that

$$\lim_{n\to\infty} |||X_n||| = |||X_\infty||| , \text{ a.e.} \tag{2}$$

since for every $x \in E$, $|||x||| = \sup_{x'\in D} |x'(X)|$. The Kadec property of $|||\cdot|||$ now shows that from (1) and (2) follows, since D is countable,

$$\lim_{n\to\infty} X_n = X_\infty , \text{ a.e..} \qquad \square$$

The second result on pramart convergence was also mentioned and is due to N.E. Frangos [1983] :

Theorem VIII.2.2 (N.E. Frangos) : Let E be isomorphic with a subspace of a separable dual Banach space. Then every pramart with an L^1_E-bounded subsequence converges strongly a.e. to an integrable function.

Proof : We can of course assume that E is the separable space F' itself. The same argument as the first part of the proof in VIII.1.20, with D now in F again yields that there exists $X_\infty \in L^1_E$ such that for every $x' \in F$

(i) $$\lim_{n\to\infty} x'(X_n(\omega)) = x'(X_\infty(\omega))$$

for every $\omega \in \Omega \setminus N$, where $P(N) = 0$.
The second part of this proof can also be used, but now using the classical Kadec-renorming theorem II.2.4.4. □

It is interesting to know that for theorem VIII.2.2 we do not need the Kadec-Klee renorming result. Indeed :

VIII.2.3 - Second proof of theorem VIII.2.2 (Frangos) : We use the elementary fact that since E is a separable space, there is a countable set $D \subset E'$ such that $\|x'\| \leq 1$ for every $x' \in D$ and such that

$$\|x\| = \sup_{x' \in D} |x'(x)|$$

for every $x \in E$ (see e.g. introduction of IV.1). The same argument as the first part of the proof in VIII.1.20 is also used here, yielding again that there exists a function X_∞ such that

$$\text{(i)} \qquad \lim_{n \to \infty} x'(X_n(\omega)) = x'(X_\infty(\omega)) \ ,$$

a.e., for every $x' \in D$. So by lemma VIII.1.15 we again have that

$$\text{(ii)} \qquad \lim_{n \to \infty} \|X_n(\omega)\| = \|X_\infty(\omega)\| \ ,$$

a.e..
We proceed now as follows, still using the original norm $\|\cdot\|$ of E (cf. also the argument in Neveu [1975], p.110).
For every $x \in E$ fixed, the same argument as above applies yielding

$$\lim_{n \to \infty} \|X_n - x\| = \|X_\infty - x\| \ ,$$

a.e.. So, if A is a countable dense subset of E, we have

$$\lim_{n \to \infty} \|X_n - x\| = \|X_\infty - x\| \ ,$$

a.e. for every $x \in A$, hence for all $x \in E$.
In particular

$$\lim_{n\to\infty} \|X_n - X_\infty\| = 0 ,$$

a.e.. □

Our next application deals with weak pramarts, W pramarts for short, as indicated in VII.3.7. It concerns a refinement of theorem VII.2.28, valid for class (B) W mils, now valid only for W pramarts, but here L^1_E-boundedness suffices, even of a subsequence. The result is of L. Sucheston (written communication) and I think is not published.

Theorem VIII.2.4. (L. Sucheston) : Let E be a reflexive Banach space and let $(X_n, F_n)_{n\in\mathbb{N}}$ be a W pramart with an L^1_E-bounded subsequence. Then $(X_n)_{n\in\mathbb{N}}$ converges weakly a.e..

Proof : We remark that in the first proof of theorem VII.2.28, we only used class (B) to have an a.e. finite supremum. For W pramarts this is true without assuming class (B) as the next lemma shows, since, by lemma VIII.1.12, $(\|X_n\|, F_n)_{n\in\mathbb{N}}$ is a subpramart with an L^1-bounded subsequence :

Lemma VIII.2.4.1 : Let $(X_n, F_n)_{n\in\mathbb{N}}$ be a real positive subpramart with an L^1-bounded subsequence. Then $\sup_{n\in\mathbb{N}} X_n < \infty$, a.e..

Proof : By lemma VIII.1.9 there exists a positive submartingale $(R_n, F_n)_{n\in\mathbb{N}}$ such that $0 \leqslant R_n \leqslant X_n$, a.e., for every $n \in \mathbb{N}$ and such that $\lim_{n\to\infty} (X_n - R_n) = 0$, a.e.. From this $\sup_{n\in\mathbb{N}} X_n < \infty$ if and only if $\sup_{n\in\mathbb{N}} R_n < \infty$. But since $0 \leqslant R_n \leqslant X_n$, a.e. for every $n \in \mathbb{N}$, $(R_n)_{n\in\mathbb{N}}$ is also an L^1-bounded subsequence. Hence $(R_n)_{n\in\mathbb{N}}$ is of class (B), being a submartingale : indeed, since $(R_n, F_n)_{n\in\mathbb{N}}$ is a submartingale we have for every $\sigma \in T$ and $\tau \in T(\sigma)$, $R_\sigma \leqslant E^{F_\sigma} R_\tau$, by definition III.2.1. So every $\int_\Omega R_\sigma$ is smaller than or equal $\int_\Omega R_{n_i}$ where R_{n_i} is one term in the L^1-bounded subsequence $(R_{n_k})_{k\in\mathbb{N}}$. Now invoke lemma II.1.5(i) or (ii) yielding $\sup_{n\in\mathbb{N}} R_n < \infty$, a.e.. Hence also $\sup_{n\in\mathbb{N}} X_n < \infty$, a.e.. □

VIII.3. Notes and remarks

VIII.3.1. It is interesting to remark that theorem VIII.1.7 on the convergence a.e. of positive superpramarts has been proved in a self-contained way : it does not use the supermartingale convergence theorem.

A second interesting point is that from this theorem VIII.1.7 the real positive submartingale convergence theorem also follows : indeed, let $(X_n, \mathcal{F}_n)_{n\in\mathbb{N}}$ be a real positive L^1-bounded submartingale. As follows from example V.2.3(2), $(X_n, \mathcal{F}_n)_{n\in\mathbb{N}}$ is an amart (in $\mathbb{R}$) and hence a uniform amart. So theorem V.1.4 shows that for every $n \in \mathbb{N}$

$$X_n = Y_n + Z_n$$

where $(X_n, \mathcal{F}_n)_{n\in\mathbb{N}}$ is a martingale. From the special case we have here, and from the proof of theorem V.1.4 we see that

$$Y_m = \lim_{n\in\mathbb{N}(m)} \uparrow E^{\mathcal{F}_m} X_n$$

So $(Y_n, \mathcal{F}_n)_{n\in\mathbb{N}}$ and $(-Z_n, \mathcal{F}_n)_{n\in\mathbb{N}}$ both are positive superpramarts, converging a.e. by theorem VIII.1.7. Hence, also $(X_n)_{n\in\mathbb{N}}$ converges a.e..

VIII.3.2. Consider a real process $(X_n, \mathcal{F}_n)_{n\in\mathbb{N}}$ with the following property :

(1) for every $\varepsilon > 0$ there is a $m_o \in \mathbb{N}$ such that for every $m \in \mathbb{N}(m_o)$ and $\tau \in T(m)$ we have

$$P(\{\omega\in\Omega \| E^{\mathcal{F}_m} X_\tau(\omega) - X_m(\omega) \geq -\varepsilon\}) \geq 1 - \varepsilon$$

This is a property between the notions subpramart and GBT. Can theorem VIII.1.11 be extended to adapted sequences of type (1) above? This is not possible as the following example of Millet and Sucheston shows : Let $(A_n)_{n\in\mathbb{N}}$ be an independent sequence of sets in a probability space $(\Omega,\mathcal{F},P)$

such that $P(A_n) = \frac{1}{n}$ for every $n \in \mathbb{N}$ and put

$$X_n = \chi_{A_n}$$

$$F_n = \sigma(X_1,\ldots,X_n)$$

for every $n \in \mathbb{N}$. Then $(X_n,F_n)_{n\in\mathbb{N}}$ satisfies (1) : for every $\varepsilon > 0$, $n \in \mathbb{N}$ and $\tau \in T(n)$ one has

$$P(\{\omega\in\Omega \| X_n(\omega) - E^{F_n} X_\tau(\omega) > \varepsilon\}) \leqslant P(A_n)$$

But the Borel-Cantelli lemma shows that

$$\liminf_{n\in\mathbb{N}} X_n = 0, \text{ a.e.} \quad \text{and} \quad \limsup_{n\in\mathbb{N}} X_n = 1, \text{ a.e..}$$

We have shown at the same time that the class of subpramarts is strictly smaller than the class of adapted sequences discussed here.

What can be shown is that adapted sequences of type (1) above do converge in probability if they satisfy the boundedness condition in theorem VIII.1.11.
This is proved exactly as in theorem VIII.1.11 and lemma VIII.1.9, where now only $\lim_{n\to\infty} (X_n - R_n) = 0$ in probability can be shown (the only change is in lemma VIII.1.9 where theorem I.3.5.5 cannot be used any more). So we repeat :

Theorem VIII.3.2.1 : Let $(X_n,F_n)_{n\in\mathbb{N}}$ be a real positive adapted sequence satisfying property (1) above and the boundedness condition

$$\liminf_{n\in\mathbb{N}} \int_\Omega X_n^+ + \liminf_{n\in\mathbb{N}} \int_\Omega X_n^- < \infty$$

Then $(X_n)_{n\in\mathbb{N}}$ converges in probability.

VIII.3.3. A special proof of the subpramart and pramart convergence theorems of Słaby, valid only in Banach spaces with an unconditional basis is given in Słaby [1983a]. The presented results in chapter VIII however are much more general. Also in this article, the following extension of theorem III.5.1, on convergence of positive submartingales in Banach lattices which do not necessarily have (RNP), is proved.

Theorem VIII.3.3.1 (Słaby [1983a]) : Let E be an order-continuous Banach lattice and let $(X_n, F_n)_{n \in \mathbb{N}}$ be a subpramart such that

$$\liminf_{n \in \mathbb{N}} \int_\Omega \|X_n\| < \infty$$

and such that

$$0 \leq X_n \leq Y_n \ , \quad \text{a.e.} \qquad (1)$$

for each $n \in \mathbb{N}$, where $(Y_n)_{n \in \mathbb{N}}$ is a sequence of measurable functions converging strongly a.e. to a function Y. Then there exists $X_\infty \in L^1_E$ such that $\lim_{n\to\infty} X_n = X_\infty$, strongly a.e..

Proof : The proof is the same as that of theorem III.5.1, now using theorem VIII.1.11, lemma VIII.1.12 and VIII.1.15. □

The condition (1) in theorem VIII.3.3.1 can be weakened to

(1') there exists an E-valued measurable function Y such that

$$\lim_{n\to\infty} x'(X_n \vee Y) = x'(Y) \ ,$$

a.e. for all $x' \in D$, where D is the countable norming subset of E'^{+}, mentioned in the lattice renorming theorem of Davis-Ghoussoub-Lindenstrauss, see theorem III.2.3.
In fact this generalization, remarked by Frangos [1983], may be trivially seen from the original proof.

VIII.3.4. Lemma VIII.1.9 can be extended to Banach lattices with unconditional basis. One uses lemma VIII.1.9 in every coordinate. For the easy proof, see Słaby [1983b] .

VIII.3.5. For GBT a convergence result of the type discussed in VII.3.8.2 and VII.3.8.3 is also valid, but in a rather restricted way, see Baek and Hong, [1982] , theorem 7 :

Theorem VIII.3.5.1 (Baek In-Soo and Hong Duk-Hun) : Let $(X_n, F_n)_{n \in \mathbb{N}}$ be a GBT and let A be an atom of the probability space (Ω, F, P). Suppose that

$$\int_{\Omega} X_{\tau}^{+} < \infty$$

for every $\tau \in T_f$, where we follow the notation in VII.3.8. Then $\lim_{n\to\infty} X_n$ exists on A and is $> -\infty$ on A.

Chapter IX : CLOSING REMARKS

IX.1. A general remark concerning scalar convergence

In the following chapters we were mainly interested in a.e. convergence properties of the studied adapted sequences. However we have not studied the properties (a) and (b) below where E is a Banach space and $(X_n, F_n)_{n \in \mathbb{N}}$ is an adapted sequence.

(a) Scalar convergence a.e. (cf. I.3.1)
There exists $X_\infty \in L_E^1$ such that for every $x' \in E'$

$$\lim_{n\to\infty} x'(X_n) = x'(X_\infty),$$

a.e..

(b) Scalar convergence in probability
There exists $X_\infty \in L_E^1$ such that for every $x' \in E'$

$$\lim_{n\to\infty} x'(X_n) = x'(X_\infty) ,$$

in probability.

The following result is very easy; however, as is seen in the corollaries, it completely solves the problem :

Theorem IX.1.1 : Let E be a Banach space with (RNP). Let $(X_n, F_n)_{n \in \mathbb{N}}$ be an adapted sequence with the following properties :

(a) There is an L_E^1-bounded sequence $(Y_n)_{n \in \mathbb{N}}$ in L_E^1 with $\lim_{n\to\infty} (x'(X_n) - x'(Y_n)) = 0$, a.e., for every $x' \in E'$, such that

$\text{w-}\lim\limits_{n\to\infty} \int_A Y_n$ exists for every $A \in \bigcup\limits_n F_n$.

(b) For every $x' \in E'$, $(x'(X_n))_{n\in\mathbb{N}}$ converges a.e. (resp. in probability). If E is a Banach lattice, the above is only supposed to be valid for every $x' \in (E')^+$.

Then there is $X_\infty \in L_E^1$ such that $(X_n)_{n\in\mathbb{N}}$ converges to X_∞ scalarly a.e. (resp. scalarly in probability).

Proof : Denote

$$\mu(A) = \text{w-}\lim_{n\to\infty} \int_A Y_n$$

for every $A \in \bigcup\limits_n F_n$. Extend μ to $\sigma(\bigcup\limits_n F_n)$ by the classical Caratheodory-Hahn-Kluvanek extension theorem yielding a measure, still denoted by μ, of bounded variation and P-continuous on $\sigma(\bigcup\limits_n F_n)$. From (RNP) of E, we derive the existence of $X_\infty \in L_E^1$ such that, for every $A \in \sigma(\bigcup\limits_n F_n)$

$$\mu(A) = \int_A X_\infty = \text{w-}\lim_{n\to\infty} \int_A Y_n$$

Fix $x' \in E'$. Let $\lim\limits_{n\to\infty} x'(X_n) = f_{x'} \in L^1$, a.e. (resp. in probability). So also $\lim\limits_{n\to\infty} x'(Y_n) = f_{x'}$, a.e. (resp. in probability). Since $\lim\limits_{n\to\infty} \int_A x'(Y_n) = \int_A x'(X_\infty)$ it follows that $x'(X_\infty) = f_{x'}$, a.e.. Indeed, this follows readily from Egorov's theorem. Hence

$$\lim_{n\to\infty} x'(X_n) = x'(X_\infty),$$

a.e. (resp. in probability). □

Corollary IX.1.2 :

1. Every W amart with an L_E^1-bounded subsequence converges scalarly a.e. to a function in L_E^1, if E has (RNP). This extends Edgar [1982a], theorem 8.

2. Every W mil which is L_E^1-bounded converges scalarly a.e. to a function in L_E^1, if E has (RNP) and is weakly sequentially complete.

3. Every subpramart in a Banach lattice E with an L^1_E-bounded subsequence converges scalarly a.e., to a function in L^1_E, if E has (RNP).

Proof :

1. Let $(X_{n_k})_{k\in\mathbb{N}}$ be L^1_E-bounded. Since $\text{w-}\lim_{n\to\infty} \int_A X_{n_k}$ exists on $\bigcup_n F_n$, and since $\lim_{n\to\infty} x'(X_n)$ exists a.e., being a scalar amart with an L^1_E-bounded subsequence, hence L^1_E-bounded itself, we see that theorem IX.1.1 applies for $(X_{n_k}, F_{n_k})_{k\in\mathbb{N}}$ and $Y_k = X_{n_k}$ for every $k \in \mathbb{N}$. Hence, there is a $X_\infty \in L^1_E$ such that $\lim_{k\to\infty} x'(X_{n_k}) = x'(X_\infty)$, a.e.. Hence also $\lim_{n\to\infty} x'(X_n) = x'(X_\infty)$, a.e..

2. Apply lemma VIII.1.17, yielding a uniformly integrable sequence $(Y_n)_{n\in\mathbb{N}}$ such that $\lim_{n\to\infty} (X_n - Y_n) = 0$, a.e.. $\lim_{n\to\infty} x'(X_n)$ exists a.e., being a scalar L^1-bounded mil and using theorem VII.2.12. Also $(Y_n)_{n\in\mathbb{N}}$ is uniformly integrable and E is weakly sequentially complete. Now it follows that theorem IX.1.1 applies for $(X_n, F_n)_{n\in\mathbb{N}}$. Hence, there is $X_\infty \in L^1_E$ such that $\lim_{n\to\infty} x'(X_n) = x'(X_\infty)$, a.e..

3. This proof is the same as the proof above, now using theorem VIII.1.11 and the fact that a (RNP) Banach lattice is always weakly sequentially complete. □

IX.2. Summary of the most important convergence results

In this section we shall repeat the most important and most general convergence results of adapted sequences known so far.

IX.2.1. We deal with the following classes and the following interrelations :

GENERAL BANACH SPACES

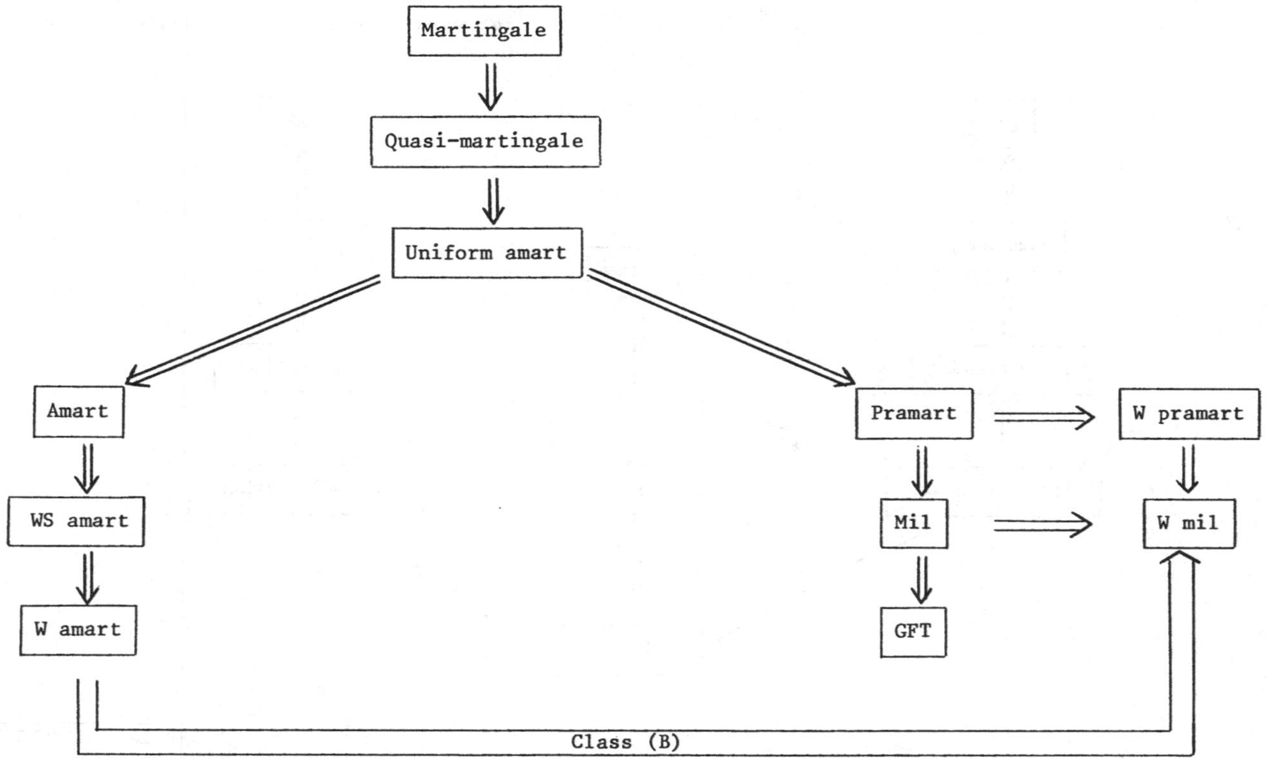

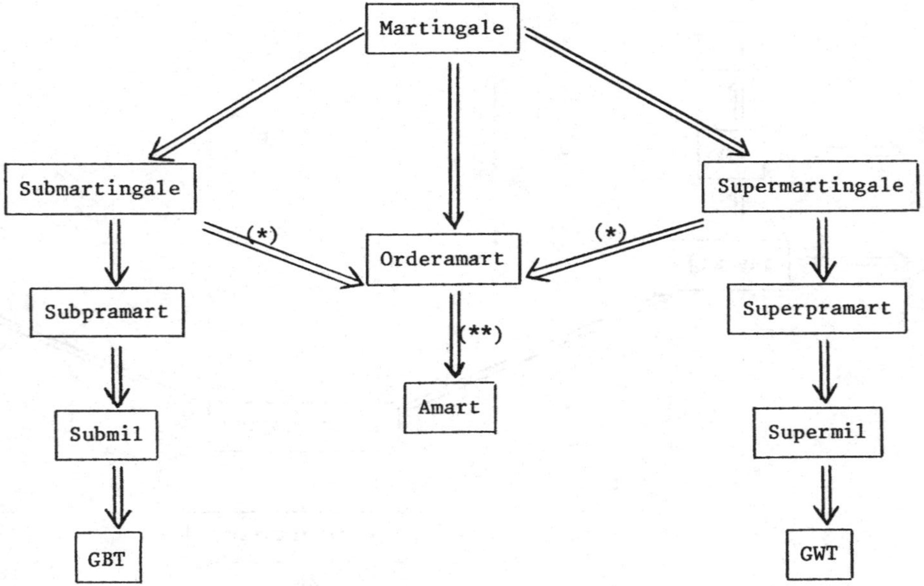

IX.2

(*) : under the condition $c_o \not\hookrightarrow E$ and class (B)

(**) : if E has ordercontinuous norm.

CASE E = $\mathbb{R}$

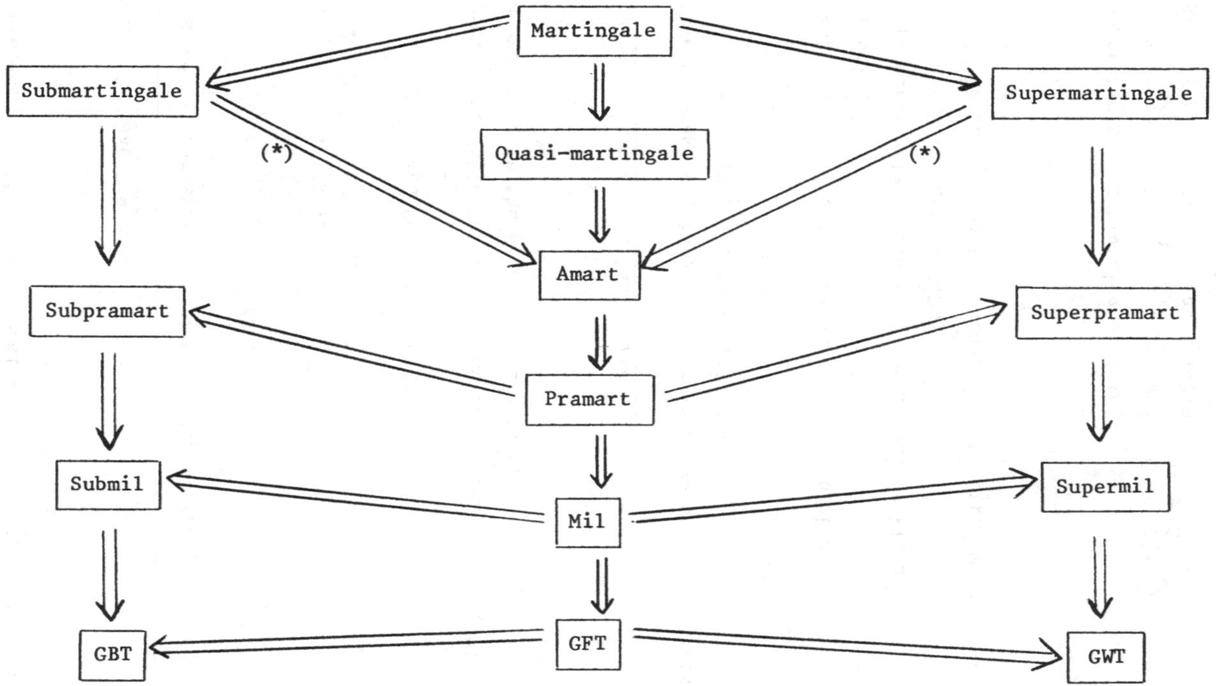

(*) : if the processes are of class (B).

IX.2.2. The results

We present only the most general results w.r.t. convergence.

IX.2.2.1. Results in case $E = \mathbb{R}$

1. Every L^1-bounded mil converges a.e. (Mucci, VII.2.12).

2. Every subpramart $(X_n, F_n)_{n\in\mathbb{N}}$ such that

$$\liminf_{n\in\mathbb{N}} \int_\Omega X_m^+ + \liminf_{n\in\mathbb{N}} \int_\Omega X_n^- < \infty$$

converges a.e. (Millet and Sucheston, VIII.1.11).

3. Every uniformly integrable GBT is L^1-convergent (Subramanian, VIII.1.22, which is an extension in the Banach lattice case).

IX.2.2.2. Results in case E is a Banach lattice

1. Every positive subpramart with an L_E^1-bounded subsequence converges strongly a.e. if E has (RNP). (Słaby, VIII.1.18).

2. (a) ⇔ (b) ⇔ (c)
   - (a) E has (RNP) and every separable sublattice F of E has a quasi-interior point in F'.
   - (b) Every supermartingale of class (B) converges weakly a.e.. (Ghoussoub and Talagrand, V.3.20).
   - (c) Every positive superpramart of class (B) converges weakly a.e.. (Egghe, VIII.1.21).

3. (a) ⇔ (b)
   - (a) E is isomorphic to a sublattice of $\ell^1(\Gamma)$.
   - (b) Every submartingale $(X_n, F_n)_{n\in\mathbb{N}}$ for which $\sup_{n\in\mathbb{N}} \int_\Omega \|X_n^+\| < \infty$ converges strongly a.e. (Szulga, III.3.8).
   - (c) Every positive supermartingale converges strongly a.e. (III.4.1).
   - (d) Every uniformly integrable GBT is L_E^1-convergent (Egghe, VIII.1.22).

4. Let E be an ordercontinuous Banach lattice and let $(X_n, F_n)_{n \in \mathbb{N}}$ be a subpramart such that

$$\liminf_{n \in \mathbb{N}} \int_{\Omega} \|X_n\| < \infty$$

and such that

$$0 \leq X_n \leq Y_n \ , \quad \text{a.e.}$$

for each $n \in \mathbb{N}$, where $(Y_n)_{n \in \mathbb{N}}$ is a sequence of measurable functions converging strongly a.e. to a function Y. Then there exists $X_\infty \in L^1_E$ such that $\lim_{n\to\infty} X_n = X_\infty$, strongly a.e. (Słaby, VIII.3.3.1).

IX.2.2.3. <u>Results in case E is a Banach space</u>

1. Every L^1_E-bounded uniform amart converges strongly a.e. if E has (RNP) (Bellow, V.1.3).

2. Every pramart with an L^1_E-bounded subsequence converges strongly a.e. if E has (RNP) and if E is weakly sequentially complete (Słaby, VIII.2.1), or if E is a subspace of a separable dual space (Frangos, VIII.2.2).

3. Every pramart of class (B) converges strongly a.e. if E has (RNP) (Millet and Sucheston, VII .2.19).

4. Every mil with a uniformly integrable subsequence converges strongly a.e. if E has (RNP) (Bellow and Dvoretzky, VII.2.18).

5. Every class (B) W mil converges weakly a.e. if E is reflexive (Egghe, VII.2.28).

6. Every W pramart with an L^1_E-bounded subsequence converges weakly a.e. if E is reflexive (Sucheston, VIII.2.4).

7. Every WS amart of class (B) converges weakly a.e. if E and E' have (RNP) (Brunel and Sucheston, V.3.5).

IX.3. Convergence of adapted sequences of Pettis integrable functions

In this book we limited ourselves to adapted sequences or nets of Bochner integrable functions. So we did not consider the case of adapted sequences of Pettis integrable functions. The reason for this is clear : virtually nothing is known about these processes. The study of it is complicated due to the fact that Pettis integrability is a difficult notion and - until recently - not well-known. Recently there has been some progress in the study of Pettis integrability. Let us give the definition first.

Definition IX.3.1 : Let $X : (\Omega,F,P) \to E$ be scalarly integrable i.e. for every $x' \in E'$, $x'(X) \in L^1$. Suppose also that, for every $A \in F$, there exists $x_A \in E$ such that

$$x'(x_A) = \int_A x'(X)$$

Then we say that X is Pettis integrable and we denote by

$$x_A = P - \int_A X$$

or shortly by $\int_A X$ if there cannot be a confusion, the Pettis integral of X over A.

It is clear that every Bochner integrable function is Pettis integrable, and, in this case, the integrals coincide. The converse is not true. In fact, we even have

Theorem IX.3.2 : The following assertions are equivalent where (Ω,F,P) is a fixed, atomless probability space :

(i) $\dim E < \infty$.

(ii) Every Pettis integrable function is Bochner integrable.

(iii) Every strongly measurable Pettis integrable function is Bochner integrable.

Proof : (i) ⇒ (ii)

The Pettis integral

$$\mu(A) = \int_A X$$

is easily seen to be of bounded variation. Furthermore X is strongly measurable. Hence, $X \in L^1_E$.

(ii) ⇒ (iii)

Is trivial.

(iii) ⇒ (i)

Apply, as we did before, the theorem of Dvorerzky and Rogers [1950], supposing dim E = ∞. Let then $(x_n)_{n \in \mathbb{N}}$ in E be such that $\Sigma\, x_n$ converges unconditionally and such that $\Sigma \|x_n\| = \infty$. Let $(A_n)_{n \in \mathbb{N}}$ be a countable measurable partition of Ω such that $P(A_n) > 0$ for every $n \in \mathbb{N}$. Now it is easy to see that

$$X = \sum_{n \in \mathbb{N}} \frac{x_n}{P(A_n)} \chi_{A_n}$$

is Pettis integrable, strongly measurable, and that $X \notin L^1_E$. □

IX.3.3.

1. The concept of Pettis integrability is a very strange one. To illustrate this we can remark that conditional expectations of Pettis integrable functions do not necessarily exist. See Musiał [1979]. Compare also with the problems concerning vector measures of σ-bounded variation, see Thomas [1974], Egghe [1984b], Musiał [1977]. Another problem with Pettis integrable functions is the following. Denote by $\mathcal{P}^1_E$ the class of all Pettis integrable functions. Put on this vector space the seminorm

$$Pe : \mathcal{P}^1_E \to \mathbb{R}^+$$

$$X \to \sup_{\substack{\|x'\| \leq 1 \\ x' \in E'}} \int_\Omega |x'(X)|$$

Consider in $\mathcal{P}^1_E$ the equivalence classes of functions which are scalarly equal a.e. : $f \sim g$ if $x'(f) = x'(g)$, a.e. for every $x' \in E'$. The quotientspace of $\mathcal{P}^1_E$ w.r.t. this equivalence relation is denoted by P^1_E and the quotientnorm by $\|\cdot\|_{Pe}$. The following negative results are to be noted :

2. $P^1_E, \|\cdot\|_{Pe}$, is a Banach space if and only if $\dim E < \infty$, see Thomas [1979].

3. Pettis integrable functions are not necessarily $\|\cdot\|_{Pe}$-limits of step-functions, see Musiał [1980].

   IX.3.3.1, IX.3.3.2 and IX.3.3.3 are bad results when studying adapted sequences of Pettis integrable functions.

Definition IX.3.4 : Let $(\mathcal{F}_n)_{n\in\mathbb{N}}$ be a stochastic basis and let $(X_n)_{n\in\mathbb{N}}$ be a sequence of Pettis integrable functions. $(X_n,\mathcal{F}_n)_{n\in\mathbb{N}}$ is called a scalarly adapted sequence (or adapted sequence, if there cannot be a confusion) if X_n is scalarly $\mathcal{F}_n$-measurable for every $n \in \mathbb{N}$. A martingale is then defined as an adapted sequence such that

$$\int_A X_n = \int_A X_{n+1}$$

for every $A \in \mathcal{F}_n$ and every $n \in \mathbb{N}$.

But even in this case, we still have problems. We have seen that a.e. convergence of martingales of Bochner integrable functions is heavily related to (RNP). In the same way the notion "weak Radon-Nikodym property" comes on the scene when studying convergence of martingales as in definition IX.3.4.

Definition IX.3.5 (Musiał [1977], [1979]) : A Banach space E is said to have the Weak Radon-Nikodym property (abbreviated (WRNP)) if for every complete probability space $(\Omega,\mathcal{F},P)$ and for every vectormeasure $F : \mathcal{F} \to E$ of σ-bounded variation such that $F \ll P$, there exists a Pettis integrable function X, such that for every $A \in \mathcal{F}$:

$$F(A) = \int_A X$$

An equivalent statement of (WRNP) is obtained if one replaces "σ-bounded variation" by "bounded variation". However, the first definition is more natural since a Pettis integral is always of σ-bounded variation, see Musiał [1977], but not necessarily of bounded variation.

IX.3.6.

Now another difficulty arises by the fact that the property (WRNP) is not hereditary for closed subspaces, as is (RNP) - see Musiał [1977].

IX.3.7.

As follows from theorem I.1.1.1 (WRNP) and (RNP) are equivalent for separable Banach spaces. So c_o, L^1 do not have (WRNP). A dual space E' has (WRNP) if and only if $\ell^1 \not\subset E$ (see Musiał [1977], Egghe [1980c]). For further results on Pettis integrability as well as on (WRNP), see Musiał [1977] and [1979], Jeurnink [1982], Geitz [1981] and [1982], Riddle [1982], Riddle and Uhl [1982], Edgar [1982b], Talagrand [1980], Sentilles and Wheeler [1982], Sentilles [1982], Thomas [1974] and [1976], Chatterji [1974], Egghe [1980c], Ghoussoub and Saab [1981] and references therein.

IX.3.8.

We have explained enough why the study of adapted sequences of Pettis integrable functions is not well-developed at the moment.
Concerning adapted sequences of general Pettis integrable functions which are not necessarily strongly measurable, in fact only one result is proved. It extends theorem 5 in Musiał [1980], p.334.
First a definition :

Definition IX.3.8.1 : An adapted sequence $(X_n, F_n)_{n \in \mathbb{N}}$ is called an amart if $(\int_\Omega X_\tau)_{\tau \in T}$ converges in E. $(X_n)_{n \in \mathbb{N}}$ is called Pettis uniformly integrable if $\sup_{n \in \mathbb{N}} \|X_n\|_{Pe} < \infty$ and

$$\lim_{P(A) \to 0} \sup_{n \in \mathbb{N}} \sup_{B \in F} \left\| \int_{A \cap B} X_n \right\| = 0 .$$

Theorem IX.3.8.2 (Egghe) : Let E have (WRNP) and let $(X_n, F_n)_{n \in \mathbb{N}}$ be an amart such that there is a disjoint sequence $(A_k)_{k \in \mathbb{N}}$ in F_1 and a strictly increasing sequence $(m_n)_{n \in \mathbb{N}}$ in $\mathbb{N}$ such that

$$\sup_{n \in \mathbb{N}} \int_{A_k} \|X_{m_n}\| < \infty$$

for each $k \in \mathbb{N}$. Suppose that $(X_n)_{n\in\mathbb{N}}$ is Pettis uniformly integrable. Then there is a function $X_\infty \in P_E$ such that $\lim_{n\to\infty} \|X_n - X_\infty\|_{Pe} = 0$.

Concerning adapted sequences of strongly measurable Pettis integrable functions a little bit more is known, although even this theory is far from complete. We only mention the martingale convergence theorem of J.J. Uhl Jr. extending theorem II.2.4.3; see Uhl [1972] , theorem 3.1.

Theorem IX.3.8.3 (Uhl) : Let $(X_n, F_n)_{n\in\mathbb{N}}$ be a martingale consisting of strongly measurable Pettis integrable functions. Suppose

(i) $\sup_{n\in\mathbb{N}} \|X_n\|_{Pe} < \infty$.

(ii) The set $\{\int_A X_n \| A \in F_n,\ n\in\mathbb{N}\}$ is weakly relatively compact.

(iii) For each $\varepsilon > 0$, there is a weakly compact set $K \subset E$ such that for each $\delta > 0$, there exists $n_o \in \mathbb{N}$ and $A_o \in F_{n_o}$ for which $P(\Omega \setminus A_o) < \varepsilon$ such that $n \in \mathbb{N}(n_o)$ implies

$$\int_A X_n \in P(A)K + \delta B_E$$

for all $A \subset A_o$ such that $A \in F_n$, where $B_E =: \{x\in E \| \|x\| \leq 1\}$.

Then $(X_n)_{n\in\mathbb{N}}$ converges strongly a.e. to a function in P_E.

Indeed this theorem extends theorem II.2.4.3 of Ionescu-Tulcea : condition (iii) is a Rieffel-type condition which is satisfied if E has (RNP) and if (i) is strengthened to $\sup_{n\in\mathbb{N}} \|X_n\|_1 < \infty$, in which case we have theorem II.2.4.3 (condition (ii) can be dropped in this case as is proved in cor.3.4 in Uhl [1972]). In the proof of theorem IX.3.8.3, Uhl did not use theorem II.2.4.3 and in fact not even the real case of it, being Doob's theorem.

For other results on convergence of adapted sequences of Pettis integrable functions, see Egghe [1984b] , Métivier [1967] , Musiał [1980] , Słaby [1982] , Uhl [1974] and [1977] .

The possibility is open of proving nicer and more applicable theorems guaranteeing a.e. convergence results for adapted sequences of Pettis integrable functions, even for strongly measurable ones.

REFERENCES

Astbury, K.A. (1978), Amarts indexed by directed sets. Ann. Prob. 6, n° 2, 267-278.

Austin, D.G., Edgar G.A. and Ionescu-Tulcea, A. (1974), Pointwise convergence in terms of expectations. Z. Wahrscheinlichkeitstheorie verw. Geb. 30, 17-26.

Baek-In-Soo and Hong Duk-Hun (1982), Convergences of games better with time. Kyungpook Math. J. 22, n° 2, 295-302.

Baxter, J.R. (1974), Pointwise in terms of weak convergence. Proc. Amer. Math. Soc. 46, n° 3, 395-398.

Bellow, A. : see also D.G. Austin, A. Ionescu-Tulcea.

Bellow, A. (1976), On vector-valued asymptotic martingales. Proc. Natl. Acad. Sc. USA, 73, n° 6, 1798-1799.

Bellow, A. (1977a), Several stability properties of the class of asymptotic martingales. Z. Wahrscheinlichkeitstheorie verw. Geb. 37, 275-290.

Bellow, A. (1977b), Les amarts uniformes. C.R. Acad. Sc. Paris, série A, t.284, 1295-1298.

Bellow, A. (1978a), Uniform amarts : a class of asymptotic martingales for which strong almost sure convergence obtains. Z. Wahrscheinlichkeitstheorie verw. Geb. 41, 177-191.

Bellow, A. (1978b), Some aspects of the theory of vector-valued amarts. Proc. conf. vector space measures and applications I, Dublin 1977. Lect. Notes in Math. 644, 57-67, Springer Verlag Berlin.

Bellow, A. (1978c), Sufficiently rich sets of stopping times, measurable cluster points and submartingales. Semin. Géom. des esp. de Banach, Ec. Polytechn., Palaiseau, 1977-1978, Apendice 1, 11 p.

Bellow, A. (1981), Martingales, amarts and related stopping time techniques. Proc. Tufts Conf. on Prob. in Banach spaces III, Medford Massachusetts USA 1980. Lect. Notes in Math. 860, 9-24, Springer Verlag Berlin.

Bellow, A., and Dvoretzky A. (1979), A characterization of almost sure convergence, Prob. in Banach spaces II. Proc. second int. conf. Oberwolfach 1978. Lect. Notes in Math. 709, 45-65, Springer Verlag Berlin.

Bellow, A. and Dvoretzky, A. (1980a), On martingales in the limit. Ann. Prob. 8, n° 3, 602-606.

Bellow, A. and Dvoretzky, A. (1980b), A note on the strong convergence of mils in Banach spaces, Preprint.

Bellow, A. et Egghe, L. (1981), Inégalités de Fatou généralisées, C.R. Acad. Sc. Paris, série I, t.292, 847-850.

Bellow, A. and Egghe, L. (1982), Generalized Fatou inequalities, Ann. Inst. H. Poincaré, 18, n° 4, 335-365.

Benyamini, Y. et Ghoussoub, N. (1978), Une caractérisation probabiliste de ℓ^1. C.R. Acad. Sc. Paris, série A, t.286, 795-797.

Bessaga, C. and Pełczynski, A. (1975), Topics in infinite dimensional topology. Monografie Mat., tome 58, PWN Warszawa.

Blake, L.H. (1970), A generalization of martingales and two consequent convergence theorems. Pac. J. Math. 35, n° 2, 279-283.

Blake, L.H. (1972), A note concerning the L_1 convergence of a class of games which become fairer with time. Glasgow Math. J. 13, 39-41.

Blake, L.H. (1975), A note concerning first order games which become fairer with time. J. London Math. Soc. 9, 589-592.

Blake, L.H. (1978), Every amart is a martingale in the limit. J. London Math. Soc. 18, n° 2, 381-384.

Blake, L.H. (1979), Weak submartingales in the limit. J. London Math. Soc. 19, n° 2, 573-575.

Blondia, C. (1981a), Integration in locally convex spaces. Simon Stevin 55, n° 3, 81-102.

Blondia, C. (1981b), A Radon-Nikodym theorem for vector valued measures. Bull. Soc. Math. Belg. 33, n° 2, série B, 231-249.

Bourgain, J. and Delbaen, F. (1980), A class of special $\mathcal{L}_\infty$ spaces. Acta Math. 145, n° 3-4, 155-176.

Bourgain, J. and Rosenthal, H.P. (1980), Martingales valued in certain subspaces of L_1. Israel J. Math. 37, n° 1-2, 54-75.

Breiman, L. (1978), Probability. Addison-Wesley Publ. Cy. Reading Massachusetts USA.

Bru, B. (1980), Espérance d'ordre. C.R. Acad. Sc. Paris série A, t.290, 111-114.

Bru, B. et Heinich, H. (1979a), Sur les suites de mesures vectorielles adaptées. C.R. Acad. Sc. Paris série A, t.288, 363-366.

Bru, B. et Heinich, H. (1979b), Sur l'espérance des variables aléatoires vectorielles. Preprint.

Brunel, A. and Sucheston, L. (1973), On B-convex Banach spaces. Math. Systems Theory, 7, n° 4, 294-299.

Brunel, A. et Sucheston, L. (1976a), Sur les amarts faibles à valeurs vectorielles. C.R. Acad. Sc. Paris série A, t.282, 1011-1014.

Brunel, A. et Sucheston, L. (1976b), Sur les amarts à valeurs vectorielles. C.R. Acad. Sc. Paris série A, t.283, 1037-1040.

Brunel, A. et Sucheston, L. (1977), Une caractérisation probabiliste de la séparabilité du dual d'un espace de Banach. C.R. Acad. Sc. Paris série A, t.284, 1469-1472.

Burkholder, D.L. (1981a), A geometrical characterization of Banach spaces in which martingale difference sequences are unconditional. Ann. Prob. 9, n° 6, 997-1011.

Burkholder, D.L. (1981b), Martingale transforms and the geometry of Banach spaces. Proc. Tufts Conf. on Prob. in Banach spaces III, Medford, Massachusetts USA 1980. Lect. Notes in Math. 860, 35-50, Springer Verlag Berlin.

Burkholder, D.L. (1983), A geometrical condition that implies the existence of certain singular integrals of Banach-space-valued functions. Proc. Conf. on Harmonic Anal. in honor of A. Zygmund, Chicago, 1981. Ed. by W. Bechner, et al., 270-286.

Burkholder, D.L. and Shintani, T. (1978), Approximation of L^1-bounded martingales by martingales of bounded variation. Proc. Amer. Math. Soc. 72, n° 1, 166-169.

Cairoli, R. (1970), Une inégalité pour martingales à indices multiples et ses applications. Sém. Prob. Univ. Strasbourg 1968/1969. Lect. Notes in Math. 124, Springer Verlag Berlin.

Chacon, R.V. (1974), A "stopped" proof of convergence. Adv. in Math. 14, 365-368.

Chacon, R.V. and Sucheston, L. (1975), On convergence of vector-valued asymptotic martingales. Z. Wahrscheinlichkeitstheorie verw. Geb. 33, 55-59.

Chatterji, S.D. (1960), Martingales of Banach space-valued random variables. Bull. Amer. Math. Soc. 66, 395-398.

Chatterji, S.D. (1964), A note on the convergence of Banach-space valued martingales. Math. Annalen, 153, 142-149.

Chatterji, S.D. (1968), Martingale convergence and the Radon-Nikodym theorem in Banach spaces. Math. Scand. 22, 21-41.

Chatterji, S.D. (1971), Differentiation along algebras. Manuscripta Math. 4, 213-224.

Chatterji, S.D. (1973), Les martingales et leurs applications analytiques. Lect. Notes in Math. 307, 27-164, Springer Verlag Berlin.

Chatterji, S.D. (1974), Sur l'intégrabilité de Pettis. Math. Z. 136, 53-55.

Chatterji, S.D. (1976), Vector-valued martingales and their applications. Proc. Conf. measure theory Oberwolfach 1975. Lect. Notes in Math. 526, 33-51, Springer Verlag Berlin.

Chen, R. (1976), Some inequalities for randomly stopped variables with applications to pointwise convergence. Z. Wahrscheinlichkeitstheorie verw. Geb. 36, 75-83.

Chi, G.Y.H. (1975), A geometric characterization of Fréchet spaces with the Radon-Nikodym-property. Proc. Amer. Math. Soc. 48, n° 2, 371-380.

Chow, Y.S. (1963), Convergence theorems of martingales. Z. Wahrscheinlichkeitstheorie verw. Geb. 1, 340-346.

Chow, Y.S. and Teicher, H. (1978), Probability theory. Independence, interchangeability, martingales. Springer Verlag Berlin.

Chung, K.L. (1974), A course in probability theory, second ed., Probab. and Math. Stat., Academic Press New York.

Davis, W.J. and Phelps, R.R. (1974), The Radon-Nikodym-property and dentable sets in Banach spaces. Proc. Amer. Math. Soc. 45, n° 1, 119-122.

Davis, W.J., Ghoussoub, N. and Lindenstrauss, J. (1981), A lattice renorming theorem and applications to vector-valued processes. Trans. Amer. Math. Soc. 263, n° 2, 531-540.

Delbaen, F. : see J. Bourgain.

Diestel, J. (1975a), The Radon-Nikodym-property and spaces of operators. Proc. conf. measure theory Oberwolfach 1975. Lect. Notes in Math. 541, 211-227, Springer Verlag Berlin.

Diestel, J. (1975b), Geometry of Banach spaces-selected topics. Lect. Notes in Math. 485, Springer Verlag Berlin.

Diestel, J. and Uhl J.J. Jr. (1977), Vector Measures. Amer. Math. Soc. Surveys, 15.

Dieudonné, J. (1950), Sur un théorème de Jessen. Fund. Math. 37, 242-248.

Dinculeanu, N. (1967), Vector measures. Pergamon Press.

Dunford, N. and Schwartz, J.T. (1957), Linear Operators, part I. Interscience publishers, New York.

Dvoretzky, A. : see also Bellow A.

Dvoretzky, A. (1961), Some results on convex bodies. Proc. on a symposium on linear spaces, Jerusalem.

Dvoretzky, A. (1976), On stopping time directed convergence. Bull. Amer. Math. Soc. 82, n° 2, 347-349.

Dvoretzky, A. and Rogers, C.A. (1950), Absolute and unconditional convergence in normed linear spaces. Proc. Natl. Acad. Sc. USA 36, 192-197.

Edgar, G.A. : see also Austin, D.G.

Edgar, G.A. (1976), Extremal integral representations. J. Funct. Anal. 23, 145-161.

Edgar, G.A. (1979), Uniform semiamarts. Ann. Inst. H. Poincaré, sect. B, 15, n° 3, 197-203.

Edgar, G.A. (1980), Asplund operators and a.e. convergence. J. Multivariate Anal. 10, 460-466.

Edgar, G.A. (1982a), Additive amarts. Ann. Prob. 10, n° 1, 199-206.

Edgar, G.A. (1982b), On pointwise-compact sets of measurable functions. Proc. conf. measure theory Oberwolfach 1982. Lect. Notes in Math. 945, 24-28, Springer Verlag Berlin.

Edgar, G.A. and Sucheston, L. (1976a), Amarts : a class of asymptotic martingales. A. Discrete parameter. J. Multivariate Anal. 6, 193-221.

Edgar, G.A. and Sucheston, L. (1976b), Amarts : a class of asymptotic martingales. B. Continuous parameter. J. Multivariate Anal. 6, 572-591.

Edgar, G.A. and Sucheston, L. (1976c), The Riesz-decomposition for vector-valued amarts. Bull. Amer. Math. Soc. 82, n° 4, 632-634.

Edgar, G.A. and Sucheston, L. (1976d), The Riesz-decomposition for vector-valued amarts. Z. Wahrscheinlichkeitstheorie verw. Geb. 36, 85-92.

Edgar, G.A. and Sucheston, L. (1977a), Martingales in the limit and amarts. Proc. Amer. Math. Soc. 67, n° 2, 315-320.

Edgar, G.A. and Sucheston, L. (1977b), On vector-valued amarts and dimension of Banach spaces. Z. Wahrscheinlichkeitstheorie verw. Geb. 39, 213-216.

Edgar, G.A. and Sucheston, L. (1981), Démonstrations de lois des grands nombres par les sous-martingales descendantes. C.R. Acad. Sc. Paris série I, t.292, 967-969.

Egghe, L. : see also Bellow, A.

Egghe, L. (1978a), On the Radon-Nikodym-property, and related topics in locally convex spaces. Vector space measures and applications II, proceedings. Dublin 1977. Lect. Notes in Math. 645, 77-90, Springer Verlag Berlin.

Egghe, L. (1978b), Caractérisations de la nucléarité dans les espaces de Fréchet. C.R. Acad. Sc. Paris série A, t.287, 9-11.

Egghe, L. (1980a), The Radon-Nikodym-Property, σ-dentability and martingales in locally convex spaces. Pac. J. Math. 87, n° 2, 213-322.

Egghe, L. (1980b), Characterizations of nuclearity in Fréchet spaces. J. Funct. Anal. 35, 207-214.

Egghe, L. (1980c), A new characterization of Banach spaces X for which every operator in $\mathcal{L}(L^1,X)$ is completely continuous. Simon Stevin 54, n° 3-4, 135-150.

Egghe, L. (1980d), Some new Chacon-Edgar type inequalities for stochastic processes and characterizations of Vitali conditions. Ann. Inst. H. Poincaré sect.B, 16, n° 4, 327-337.

Egghe, L. (1981), Strong convergence of pramarts in Banach spaces. Can. J. Math. 33, n° 2, 357-361.

Egghe, L. (1982a), Weak and strong convergence of amarts in Fréchet spaces. J. of Multivariate Anal. 12, n° 2, 291-305.

Egghe, L. (1982b), Convergence of adapted sequences with values in a Banach or Fréchet space. Proc. Conf. Soc. Math. Belg., Bull. Soc. Math. Belg., 34, 133-146.

Egghe, L. (1982c), On sub- and superpramarts with values in a Banach lattice. Proc. Conf. measure theory Oberwolfach 1981. Lect. Notes in Math. 945, 353-365, Springer Verlag Berlin.

Egghe, L. (1982d), An extension of a theorem of Brunel-Sucheston. Bull. Acad. Pol. Sci., 30, n° 11-12, 561-564.

Egghe, L. (1984a), Strong convergence of positive subpramarts in Banach lattices. To appear in the Bull. Polish. Acad. Sci. (formerly Bull. Acad. Pol. Sci.).

Egghe, L. (1984b), Convergence of adapted sequences of Pettis integrable functions. Pac. J. Math., 114, n° 2 (to appear).

Engelbert, A. and Engelbert, H.J. (1979), Optional stopping and almost sure convergence of random sequences. Z. Wahrscheinlichkeitstheorie verw. Geb. 48, 309-325.

Fouque, J.P. (1980a), Régularité des trajectoires des amarts et hyperamarts réels. C.R. Acad. Sc. Paris série A, t.290, 107-110.

Fouque, J.P. (1980b), Enveloppe de Snell et théorie générale des processus [1]. C.R. Acad. Sc. Paris série A, t.290, 285-288.

Fouque, J.P. et Millet, A. (1980), Régularité à gauche des martingales fortes à plusieurs indices. C.R. Acad. Sc. Paris série A, t.290, 773-776.

Frangos, N.E. (1983), On convergence of vector valued pramarts and subpramarts. To appear in Can. J. Math.

Gabriel, J.P. (1977a), An inequality for sums of independent random variables indexed by finite dimensional filtering sets, and its applications to the convergence of series. Ann. Prob. 5, 779-786.

Gabriel, J.P. (1977b), Martingales with countable filtering index set. Ann. Prob. 5, n° 6, 888-898.

Geitz, R.F. (1981), Pettis integration. Preprint.

Geitz, R.F. (1982), Geometry and the Pettis integral. Trans. Amer. Math. Soc. 269, n° 2, 535-548.

Ghoussoub, N. : see also Benyamini, Y., Davis, W.J.

Ghoussoub, N. (1977), Banach lattices valued amarts. Ann. Inst. H. Poincaré sect.B, 13, n° 2, 159-169.

Ghoussoub, N. (1979a), Orderamarts : a class of asymptotic martingales. J. Multivariate anal. 9, 165-172.

Ghoussoub, N. (1979b), Summability and vector amarts. J. Multivariate Anal. 9, 173-178.

Ghoussoub, N. (1982), Riesz space valued measures and processus. Bull. Soc. Math. France, 110, 233-257.

Ghoussoub, N. and Saab, E. (1981), On the weak Radon-Nikodym Property. Proc. Amer. Math. Soc. 81, 81-84.

Ghoussoub, N. and Sucheston, L. (1978), A refinement of the Riesz-decomposition for amarts and semiamarts. J. Multivariate Anal. 8, n° 1, 146-150.

Ghoussoub, N. and Talagrand, M. (1979a), Convergence faible des potentiels de Doob vectoriels. C.R. Acad. Sc. Paris série A, t.288, 599-602.

Ghoussoub, N. and Talagrand, M. (1979b), Order dentability and the Radon-Nikodym Property in Banach lattices. Math. Annalen 243, 217-225.

Gilliam, D. (1976), On integration and the Radon-Nikodym theorem in quasi-complete locally convex topological vector spaces. Preprint.

Gut, A. (1982), A contribution to the theory of asymptotic martingales. Glasgow Math. J. 23, 177-186.
Heinich, H. : see also Bru, B.
Heinich, H. (1978a), Convergence des sous-martingales positives dans un Banach réticulé. C.R. Acad. Sc. Paris série A, t.286, 279-280.
Heinich, H. (1978b), Martingales asymptotiques pour l'ordre. Ann. Inst. H. Poincaré, sect.B, 14, n° 3, 315-333.
Heinich, H. (1979), Sur les mesures vectorielles signées. C.R. Acad. Sc. Paris série A, t.289, 285-286.
Ho, A. (1979), The Radon-Nikodym property and weighted trees in Banach spaces. Israel J. Math. 32, n° 1, 59-66.
Hoffmann-Jørgensen,J. (1971), Vector measures.Math. Scand. 28, 5-32.
Hoffmann-Jørgensen,J. (1974), Sums of independent Banach space valued random variables. Studia Math. 52, 159-186.
Huff, R.E. (1974), Dentability and the Radon-Nikodym property. Duke Math. J. 41, 111-114.
Ionescu-Tulcea, A. : see also Bellow, A.
Ionescu-Tulcea, A. and C. (1963), Abstract ergodic theorems. Trans. Amer. Math. Soc. 107, 107-124.
Isaac, R. (1965), A proof of the martingale convergence theorem. Proc. Amer. Math. Soc. 16, 842-844.
Janicka, L. (1977), Vector measures of infinite variation. Bull. l'acad. Polon. Sc. 25, n° 3, 239-241.
Jeurnink, G.A.M. (1982), Integration of functions with values in a Banach lattice. Ph. Doct. Diss. Kath. Univ. Nijmegen, the Netherlands.
Korzeniowski, A. (1978a), Martingales in Banach spaces for which the convergence with probability one, in probability and in law coincide. Coll. Math. 39, n° 1, 153-159.
Korzeniowski, A. (1978b), A proof of the ergodic theorem in Banach spaces via asymptotic martingales. Bull. l'acad. Polon. Sc. 26, n° 12, 1041-1044.
Krengel, U. and Sucheston, L. (1977), Semiamarts and finite values. Bull. Amer. Math. Soc. 83, n° 4, 745-747.
Krengel, U. and Sucheston, L. (1978), On semiamarts, amarts, and processes with finite value. Adv. Prob. 4, 197-266.
Krickeberg, K. (1956), Convergence of martingales with a directed index set. Trans. Amer. Math. Soc. 83, 313-337.
Kubacki, K.S. and Szynal, D. (1980), Bernstein's inequality for a class of asymptotic martingales. Bull. l'acad. Polon Sc. 28, 1-2, 77-80.
Kunen, K. and Rosenthal, H. (1982), Martingale proofs of some geometrical results in Banach space theory. Pac. J. Math. 100, n° 1, 153-175.
Lamb, C.W. (1973), A short proof of the martingale convergence theorem. Proc. Amer. Math. Soc. 38, n° 1, 215-217.
Landers, D. and Rogge, L. (1980), A short proof for a.e. convergence of generalized conditional expectations. Proc. Amer. Math. Soc. 79, n° 3, 471-473.
Landers, D. and Rogge, L. (1981), Characterizations of conditional expectations operators for Banach-valued functions. Proc. Amer. Math. Soc. 81, n° 1, 107-110.
Lewis, D.R. (1972), A vector measure with no derivative. Proc. Amer. Math. Soc. 32, 535-536.
Lindenstrauss, J. : see also Davis, W.J.

Lindenstrauss, J. and Stegall, C. (1975), Examples of separable spaces which do not contain ℓ^1, and whose duals are not separable. Studia Math. 54, 81-105.

Lindenstrauss, J. and Tzafriri, L. (1977), Classical Banach spaces I. Sequence spaces. Ergebnisse der Math. und ihrer Grensgeb. 92, Springer Verlag Berlin.

Lindenstrauss, J. and Tzafriri, L. (1979), Classical Banach spaces II. Function spaces. Ergebnisse der Math. und ihrer Grensgeb. 97, Springer Verlag Berlin.

Maisonneuve, B. (1981), Submartingales-mesures. Sém. Prob. univ. Strassbourg XV, 1979/80. Lect. Notes in Math. 850, 347-350, Springer Verlag Berlin.

Maurey, B. et Pisier, G. (1976), Séries de variables aléatoires indépendantes et propriétés géométriques des espaces de Banach. Studia Math. 58, 45-90.

Maynard, H.B. (1973), A geometric characterization of Banach spaces possessing the Radon-Nikodym Property. Trans. Amer. Math. Soc. 185, 493-500.

Mc Cartney, P.W. (1980), Neighborly bushes and the Radon-Nikodym property for Banach spaces. Pac. J. Math. 87, n° 1, 157-168.

Mc Cartney, P.W. and O'Brien, R.C. (1980), A separable Banach space with the Radon-Nikodym property which is not isomorphic to a subspace of a separable dual. Proc. Amer. Math. Soc. 78, 40-43.

Métivier, M. (1967), Martingales à valeurs vectorielles. Applications à la dérivation des mesures vectorielles. Ann. Inst. Fourier (Grenoble) 17, n° 2, 175-208.

Meyer, P.E. (1972), Martingales and stochastic integrals I. Lect. Notes in Math. 284, Springer Verlag Berlin.

Millet, A. : see also Fouque, J.P.

Millet, A. (1980), Convergence and regularity of strong submartingales. Proc. of the Coll. ENST CNET sur le processus a deux indices.

Millet, A. (1982), On convergence and regularity of two-parameter 1-submartingales. Preprint.

Millet, A. and Sucheston, L. (1978a), Classes d'amarts filtrants et conditions de Vitali. C.R. Acad. Sc. Paris série A, t.286, 835-837.

Millet, A. and Sucheston, L. (1978b), Sur la caractérisation des conditions de Vitali par la convergence essentielle de classes d'amarts. C.R. Acad. Sc. Paris série A, t.286, 1015-1017.

Millet, A. and Sucheston, L. (1979a), Characterizations of Vitali conditions with overlap in terms of convergence of classes of amarts. Can. J. Math. 31, n° 5, 1033-1046.

Millet, A. and Sucheston, L. (1979b), La convergence essentielle des martingales bornées dans L^1 n'implique pas la condition de Vitali V. C.R. Acad. Sc. Paris série A, t.288, 595-598.

Millet, A. and Sucheston, L. (1980a), Convergence of classes of amarts indexed by directed sets. Can. J. Math. 32, n° 1, 86-125.

Millet, A. and Sucheston, L. (1980b), A characterization of Vitali conditions in terms of maximal inequalities. Ann. Prob. 5, n° 2, 339-349.

Millet, A. and Sucheston, L. (1980c), On covering conditions and convergence. Proc. conf. Measure theory Oberwolfach 1979. Lect. Notes in Math. 794, 432-454, Springer Verlag Berlin.

Millet, A. and Sucheston, L. (1980d), Convergence et régularité des martingales à indices multiples. C.R. Acad. Sc. Paris série A, t.291, 147-150.

Millet, A. and Sucheston, L. (1981a), On convergence of L_1-bounded martingales indexed by directed sets. Preprint.
Millet, A. and Sucheston, L. (1981b), On regularity of multiparameter amarts and martingales. Z. Wahrscheinlichkeitstheorie verw. Geb. 56, 21-45.
Millet, A. and Sucheston, L. (1981c), Demi-convergence des processus à deux indices. C.R. Acad. Sc. Paris série I, t.293, 435-438.
Moedomo, S. and Uhl, J.J. Jr. (1971), Radon-Nikodym theorems for the Bochner and Pettis integrals. Pac. J. Math. 38, n° 2, 531-536.
Mourier, E. (1977), Eléments aléatoires dans un espace de Banach. Ann. Inst. H. Poincaré 13, n° 3, 161-244.
Mucci, A.G. (1973), Limits for martingale like sequences. Pac. J. Math. 48, 197-202.
Mucci, A.G. (1976), Another martingale convergence theorem. Pac. J. Math. 62, n° 2, 539-541.
Musiał, K. (1977a), An operator characterization of Banach spaces which do not contain any isomorphic copy of ℓ^1. Preprint series n° 28. Aarhus Matematisk Institut, Aarhus Universitet, 8 p.
Musiał, K. (1977b), The weak Radon-Nikodym property in Banach spaces. Preprint series n° 30. Aarhus Matematisk Institut,Aarhus Universitet, 53 p.
Musiał, K. (1979), The weak Radon-Nikodym Property in Banach spaces. Studia Math. 64, 151-173.
Musiał, K. (1980), Martingales of Pettis integrable functions. Proc. conf. measure theory Oberwolfach 1979. Lect. Notes in Math. 794, 324-339, Springer Verlag Berlin.
Neveu, J. (1965), Relations entre la théorie des martingales et la théorie ergodique. Ann. Inst. Fourier (Grenoble) 15, n° 1, 31-42.
Neveu, J. (1975), Discrete parameter martingales. North-Holland Publ. Cy New York.
Namioka, I. and Phelps, R. (1975), Banach spaces which are Asplund spaces. Duke Math. J. 42, 735-750.
O'Brien, R. : see Mc Cartney, P.
Parthasarathy, K.R. (1967), Probability measures on metric spaces. Acad. Press. New York.
Parthasarathy, K.R. (1972), Selection theorems and their applications. Lect. Notes in Math. 263, Springer Verlag Berlin.
Pełczynski, A. : see Bessaga, C.
Peligrad, M. (1976), A limit theorem for martingale-like sequences. Rev. Roum. Math. Pures et Appl. 21, n° 6, 733-736.
Phelps, R.R. : see Davis, W.J., Namioka, I.
Pestman, W.R. (1981), Convergence of Martingales with Values in Locally Convex Suslinspaces. Preprint.
Phillips, R.S. (1943), On weakly compact subsets of a Banach space. Amer. J. Math. 65, 108-136.
Pietsch, A. (1972), Nuclear locally convex spaces. Ergebnisse der Math. und ihrer Grenzgeb. 66, Springer Verlag Berlin.
Pisier, C. (1975), Martingales with values in uniformly convex spaces. Israel J. Math. 20, 326-350.
Riddle, L.H. (1982), The geometry of weak Radon-Nikodym sets in dual Banach spaces. Proc. Amer. Math. Soc. 86, n° 3, 433-438.
Riddle, L.H. and Uhl, J.J. Jr. (1982), The fine line between Asplund spaces and spaces not containing a copy of ℓ^1. Proc. Conf. Martingale theory in harmonic anal. and Banach spaces, NSF-CBMS Conf., Cleveland, Ohio, 1981. Lect Notes in Math. 939, Springer Verlag Berlin, 145-156.

Rieffel, M.A. (1967), Dentable subsets of Banach spaces, with applications to a Radon-Nikodym theorem. Funct. Anal., Proc. Conf. Irving California USA 1966. B.R. Gelbaum ed., Acad. Press London, Thompson Washington D.C., 71-77.
Rodriguez-Salinas, B. (1980), La propriedad de Radon-Nikodym, σ-dentabilidad y martingalas en espacios localmente convexos. Rev. Real Acad. Cienc. exact. Fis. Natur. Madrid, 74, n° 1, 65-89.
Rogers, C.A. : see Dvoretzky, A.
Rogge, L. : see Landers, D.
Rosenthal, H.P. : see also Bourgain, J., Kunen, E.
Rosenthal, H.P. (1974), A characterization of Banach spaces containing ℓ^1. Proc. Natl. Acad. Sc. USA 71, 2411-2413.
Saab, E. (1978), On the Radon-Nikodym-Property in a class of locally convex spaces. Pac. J. Math. 75, n° 1, 281-291.
Scalora, F.S. (1961), Abstract martingale convergence theorems. Pac. J. Math. 11, 347-374.
Schaefer, H.H. (1971), Topological Vector Spaces. Graduate Texts in Math. 3, Springer Verlag Berlin.
Schaefer, H.H. (1974), Banach lattices and positive operators. Grundlehren der Math. Wiss. 215, Springer Verlag Berlin.
Schmidt, K.D. (1979a), Espaces vectoriels réticulés, décompositions de Riesz et caractérisations de certains processus de fonctions d'ensembles. C.R. Acad. Sc. Paris série A, t.289, 75-78.
Schmidt, K.D. (1979b), Sur la convergence d'une amartingale bornée et un théorème de Chatterji. C.R. Acad. Sc. Paris série A, t.289, 181-183.
Schmidt, K.D. (1980a), Théorèmes de structure pour les amartingales en processus de fonctions d'ensembles à valeurs dans un espace de Banach. C.R. Acad. Sc. Paris série A, t.290, 1069-1072.
Schmidt, K.D. (1980b), Théorèmes de convergence pour les amartingales en processus de fonctions d'ensembles à valeurs dans un espace de Banach. C.R. Acad. Sc. Paris série A, t.290, 1103-1106.
Schmidt, K.D. (1980c), On the value of a stopped set function process. J. Multivariate Anal. 10, n° 1, 123-134.
Schmidt, K.D. (1981a), On the convergence of a bounded amart and a conjecture of Chatterji. J. Multivariate Anal. 11, n° 1, 58-68.
Schmidt, K.D. (1981b), A measure theoretic approach to vector valued amarts. Preprint.
Schmidt, K.D. (1982), Generalized martingales and set function processes. Preprint.
Schwartz, J.T. : see Dunford, N.
Schwartz, L. (1973), Radon Measures on arbitrary topological spaces. Tata Institute of Fund. Res., Bombay.
Schwartz, L. (1981), Geometry and probability in Banach spaces. Lect. Notes in Math. 852, Springer Verlag Berlin.
Sentilles, F.D. (1982), Decomposition of weakly measurable functions. Preprint.
Sentilles, F.D. and Wheeler, R.F. (1982), Pettis integration via the Stonian transform. Preprint.
Shintani, T. : see Burkholder, D.L.
Sion, M. (1960), Topological and measure theoretic properties of analytic sets. Proc. Amer. Math. Soc. 11, 769-776.
Słaby, M. (1982), Convergence of submartingales and amarts in Banach lattices. Bull. Acad. Pol. Sc., sér. Sc. Math. 30, n° 5-6, 291-299.

Słaby, M. (1983a), Convergence of positive subpramarts and pramarts in Banach spaces with unconditional basis. To appear in Bull. l'acad. Pol. Sc.
Słaby, M. (1983b), Strong convergence of vector-valued pramarts and subpramarts. To appear in Prob. Math. Stat.
Stegall, C. : see also Lindenstrauss, J.
Stegall, C. (1978), The duality between Asplund spaces and spaces with the Radon-Nikodym property. Israel J. Math. 29, n° 4, 408-412.
Stern, J. (1978), A Ramsey theorem for trees, with an application to Banach spaces. Israel J. Math. 29, n° 2-3, 179-188.
Subramanian, R. (1973), On a generalization of martingales due to Blake. Pac. J. Math. 48, n° 1, 275-278.
Sucheston, L. : see also Brunel, A., Chacon, R.V., Edgar, G.A., Ghoussoub, N., Krengel, U., Millet, A.
Sucheston, L. (1967), Banach limits. Amer. Math. Monthly, march 1967, 308-311.
Sucheston, L. (1980), Math. Rev. # 80 g : 60052.
Szulga, J. (1978a), Boundedness and convergence of Banach lattice valued submartingales. Proc. conf. "Probability on Vector spaces". Trebienowice 1977, Lect. Notes in Math. 656, 251-256, Springer Verlag Berlin.
Szulga, J. (1978b), On the submartingale characterization of Banach lattices isomorphic to ℓ_1. Bull. l'acad. Polon. Sc. 26, n° 1, 65-68.
Szulga, J. (1979), Regularity of Banach lattice valued martingales. Coll. Math. 41, n° 2, 303-312.
Szulga, J. and Woyczynski, W.A. (1976), Convergence of submartingales in Banach lattices. Ann. Prob. 4, n° 3, 464-469.
Szynal, D. : see Kubacki, K.S.
Talagrand, M. : see also Ghoussoub, N.
Talagrand, M. (1980), Sur les mesures vectorielles définies par une application Pettis-intégrable. Bull. Soc. Math. France 108, 475-483.
Talagrand, M. (1981), Dual Banach lattices and Banach lattices with the Radon-Nikodym-Property. Israel J. Math. 38, 46-50.
Teicher, H. : see Chow, Y.S.
Thomas, G.E.F. (1974), The Lebesgue-Nikodym theorem for vector-valued Radon measures. Mem. Amer. Math. Soc. 139.
Thomas, G.E.F. (1976), On some negative properties of the Pettis-integral (addendum). Lect. Notes in Math. 541, 131, Springer Verlag Berlin.
Tuyên, D.Q. (1981), On the asymptotic behaviour of sequences of random variables and of their previsible compensators. Ann. Inst. H. Poincaré sect.B, 17, n° 1, 63-73.
Tzafriri, L. : see Lindenstrauss, J.
Uhl, J.J. Jr. : see also Diestel, J., Moedomo, S., Riddle, L.H.
Uhl, J.J. Jr. (1969a), The Radon-Nikodym theorem and the mean convergence of Banach space valued martingales. Proc. Amer. Math. Soc. 21, 139-144.
Uhl, J.J. Jr. (1969b), Martingales of vector valued set functions. Pac. J. Math. 30, n° 2, 533-548.
Uhl, J.J. Jr. (1969c), Applications of Radon-Nikodym theorems to martingale convergence. Trans. Amer. Math. Soc. 145, 271-285.
Uhl, J.J. Jr. (1972), Martingales of strongly measurable Pettis integrable functions. Trans. Amer. Math. Soc. 167, 369-378. Erratum 181, 507, 1973.

Uhl, J.J. Jr. (1977), Pettis mean convergence of vector-valued asymptotic martingales. Z. Wahrscheinlichkeitstheorie verw. Geb. 37, 291-295.

Van Dulst, D. (1982), Unpublished seminar notes.

Walsh, J.B. (1979), Convergence and regularity of multiparameter strong martingales. Z. Wahrscheinlichkeitstheorie verw. Geb. 46, 177-192.

Woyczynski, W. : see also Szulga, J.

Woyczynski, W. (1975), Geometry and martingales in Banach spaces. Winter School on probability, Karpacz 1975. Lect. Notes in Math. 472, 229-275, Springer Verlag Berlin.

Woyczynski, W. (1978), Geometry and martingales in Banach spaces, part II : independent increments. Adv. Prob., 4 (probab. in Banach spaces, J. Kuelbs, ed.), M. Dekker, New York, 267-518.

Yamasaki, Y. (1981), Another convergence theorem of martingales in the limit. Tôhoku Math. J. 33, 555-559.

LIST OF NOTATIONS

E	I.1
Ω	I.1
$\mathcal{F}$	I.1
P	I.1
χ_A	I.1.1
$\int$	I.1.2
$\mathcal{L}^1_E$	I.1.2
L^p_E $(1 \leq p \leq \infty)$	I.1.2
$\|\cdot\|_p$ $(1 \leq p \leq \infty)$	I.1.2
$\|\cdot\|_{Pe}$	I.1.2
$(\mathcal{F}_n)_{n\in\mathbb{N}}$	I.2.1
$(X_n,\mathcal{F}_n)_{n\in\mathbb{N}}$	I.2.1
$\sigma(X_1,\ldots,X_n)$	I.2.1
T^*	I.2.1
T	I.2.1
$\mathbb{N}(\sigma), T(\sigma)$	I.2.1
$\mathcal{F}_\tau$	I.2.1
X_τ	I.2.1
$\mathcal{F}_\infty$	I.2.1
$E^{\mathcal{G}}$	I.2.2.1
$P(X > \lambda)$	I.3.5.2
$(\mu_n,\mathcal{F}_n)_{n\in\mathbb{N}}$	I.4.1
μ_τ	I.4.1
$\mu \ll P$	II.2.1.1
(RNP)	II.2.1.1
Π	II.2.2.1
Co	II.2.3.1
$\mathcal{F}^+$	II.2.3.2
$\mathcal{A}_A(\mu)$	II.2.3.2
$\#$	II.2.3.6
$\mathcal{B}(E)$	II.3.1.1
$C_b(T)$	II.3.1.5
$M^1_E, \|\cdot\|_1,$	II.3.2
d_n	II.3.2.1
BV	II.3.2.1
AE	II.3.2.1
(ANP)	II.4.7.2
E^+	III.1.4
$\mathcal{T}$	IV.1
$Q_{\mathcal{T}}$	IV.1
(CRP)	V.2.20
H_A	V.3
dens (E)	V.3
t.o.s.	V.3.19.2
L, L^*, L_*	V.5.7-V.5.8
$(X_i,\mathcal{F}_i)_{i\in I}$	VI
e sup	VI.1.1
e inf	VI.1.1
e lim sup	VI.1.1
e lim inf	VI.1.1
e lim	VI.1.1
s lim sup	VI.1.1
s lim inf	VI.1.1
s lim	VI.1.1

SUBJECT INDEX

S

T

Printed in the United States
By Bookmasters